P9-BEE-601

Purchased with funds from a
Library Services & Technology
Act Grant, 1998-99

GROWING AND

selling

Fresh-Cut Herbs

GROWING AND

selling

Fresh-Cut Herbs

SANDIE SHORES

**STOREY
BOOKS**

The mission of Storey Communications is to serve our customers
by publishing practical information that encourages personal independence
in harmony with the environment.

Edited by Deborah Balmuth and Robin Catalano
Cover design by Eva Weymouth and Meredith Maker
Cover photograph by A. Blake Gardener
Text design by Mark Tomasi
Text production by design+FORMAT
Photographs by Agra Tech, Inc.; Poly-tex, Inc.; Schaefer Fan Company, Inc.;
  P.L. Light Systems Canada, Inc.; Jack Huhnerkoch, and Deb House-Finlay
Line drawings by Beverly Duncan, LaVonne Francis, Brian Whitehurst,
  Judy Eliason, Elayne Sears, Charles Joslin, Alison Kolesar, Doug Paisley,
  Cathy Baker, and Brigita Fuhrmann
Indexed by Hagerty & Holloway

Copyright © 1999 by Sandie Shores

The information in this book is true and complete to the best of our knowledge. All recommendations are made without guarantee on the part of the author or Storey Books.
The author and publisher disclaim any liability in connection with the use of this information. For additional information, please contact Storey Books, Schoolhouse Road,
Pownal, Vermont 05261.

Storey Books are available for special premium and promotional uses and for customized
editions. For further information, please call Storey's Custom Publishing Department at
1-800-793-9396.

Printed in Canada by Webcom Limited
10 9 8 7 6 5 4 3 2 1

Library of Congress Cataloging-in-Publication Data

Shores, Sandie, 1944–
      Growing and selling fresh-cut herbs / Sandie Shores.
          p.   cm.   — (Making a living naturally)
      Includes bibliographical references and index.
      ISBN 1-58017-128-1   (HC : alk. paper)
      1. Herb farming.   2. Herbs—Marketing.   3. Herbs.
      I. Title   II. Series
      SB351.H5S538   1999
      635'.7—dc21   98-30053
                                                                              CIP

contents

contents *(cont.)*

*Dedicated to the memories of my son, Ken Kranz,
and my mother, Vi Shores,
whose untimely passings gave me the courage
and strength to write this book.*

ACKNOWLEDGMENTS

From the bottom of my heart I want to thank the many people who gave the love, friendship, and encouragement that helped me to get this book to you:

Finn Halvorsen for his love, handyman skills, and helping to build my greenhouses; Cathy Osborne and Katie Bleed, for their great herb-picking abilities and much more; Alberta Roberson (Dirty Thumb Greenhouses), for her energy and inspiration; Beth Tidwell (Perennial Design), for her knowledge and insights; Deb House, for her solace and sanctuary; Bob (NA) Boyer, for the traditional spiritual lessons; Ray Howe (Lone Oak Press), for his wit and insights into the publishing business; Cathryn (The Wicker Woman) Peters, for the promotion lessons; Shirley Fite, for lessons and friendship; and Pauline Mansour and LaVonne Francis, for some of the artwork and longtime friendship.

Thanks to Bob, Dan, and Susan Milam for buying my business (now the Zumbro Herb Farm) and farm and allowing me to take photos there. Thanks to Rochester Restaurant Suppy, Maureen Rogers from The Herb Growing & Marketing Network, and all the great people at Storey Books.

Thanks to my family, Keith, Mary, Josh, Jeremy Kranz, Barb Bannitt, and Brandi Kranz, for their understanding and forgiveness while I was much too busy being a commercial grower and then an author. And thanks to the departed souls of my dad and Grandpa Herb for the "growing genes" and to the Creator for giving me all my blessings and the time and ability to love and care for our Mother Earth.

One of the most frustrating things in life is to have a great idea for a business and then have difficulty finding information to help you work toward that goal. When I started Herb's Herbs in 1985 there was very little information out there about how to grow fresh-cut herbs on a commercial scale. Crop scheduling and packaging information was nonexistent. There were many businesses, large and small, growing and selling fresh-cut herbs, but they were closely guarding their secrets on how to do it.

Certainly there were many beautiful books with pretty pictures about growing herbs in a home garden setting. While those books only made me want to grow the pretty and fragrant herbs even more, they hardly served the needs of someone wanting to start a business. The only book I could find that had any information about commercial production was published by this very same publisher, Storey Books. *Growing and Using Herbs Successfully* by Betty Jacobs is still one of their top-selling books.

The industry has changed a great deal since Betty's book was written. The market for fresh-cut herbs has expanded to even the smallest Mom and Pop grocery stores in the tiniest of towns across the country. People everywhere are cooking with fresh herbs, and not just with the basic parsley and chives. Many view cooking with fresh herbs as a way to express their creativity. People in all walks of life and of all ages are cooking more healthful and flavorful meals. Customer demand is growing and will continue to do so; this paves the way for fresh-cut herb growers of any size to operate successfully all around the country. You can get your herbs to the consumer fresh — really fresh! That is one of the best selling points you have.

Other than Betty Jacobs' book, there still has not been a comprehensive book on the market to help the aspiring commercial herb grower get started — until now. I have written, for you, the book that will get you on your way to growing and selling fresh-cut herbs successfully and profitably. These are my experiences, learned at the school of hard knocks, from years of research and prosperous operation. Other commercial growers may do things just a bit differently, but the basics are all here for you to learn from. For those of you who are serious home gardeners and want to learn some "insider" secrets, there is plenty here for you too!

It is my hope that this book will give you the tools you need to operate a lucrative and personally rewarding herb business. I have received hundreds of letters from people all over saying "Help! How do I do this?" I have tried to answer all of these questions and more. But, always remember to run your business and grow your herbs with love and joy in your heart — both you and your herbs will prosper. Enjoy!

the herb's herbs history: a voice of experience

Starting a business can sometimes seem like an insurmountable task, even when there is lots of information to guide you. This is the story of how I started Herb's Herbs, and some of the obstacles I faced and how they were overcome. I hope this will help you see that you too can make this business rewarding and profitable.

Making Choices

Prior to starting my business, I had had a long and varied career. My choices were eclectic and they ranged from a short stint as a nurse, to owning a retail store, to 15 years in the music industry as a booking agent and artist manager. In the early 1980s I was a Realtor and an insurance agent. From each of these careers I acquired skills that prepared me for running a successful herb business.

The common thread was my hobby farm and my love for growing herbs and vegetables. I have always wanted to earn a living by growing and selling herbs, vegetables, or flowers. I would grow anything, as long as I could work at home! So, being entrepreneurial in nature, and with a love for marketing and dealing with people, I started looking for a niche for my business.

Finding Information

I couldn't find much written about fresh-cut herbs when I started my enterprise. There were many growers operating this kind of business, but they were competitive and not willing to share the "how-to" information. Nonetheless, I talked to everyone who knew anything about the vegetable- and herb-growing, restaurant, and supermarket businesses. Eventually, through a friend, I made contact with a chef who agreed to give me an overview of the industry. After talking with him, and many others, I learned that there was definitely a place for a grower who could supply really fresh herbs.

Sure there were "fresh" herbs coming from distributors to the restaurants, hotels, and supermarkets. But they had been warehoused for anywhere from one to four days; by the time they got to the end user, they were hardly fresh. I knew, however, that I would have to have more experience in growing large amounts of herbs before I could contact anyone to try to sell them.

Starting Up

So, bullheadedly but blindly, in 1985 I started growing herbs, both perennials and annuals. Soon my little house was filled with card tables overflowing with flats of herbs, lamps, and fluorescent lights. I couldn't cook for weeks because on my stove top, with its pilot lights, were flats of seeds that needed bottom heat to germinate. Every available space was taken up by plants. I bet some of you know just what I mean!

I planted all these herbs on about ½ acre of land that had previously been pasture. It didn't take long to learn about one of the most pesky problems that commercial growers face — weeds! Many hours were spent sitting on the soil trying to pull the clover from the thyme beds and hoeing or tilling the rest. Just as much time was spent working an outside job, because the bills still had to be paid.

In a few months I had a fine crop of basil. How exciting it was to go out and try to sell it to a few chefs! My very first order was for ½ pound of basil. That morning I picked a bucketful and took it into the house, figuring that it should be washed first. But I didn't know how to package it! It seemed to me that the chef would use only the leaves, so I stripped them all from the stems! I carefully washed the leaves in the kitchen sink and cut off any brown or discolored spots. I didn't have a scale, so after the leaves were dry I just packed them up in a plastic bag.

When I proudly delivered the basil, the chef asked, wide-eyed, "This is half a pound?" "Well," I said, "it may be a little more, but that's okay!" Looking back, I now know that what I delivered to him was probably more like 3 pounds, with no stems! I sure had much to learn, but I knew that I could succeed.

Expanding

I set about doing careful and thorough market research. I discovered many possible outlets for my herbs. In my area, Rochester, Minnesota, there are many fine restaurants and hotels catering to the constant flow of visitors from around the world to the Mayo Clinic. I wrote up a business plan setting out just what my goals were and the financial information necessary to achieve them. This business plan was a great asset when I applied for a loan for my first greenhouse.

Early in the start-up phase of Herb's Herbs I was able to finance the necessary supplies and seeds myself. But there was much to learn yet about every aspect of the business: packaging, harvesting, crop scheduling, growing in volume — the list is endless! Now, looking back, it's embarrassing to admit how little I knew about my own business in the first few years.

During the second year of growing herbs for sale, great luck came my way. My Service Corps of Retired Executives (SCORE) counselor was a past employee of the largest and most popular hotel in Rochester. He provided me with an introduction to the produce buyer at that hotel, who then set up a meeting with the new chef.

That chef was Arnie Crenin, a well-known culinary artist with impressive credentials. I will always be indebted to him, for he spent hours showing me the herbs and edible flowers he was getting and how they were packaged. He told me what restaurants expect from growers and then gave me a list of the most popular herbs that I should grow. Later, whenever articles were written about him and the hotel, he would make sure my name was mentioned as the supplier of fresh herbs. I hope that as you begin your business, you can find a chef like Arnie to help you.

The second year I grew enough to supply this hotel and several other restaurants with fresh-cut herbs in season. They all wanted me to continue to supply them through the winter because my herbs were superior in quality, flavor, and freshness to any herbs that they could buy from their produce distributors.

This is also something I wanted, partly so I wouldn't have to work at another job to pay the bills and partly because this was a component of my ultimate goal.

Growing Year-Round

Growing year-round, of course, means having a greenhouse. But I knew nothing about them and little information was available at the library or at the local university extension office. I started visiting greenhouse nurseries just to look at them and try to figure out how they were built and operated. I was often met with hostility by the owners as soon as they discovered I was going to build a greenhouse. It didn't seem to matter that I wouldn't be competing with them!

Luck again prevailed and I met Alberta Roberson, who owns the Dirty Thumb Greenhouses nearby. Alberta took me under her wing and taught me much, not only about greenhouses but about growing and doing business, too. This savvy and energetic woman became my good friend, and we would often combine our supply orders to get the volume discounts. A mentor such as Alberta is invaluable.

Making Mistakes

After reading my business plan, the local bank gave me a small loan for my first greenhouse, which was built in the winter of 1987–1988. Herb's Herbs was really on its way! Little did I know how much more there was to learn.

The manufacturer that I bought the greenhouse from tried to guide me into buying the proper-size furnace and ventilation fans. However, in an effort to save money, I skimped on getting the proper equipment and putting in a water supply. Although I was able to grow herbs the first year without that equipment, it was a great deal of extra work. The water had to be brought into the greenhouse through almost 500 feet of hose. By the time the water arrived, there was not enough pressure to spray it, so I had to fill watering cans and use them to give the plants a drink. In the winter I had to carry water over to the greenhouses in my little four-wheel-drive car.

The furnace was much too small to heat the entire greenhouse. A wood-burning furnace was also used, but that meant trudging a long way through deep snow to reload it several times during the night. There were many below-zero nights when I slept on a cot in the greenhouse to keep the fire going!

I planted most of the herbs in the ground, and many in hanging baskets. This way of growing is fraught with problems. The biggest problem was that the herbs didn't provide a good yield when grown in containers. Crickets, mice, snakes, and all manner of other problems plagued my herb beds. By some quirk of fate I did not have any serious disease problems. There were troubles with nutritional deficiency, though, because the soil had not been properly prepared.

Eventually — mostly through trial and error — I found what worked and what didn't. Often a problem would develop and I would have to scramble to search for the reasons and remedies. Remember, in those early years in-depth information was not as accessible as it is today. Even the trade magazines and newsletters that I subscribed to offered only superficial information.

Meanwhile, through this learning process I was still managing to grow good-quality herbs. There were many times, however, when I was out of an herb and unable to supply my accounts. Luckily, my customers were loyal, and when I was out of something they would order that herb from the wholesale distributor and buy from me whatever I had on hand.

Crop scheduling to maintain continuous supplies was a most complex thing to master. That, along with estimating the number of plants to grow to provide the needed volume of herbs, was the biggest problem I faced. Both of these difficulties were magnified because my business was increasing constantly. I was always playing "catch up and grow"!

Meeting Growing Demand

Through word of mouth and my own marketing efforts, more and more accounts wanted my herbs. Soon it became necessary to build another greenhouse to keep up with the demand. My long-range plan, however, did not allow for the possibility of another greenhouse. The property I owned was in a narrow valley, which was also a floodplain, and another greenhouse would not receive full sun.

I mention the floodplain because, yes, there was a flood! After a 9-inch rainfall, a wall of water came racing down the valley and washed away several of my buildings, 60 chickens, and several years of hard work. The greenhouse was not severely damaged because the large doors were left open and the water flowed through it rather than moving the structure.

It is impossible to describe the sorrow and hard work you must endure after a flood unless you have lived through it yourself.

For these reasons I purchased the farm across the road. It had much more land and it was out of the floodplain. My greenhouse had to be taken apart, moved, and reassembled. Many hundreds of mature perennial plants had to be dug up and transplanted. This move was not an easy task and is a shining example of why you should have long-range plans in place before starting your business.

We built raised beds in the newly moved greenhouse with used barn lumber that was available in abundance on the new farm. It took truck-loads of topsoil, gravel, and sand and hundreds of bags of peat mix to fill these beds. I used an electric cement mixer to blend the components. That old barn lumber was free, but in less than a year the long screws were pulling loose and soil and rocks would spill onto the floor.

My business continued to grow and another greenhouse was put up. By now both greenhouses had all the necessary equipment, including supplemental lighting. I was able to grow everything I needed to supply my accounts with the exception of basil during the winter, a difficult problem indeed.

I had basically all the restaurant and supermarket accounts in my region that used fresh-cut herbs. Consistently I turned down business from outside a 40-mile delivery area: Delivery costs were simply too high. Instead of delivering herbs out of my area, I expanded the business by selling at the farmer's market Saturday mornings. This was an excellent source of "instant" money and a fun experience chatting with customers and developing a camaraderie with growers of all kinds. Selling at the farmer's market is a lot of work, but well worth it.

Business expansion also included growing and selling potted herbs wholesale to local nurseries. This requires different techniques from those needed to grow herbs for fresh-cut use. Yet more lessons to be learned. Soon I was growing and selling 10,000 potted herb plants per season.

This expansion was wonderful, but I could no longer do it alone. There were not enough hours in the day to do everything, and certainly not enough stamina. Soon I found Cathy, a great gardener who lived nearby. She also worked for Alberta at the Dirty Thumb Greenhouses part time, so she became an independent contractor and also worked for me part time. Soon another lady, Katy, came to work. I could not have done this all without them! They were great, not only at picking and packaging herbs, but at many other tasks as well. They planted, transplanted, weeded, and sometimes made deliveries. Luck had come my way again!

Changing Courses

Herb's Herbs bounded along, practically at breakneck speed, for several years. Every day I learned more and more about this business. I taught herb growing and cooking with fresh herbs for a community education program, taught classes at local nurseries, and spoke to garden clubs. Teaching and sharing my knowledge was sure fun!

I joined herb organizations and received 10 monthly grower and herb trade magazines and newsletters. Some of the newsletters featured articles about the lack of information about growing and selling fresh-cut herbs. These stories rang true; I had lived through operating without many written resources.

In 1994 a profile of my business appeared in *Growing Your Herb Business,* by Bertha Reppert. Since that book was published, I have received hundreds of letters from readers asking all types of questions about growing and selling fresh-cut herbs. The demand for information, both from individuals and the herb press, was constantly on my mind. In late 1994 I experienced some life changes and decided to sell the business and farm so I could concentrate on filling this information void.

From this introduction you have learned only a few of the problems that I faced. Some are humorous, certainly, but most of them were caused simply because there was no information on the subject. You can also see from these experiences why I offer some of the advice I do in the following chapters.

Perhaps all this sounds like it was a great deal of work and a comedy of many errors. It was, of course, but it was well worth it. I hope you can learn from this voice of experience.

A Reader's Guide to Using This Book

In an effort to keep similar subjects together, this book is divided into four parts. Part I, Starting Your Herb Business, contains information on business planning and conducting market research, successfully marketing your business to different types of clients, and managing day-to-day business details. For inspiration, I've included profiles of some of the many successful herb-growing businesses from around the country. The experience and advice of their owners and managers can help you find the right niche for your enterprise. This is the section to read first if you are really serious about growing fresh-cut herbs as a profession; the chapters are designed to take you through every step of the business end of the process, and no detail is spared.

Building and Maintaining a Greenhouse is the second major section of the book, and it covers all aspects of using a greenhouse to grow your herbs. You'll learn how to select and build the appropriate greenhouse for your type of business, property, and budget. Interior greenhouse design gets in-depth treatment, as do everyday operation and maintenance.

Of course, you might like to skip to Part III, Growing and Nurturing Your Plants, if you want to learn the secrets of the pros! The dos and don'ts of starting and growing herbs and controlling insect pests and diseases get full attention here. Finally, this section wraps up with a chapter on the proper methods of harvesting, handling, and packaging your herbs for sale.

I also include 16 separate, detailed chapters on growing specific herbs and edible flowers in Successfully Growing More Than 20 Herbs and Flowers, which is the fourth and final part of the book. In these chapters you'll learn how to grow and sell popular herbs such as basil, oregano, and rosemary, while also getting indispensible information on growing and selling techniques for edible flowers and less common but promising herbs like epazote and salad burnet. This is the part of the book to keep coming back to whenever you need information on your particular plants.

The book finishes off with a comprehensive glossary, so if you find an unfamiliar term in the text be sure to check for the definition. Also, use Resources to locate grower supplies, seed suppliers, grower publications, and information about important government organizations that you'll need to contact when you begin your business. Above all, enjoy the book, your work, and your business!

starting
your herb
business

planning your business and market research

Would you like to have your own herb business? Growing herbs and doing what you love — gardening — are a wonderful way to make a living.

This type of business has been romanticized by lovely pictures and articles in magazines and by television. Few people understand how much hard work it takes to have a successful business. While you explore your ideas and feelings for starting an herb business, it is important to

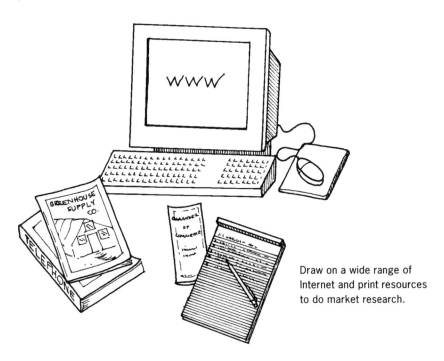

Draw on a wide range of Internet and print resources to do market research.

understand that this really is a business. It requires planning and perseverance; without a solid business plan and good market research, your chances for success are limited.

Defining Your Objectives

Before doing any market research, figure out what it is that you really want to do. Do you want a small backyard business so you can make a little money at home? Do you want a full-scale commercial operation with a nationwide clientele? Define how large you want to be, what your financial resources are, and what your goals are.

Take an honest look at yourself and ask some serious questions about your strengths and weaknesses. Do you have the stamina to put in long hours day in and day out? Will your family support the amount of work you do? Do you have the knowledge and marketing skills necessary to sell your herbs and maintain customer loyalty?

Physical location is extremely important in determining how you begin this business. Do you have to buy a farm or can you start this business in your present location? Do you have enough room for a greenhouse? Consider that five years from now you may want more greenhouses. Is there enough space for them and 20 acres of outdoor growing range?

The answers to these questions are important before you begin your market research and start this business. You must also remember that as your business grows and you learn more, your objectives may change.

 finding a market

An important step in starting any business is to discover where your market is. More than likely there will always be someone to buy your fresh-cut herbs. It is your job to learn who these people are, how many of them there are, and what your competition is. It will pay you many times over to evaluate your prospective markets before planting a single crop.

Organizing Your Research Plan

Take the time to organize your research plan. This will save you much time, money, and wasted effort. Use a notebook to keep track of all your questions and the information you learn. Here are some guidelines to help you.

1. Make a list of people who can help you with your business. Include greenhouse-supply companies, government agencies, university Extension agents, the chamber of commerce, seed companies, bankers, commercial vegetable growers, nurseries, farmer's market growers, and other herb growers.

2. Develop a list of sources of information. Now there are many excellent places to find out what you want to know. Write down any ideas that you have as they come to you; these will undoubtedly lead to many more.

3. List the people and businesses that you could possibly sell your herbs to. Seek out supermarket produce managers, hotel purchasing agents, chefs, produce buyers for distributors, and anyone else who can possibly give you information about your business.

4. Make a list of questions to ask the people you will talk to. The questions will vary according to each person's occupation. Sample questions are listed in chapter 2.

5. Set up a target date to complete your primary research. Establishing a definite time frame for your initial research will enable you to move quickly toward your goal.

Sources of Information

There is a wide variety of sources from which you can obtain marketing information. Listed here are but a few places to start. As you explore these areas, you probably will find even more resources.

Networking

Networking can be one of the best and most enjoyable ways of finding potential markets. Ask questions of everyone you can — family, friends, even bare acquaintances. Many people will have ideas and contacts for you to follow up on.

Public Library

This is *the* information spot. Go to the reference department and talk to the librarian. These people are trained to help you find the information you need. Here you will find books that can help you with your business plan, growing, potential markets, and more. Two of the most important sources are:

Telephone books. Look in the Yellow Pages under RESTAURANTS and write down the name, address, and number of each. Many restaurants, even some ethnic fast-food places, use fresh herbs. Make a list of all businesses that might buy your products. Look under the categories for caterers, supermarkets, health food stores, hotels, farmer's markets, food, herbs, specialty and ethnic markets, food products, food processing, produce companies, and food service companies. Also make note of listings for greenhouses, greenhouse supplies, nurseries, and vegetables.

If your area does not have local wholesale produce distributors, look in the telephone book of the largest city closest to you. This area could be a source of business as well. Write all these down in a notebook and leave plenty of space between names for your notes.

Reader's Guide to Periodical Literature. This is an index of newspaper and magazine articles about the subjects listed. Look under restaurants, hotels, herbs, chefs, and any other subject that might hold interest for you. There are other magazine and news indexes available. Ask the librarian for help in finding them.

The World Wide Web

If you have Internet access, much of the information you are seeking can be found there. Using a computer can decrease the amount of time it takes to search for information. Many libraries now have computers that are on-line for their patrons.

Most government agencies have Web sites that provide much of the information you will need. On-line telephone directories make it easy to find the names and numbers of those companies you want to contact. In many cases, the indexes and guides you would look for at the library are also on-line.

Small Business Administration

The Small Business Administration (SBA) has many programs and lots of information for those wanting to start a business. *The Directory of*

Business Development Publications is available free by contacting your local SBA office. If you don't have a local office, call 1-800-368-5855. There are many useful publications listed here that are very inexpensive. The SBA can provide you with free books that are designed to take you step-by-step through writing your business plan.

Service Corps of Retired Executives (SCORE)

This is an organization sponsored by the SBA. In cities with an SBA office, you can find SCORE by calling the SBA. In smaller cities, locate the group through the chamber of commerce. These people are an excellent source of help in setting up your business. They may not know anything about herbs, but the business knowledge they offer will be invaluable. Most of the counselors in SCORE have been in business all of their lives, in your area, and can be a good source of contacts.

Call the chamber of commerce or the SBA in your target city and ask for the SCORE office. Tell them that you would like to set up an appointment with a counselor. They will send you some information and a questionnaire to fill out. In this way they can match you with the right counselor.

Small Business Development Center

Most colleges have this or a similar program to help people start up a small business. These counselors, who have knowledge of city, county, or state economic development programs, will meet with you. Some of these programs even have grants or low-interest loans for business expansion or start-up, particularly if you have or plan to hire employees. The Small Business Development counselors are also knowledgeable about local licenses and start-up requirements.

Scouting for a Location

Many people who start an herb business already have a farm or acreage and are looking for a way to earn money with it. This can be a real advantage if it is located close to a metropolitan area with many places to sell herbs.

If you have yet to find a farm, you may be in a better position than many who start their business where they live. You have the opportunity to make some very important choices. These choices could determine the success of your business.

Investigating an Area for Markets

The most important aspect of your location search is proximity to places to sell your products. Does this location have an attainable market-place nearby? Are there plenty of upscale restaurants and hotels in the city? What is the population of the city and surrounding area? What is the makeup of the consumer base? How far would you have to drive to make deliveries? Is there a weekly farmer's market? What kind of super-markets does the city have? How far are the wholesale distributors? Would you like to live close to this city? Are there people selling fresh-cut herbs in the area? How do they do business? Where are they located?

You can start your investigation into this community by visiting the chamber of commerce. It offers real insight into the workings of the city. The local Extension agent can be a good source of information as well. Ask him or her about other herb growers in the area, possible areas to look for farms, soil types, and climate patterns.

If it looks favorable this far into your investigation, visit some super-markets and restaurants. Are there good opportunities for doing business? If so, proceed with looking for a physical location.

Evaluating Farm Sites

When looking for property, it is best to work with a real estate agent. Ask someone you trust to recommend one who is familiar with local farms and businesses. Agents usually belong to a Multiple Listing Service and have access to property in a wide area around your target city. Working with an agent does not usually cost the buyer any money.

There are many things to consider when looking for a farm. The first is to allow yourself room to expand. Try to buy more acres than you think you will need. Your plan now may be for one greenhouse and two acres of outdoor growing space. After five years in business, though, you may want to expand with four more greenhouses and 40 acres in tillable land. After four years I found that in order to expand my business to keep up with the demand, I had to buy another farm. It is not an easy task to dismantle and move a 30-foot by 60-foot greenhouse along with hundreds of stock plants!

Look at the lay of the land. If it is in a valley, how much room do you have for greenhouses? Will they receive the benefit of direct sun all day? Which way does the wind blow? Must the greenhouses run east-west or north-south? Where would the drainage from a 10-inch rainfall go? Where would runoff from the greenhouses fall?

Are there trees that could topple onto the greenhouse in a storm? Would there be room for a work building adjacent to the greenhouse? Are the living quarters acceptable? Who are the neighbors, and would they object to an agribusiness? Are there any plans for housing subdivisions in the next 15 years?

How far would the greenhouse be from utilities? Most power companies require you to pay for additional power poles, usually $250 apiece. Wire and transformers are costly, too. Running power and water lines underground can be a major expense.

How much earth would have to be moved to make a level pad for a greenhouse? Do you need to install a driveway for access to the greenhouse? Excavating contractors usually charge between $600 and $900 an hour.

How far would you have to drive on gravel roads three or four days a week for deliveries? How easy would it be for the public to find your farm? How far away would possible employees be?

Determining Water and Soil Quality

Water quality is of great importance in the production of fresh-cut herbs. Water-treatment systems are expensive to buy and to operate. Ask some questions about the water supply. If there is a well, how old is it? Is there any history of the well? How deep is it and how far down is the pump? What is the output of the well and pump? Are there any regulations regarding groundwater withdrawal? Is city water available? If so, what would be the cost per 1,000 gallons?

When you find a location that seems right for your plans, have the water and soil checked before making a final decision. Most lenders require testing the water for coliform bacteria and nitrates before making real estate loans.

According to Jack McGill, a water quality specialist from the University of Minnesota Extension Service, it's also wise to test for atrazine, pH, and anions (nitrates, sulfates, and chlorides). Tests are now being developed for the reaction of water to some of the newer chemicals being used in farming, he says.

Soil testing is important as well. A complete assay of the soil on different parts of the farm will give you an idea of how much work must be done to put the fields in good condition for growing herbs. You can tell much by digging a foot down to examine the soil.

How many acres have been tilled, how long ago, and what was planted there? If it was planted with conventional crops, chances are it was sprayed with herbicides. If so, you may not want to risk planting herbs in that area for several years.

It may take a while to find your ideal location. Take your time to be sure that it is right for you; this decision is a vital one for your business.

Business Laws

Each state has different laws and regulations for operating a business. Municipalities and townships differ in their regulations as well. You must check what the laws are regarding permits in your own location. The following are some places that can help you determine what permits and licenses you will need.

Licensing

Many cities, towns, and counties require you to have a business license in order to operate within that municipality even if you sell strictly wholesale. Usually these are inexpensive. Check with the city clerk, zoning administrator, or the chamber of commerce. If you are located within the boundaries of a small town, a call to the mayor may be all that is necessary. In rural areas, call someone from the township board for information about licenses and permits.

Filing for a Tax Number

Most states require that you have a sales tax number or resale number. If you sell strictly wholesale, you may not need a tax number. The retailer charges and pays the sales tax. If you sell fresh-cut herbs, a food product, you also may not need a tax number.

Many of your suppliers will want to have your tax number on file or they must charge you sales tax. In some states there is an alternate form, called a Resale Exemption Certificate, that you can use for your suppliers so they won't have to charge you sales tax. Charging, filing, and paying sales tax, either monthly or quarterly, is a nuisance but you may have to do it if you sell anything, except fresh produce, retail in most states.

Contact your State Department of Revenue to obtain the proper forms for sales tax identification numbers. They will provide you with all the information you need to begin paying sales taxes.

Building Codes and Zoning

If you intend to build a greenhouse on your property, you will need to know what building codes and zoning restrictions are in place. Even in rural farming areas, the township or county may have regulations regarding building or doing business in an agricultural zone. Know what these are before committing yourself to business in the area. In cities and counties, call the planning, zoning, and building codes departments. In rural areas, check with a township board member.

Financing

Many people start their herb business on a small scale. They operate only during the growing season and gradually build their account base and knowledge. They infuse money into the business as needed by borrowing from friends, family, and credit cards, or by using savings.

Others begin with a start-up plan that requires bank financing to purchase land and equipment. Even without a loan, the time may come when it is necessary to secure capital from a bank or some other source. It is important that you have the ability to obtain funding if needed.

Bank Financing

Banks are the first place that we think of when looking for a loan. Large financial institutions have money to lend, but many are not interested in making loans for small business start-ups. You may have better luck securing a loan from a smaller, locally owned bank situated in a farming community near you. These banks usually make operating loans to farmers and have a better understanding of agriculture.

When you find a bank, give it all of your business. Build rapport with the officers of the bank. Try to convey the nuances of your business to the banker before looking for a loan. Guard your credit rating with vigor.

If you are looking for a loan to purchase a farm, be aware that most lenders require you to have a 30 percent to 40 percent down payment. The Farmers Home Administration may have low-interest loans available whereby a beginning farmer finances an additional 30 percent of the purchase price. There may be some requirements that do not suit your situation, but it is worth checking into. Contact the office closest to you for information.

Support from Investors

Loans can sometimes be obtained from a venture capitalist. This is a person who has money to lend to businesses or for the start-up of a business. Venture capitalists usually want a percentage of the business and sometimes a say in how the business is operated. Look for these lenders in the classified section of newspapers in the nearest large city.

Many times farms can be purchased with a contract for deed (or contract for title) between the sellers and buyers. Your real estate agent can help you look for sellers who will offer a contract for deed. A contract for deed is an agreement where the seller sells the land to the buyer but does not give the title to the buyer until all, or a portion, of the purchase price is paid to the seller. This can be beneficial to buyers who do not have the full purchase price, or who can't obtain a loan for it.

This type of sale allows the buyer to use the property, but there may be restrictions on the total dollar amount of improvements that the buyer may make to the property. These restrictions are used to protect the seller if the buyer can no longer pay his installment payments. The seller may be forced to pay unpaid debts for improvements to the property or permanent structures. (In most areas greenhouses are not considered permanent structures because they can be taken apart and moved.)

Other Sources of Financing

There are other areas to explore for financing. Many localities have small business development centers that make low-interest loans for business start-up or expansion. Contact the Small Business Administration and inquire about any loan programs they may have. Check with the resource librarian at a large library. Look for *The Catalog of Federal Domestic Assistance;* this book describes all of the federal programs that help finance small businesses. *The Government Assistance Almanac* helps you sort through the catalog. Look for *The Foundation Center Collection,* which lists grants and monies available to individuals and organizations. It may be a long shot, but if there are any grants available for start-up businesses operated by women or minorities, they would be listed in this book.

maggie's herbs

St. Augustine, Florida

Maggie Ouellette's business started as a backyard hobby and turned into a livelihood 12 years ago. Maggie sells potted herb plants (200 varieties) wholesale to nurseries, and she has a small amount of retail trade at the farm. While half her fresh-cut herbs and edible flowers are sold to wholesale distributors, the rest go directly to restaurants. She does not buy in herbs to repackage when she is temporarily out. Of her 10 accounts, the farthest is 50 miles away. She prefers to make the deliveries herself, even though she has three full-time employees.

The biggest sellers for Maggie are basil, cilantro, rosemary, dill, chives, mint, sage, and edible flowers. The herbs are sold by the approximate weight in 1-ounce, ¼-pound, and 1-pound bags. She also grows and sells hot peppers and some perennial plants.

With a half acre of outdoor production space in zone 9, she is able to grow and harvest most herbs year-round without using her greenhouse. According to Maggie, the soil in her area is mostly sand. She grows her herbs in raised beds that are augmented with compost and peat. Fertilization is done with cow manure, bonemeal, and a 20-20-20 fertilizer.

She devotes 1,200 square feet of greenhouse space to fresh-cut herbs. The greenhouse is heated, when necessary, with gas. The herbs are grown indoors in raised beds with basically the same soil mix as outdoors.

The biggest problem Maggie has is keeping basil in production in the greenhouse from the end of December until the end of February. Even in her warm climate, this herb needs supplemental lighting. During the summer the humidity from afternoon rains can cause disease problems, especially with rosemary, sage, and thyme. To remedy this in the greenhouse, she grows her herbs in raised beds or on benches. The plants are meticulously weeded and mulched with crushed oyster shells. She simply throws away plants that are severely infected. Greenhouse insects are controlled with insecticidal soaps.

With the help of her employees, Maggie has been able to cut down her hours of work to 7 to 11 hours a day, five days a week. This is much appreciated by her two young children and her husband, who is a teacher at a local community college. Maggie loves this profitable business and offers these words: "You are probably not going to get rich from this business, but I bet you will love it!"

Creating a Business Plan

In order to obtain any type of financing, you must have a business plan. Lenders, even private individuals, want to see exactly how you plan to repay this loan. It is also a valuable tool for the entrepreneur to identify strengths and weaknesses of a business. It will enable you to see clearly and evaluate whether any changes need to be made in order to operate profitably, and it can help you avoid the fear of the unknown.

Most people view the writing of a business plan as an unpleasant chore. This process can actually be fun. The planning of a new business is exciting. The inspiration you experience in the planning stage need only be written down in an organized manner.

There are many different methods of writing such a plan. The trick is to supply as much information as needed to secure the funding that you want. Gather as much information and as many numbers and facts as possible. Then seek some help in finding the best system to use in writing your business plan.

Getting Help for Writing Your Plan

The Small Business Administration has a very good workbook available at no cost. You can get it from any SBA office, your SCORE counselor, or Small Business Development Center. The library has many excellent books that can guide you through the writing process.

For personal help with writing your business plan, contact the SCORE office through the local chamber of commerce or Small Business Administration. Ask for a counselor who will help you with the process. The Small Business Development Center can also provide you with a counselor knowledgeable in writing business plans.

You may wish to seek help from an accountant. A large part of your plan should include financial statements and projections. If you are not comfortable doing financial planning, do get some help. A good accountant can be an invaluable ally, now and in the years to come.

Naming Your Business

Every business must have a trade name. It is necessary for doing business in every state and for the public to identify who you are. Take care in choosing a name because it greatly influences the public's perception of your product.

Whatever name you decide on, be sure that it conveys a positive image. It must be one that you are comfortable with — you will have it a long time.

You can name your farm (a trade name), name your product (the trade-mark), and name your service (a service mark; that is, herb garden design service). For most purposes, just naming your business will suffice. Your business cards, labels, and price lists will describe your products or services.

Use imagination when naming your business. It conveys to the public information that you want them to remember. You can choose a geo-graphic name — a road, river, town, location, or place with special mean-ing to you. Many people use their own name, provided it is easy to pronounce and spell. However, a personal name can present a problem if you ever sell your business. Many buyers prefer to buy the name also because of name recognition.

If you use humor or a pun in your name, it will probably stay in a customer's mind. My own business name, Herb's Herbs, was chosen using a combination of ideas. Herb was the name of my grandfather, who was a Master Gardener. While making deliveries, people would pull up next to me, point at the signs on my truck, and laugh and smile. I knew they would not forget where to buy herbs!

Once you have settled on a business name, you may have to register it in the state where you will be operating. Each state has different laws regarding registration of business names. Many states require you to file a Certificate of Assumed Name. This is to register your business name in the state you are doing business, and to ensure that someone else is not already using that name (within your state). They may require you to run an ad in the legal column of your local newspaper for several weeks. The newspapers have the necessary forms for this ad and will ask some questions about you and your business name. Call the Secretary of State's office in your state to find out if this is necessary and what the procedures are.

Commercial Grower
Fresh Cut and Potted Herbs
Wholesale and Retail

Herb's Herbs ...
... and such

Route 11, Box 40   Sandie Shores
Zumbro Falls, MN 55991   507-555-1721

Insurance

For most of us, insurance is paying for something we hope we will never use. Most people must insure their living quarters and cars, and have some liability on the grounds. What about your business? You need only talk to a person who has lost his business and was uninsured or even under-insured to learn why we need insurance.

Think about the items you have that are important to you. What would happen if you lost your client lists, reference books, automobile, invoice numbers, computer? What would happen if you lost your green-house, stock plants, and income? As your business is growing, it may be difficult to think of spending money on insurance. I urge you to be pre-pared. Keep backup copies of computer files, paperwork, and any other important items away from your primary workplace.

Insure your greenhouses, with or without plants in them, and all of the permanent equipment installed. Make sure you have enough liability insurance on your delivery vehicles. Insure your stock plants. It takes two or three years for a rosemary plant to grow large enough to provide a good yield. How much would it cost to buy hundreds of mature rose-mary plants?

If you grow acres of vegetables and herbs, look into crop insurance. Your crops are your livelihood and your future. You can get crop insur-ance through a private insurer or through the USDA's Federal Crop Insurance Corporation, or from both.

If your main income is from your fresh-cut-herb business, check into business interruption (or loss of income) insurance. In the event of a loss, your bills and expenses continue. This type of insurance gives you income while you rebuild your business and also provides for lost profit.

Local insurance agencies may not be able to help you with your needs — the costs may just be too high. Look to companies that special-ize in nursery and crop insurance. (See Resources.)

Suppliers

It's a good idea to locate suppliers for all your business needs before you're ready to place orders. You will need to know how much your sup-plies will cost before you order them. There are many companies willing to compete for your business. You can benefit from this competition by getting better prices, service, and quality.

When looking for suppliers, interview the salespeople. They should be knowledgeable about their products and willing to answer any of your questions. Remember, they may have less time to spend with you during their busy season, which is late winter through early spring.

The best supplier can be an invaluable source of information. Finding a supplier with a toll-free number can be a money saver. Most large metropolitan areas have several greenhouse-supply companies. These companies usually carry a complete line of supplies including pots, peat/soil mixes, chemicals, greenhouse equipment, greenhouses, and even seeds. Call or write for catalogs from as many suppliers as you can. Compare prices and delivery costs.

Equipment Suppliers

When shopping for a greenhouse manufacturer, write or call for information from many of the largest suppliers. This will help you learn about the features and benefits of different types of greenhouses. When you have settled on a size and type, talk to your closest greenhouse-supply company. Some of these companies also manufacture greenhouses that may suit your needs nicely. The closer location will save you quite a bit in shipping costs, too.

Look locally for a vendor for your packaging supplies. Try to find a company that carries all the supplies that you will need. Poly bags, labels, and trays can be some of your biggest ongoing expenses, and you can save money in shipping costs by picking up your order yourself.

Form a buying partnership with other growers. Nurseries and bedding-plant growers buy supplies in large quantities. By ordering together, you may be able to save money in volume discounts and shipping costs.

You can sometimes save money on heaters and other equipment by shopping locally. The local liquid propane (LP) or natural-gas supplier may sell these items. Usually they will install them at a reduced price if you buy from them.

Major manufacturers and suppliers often advertise in national trade magazines. You can send for their catalogs and information by using the reader service cards found in these publications. These same trade magazines publish yearly buyer resource guides that list most major suppliers. (See Resources for a list of suppliers and magazines.)

Seed Suppliers

Many seed companies now carry a line of herb seeds along with flowers and vegetables. Not all the seed companies will have the varieties, size of packets, or quality of seeds that you want. Compare prices and quantities. Many seed outlets offer discount pricing for large-volume orders or for cash payment. Include this in your comparison.

Get seed catalogs from as many seed houses as you can to keep informed of the latest varieties being introduced. Try new varieties — perhaps they will perform better than your current variety. Some are better than others for fresh-cut sales. See Part IV: Successfully Growing More Than 20 Herbs and Flowers for the varieties that are best for fresh-cut use.

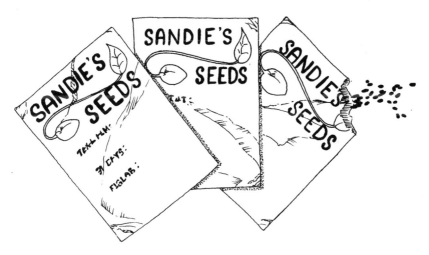

Try out a variety of seed suppliers to see which ones give you the best results.

finding
potential markets

There should be many places to sell your fresh-cut herbs. You can specialize in one type of account or diversify. Much of what you choose has to do with your market area.

If you are living near a large metropolitan area, you could sell only to restaurants and make a fine living. If, however, you are near a medium-size city, you may have to sell to all types of accounts to be profitable. Your own situation and goals will determine how you sell your herbs.

Selling to Restaurants

If your goal is to sell high-quality, fresh-cut herbs at a premium price, restaurants and hotels should be the best outlet for your products. Selling directly to restaurants provides a higher income with less packaging cost. It can be challenging, educational, and fun, too.

The herbs themselves are the best sales tool for showing what you can offer potential markets.

Providing Specialized Service

You must be able to provide specialized service to these accounts. Chefs are very busy people and sometimes have a reputation for being temperamental. Most are very nice and understanding. These accounts require more patience, care, and service than do some other types of markets. If you can give them this special treatment, direct selling can be profitable for you.

Sometimes restaurant buyers or chefs are not as organized as they should be, or they may get last-minute banquet or party bookings. This can result in "emergency" orders for you with an immediate delivery to be made. Would you be willing to drop everything to pick and deliver these orders? Yes, you probably would with a $20 service charge added in for a rush order.

One downside of selling to restaurants is that many chefs stay with a restaurant for only three to five years. If you are selling large quantities of a certain herb to a restaurant and the chef moves on, you may be left with no market for that herb. Often, the new chef will change the menu.

If a change of chefs occurs in one of your accounts, try to meet with the new chef as soon as possible to determine what herbs she will be using. Explain that there is a lead time for growing and you want to be sure to be able to supply her with everything she needs. With luck, she may use even more than the last chef!

Managing Seasonal Changes

Restaurant sales have peaks and valleys. The busiest time for most restaurants is usually May through October, with the peak in August. The winter months are slowest, with February lagging behind. This, of course, depends on your local market. If your area is a big convention center or a large city, your business may not be affected by dips in business.

If you intend to supply your accounts year-round, these dips may coincide with your most difficult time to grow. If you grow in the greenhouse during the winter months, your expenses will be high. Just when you need the business the most, your sales may be down. You will most likely make up for this during the summer. Ideally, you can lay away enough money over the summer months to cover your heating and electric bills in winter.

You could choose to sell to restaurants only during the growing season. This way, however, you run the risk of not being able to secure these accounts in the spring. If you are supplying specialized items such as fresh herb garnish and edible flowers, your accounts may not be able to find these items elsewhere. More than likely the chef will not find the kind of freshly picked quality that you have been providing. This makes for a very unhappy chef, and he may decide to buy from the "other guys" permanently.

Approaching Restaurants

If you feel that direct-to-restaurant sales are best for you, the following guidelines should help you build accounts.

1. Determine how far you are willing to drive to deliver your product twice or three times a week. Locate as many restaurants and hotels within a single delivery route. If possible, obtain a menu or talk to people who have eaten at these places to find out if they use fresh herbs. Many times you will not know if they use herbs by reading the menu. The higher prices charged in upscale restaurants usually indicate that they do. These places frequently can afford to pay more for high-quality and individualized service.

2. Call the restaurant. Ask who the executive chef is and when the best time to reach him or her is. In all but the very busiest of restaurants and hotels, the chef makes the buying decisions about fresh herbs.

3. Call the chef. There are two choices in this step, depending on how comfortable you are with making "cold calls." *Note:* Most chefs like to be addressed as "Chef," such as "Chef Smith."

You may simply call the chef on a day when he is in. Always call between 9:00 and 10:30 A.M. or from 2:00 to 3:30 P.M. Executive chefs in large restaurants are usually not cooking on "the line" during mealtimes, but they do expect sales people to be sensitive to their schedules.

When you talk to the chef, do so with confidence. Ask him if he is using fresh-cut herbs, if he is happy with the herbs he is getting, what he would change if he could, and what his needs are. Tell him that you are a commercial grower, or if you are in the start-up phase, tell him that you are building your business around the restaurant trade.

If he responds by saying that he does not use fresh-cut herbs, ask him whether he would consider using them if he could get a better product and specialized service. If he says no, don't give up. Politely thank him for his time and call him back in a few months.

You could also ask him if you could send him your card and product list in case he would like to see what you have to offer. Most chefs who use herbs would be interested in using herbs that are truly fresh, and are most willing to discuss further what you have to offer. Ask what herbs and quantities he is using. Try to set a time for a meeting so you can give him some samples and discuss your products. If you don't have herbs to show, ask for a meeting to discuss how you could best serve his needs. Most chefs will be happy to make an appointment with you. Be sure to let *him* set this appointment: He is likely to be busier than you are at this time.

If you would prefer to make your first contact with the chef by mail, do so. But remember that chefs in busy upscale restaurants are often besieged with solicitations, and your letter may never be read.

An introduction should be written on your own letterhead. The more businesslike you are, the better. Address the chef as "Chef Jane Doe." Tell her that you are writing to "introduce her to a new service available to her." Explain who you are, what you grow (or will be growing), and that you can supply her with herbs that are fresher, less expensive, and cut to order and with personalized service. Keep this letter short and to the point. Explain that you will be calling her in the near future to set up an appointment so that you may meet for a few minutes to discuss your service.

In a week, call the chef between the proper times. Build rapport by first asking if she received your letter. Then ask her the necessary questions.

If she hesitates to set an appointment, keep trying. Tell her you will send her a product/price list. Follow up with a call after she has had a chance to look it over. Remind her again that you can provide personalized product and service. Offer to drop off some free samples for her experimentation or to supply her in an "emergency."

When you make the appointment, be sure to find out the directions to her office. Repeating the date and time will help it stick in her mind. Thank the chef and tell her you look forward to meeting her. If the appointment date is a week or more away, it's a good idea to send a thank-you note and a reminder of the date, especially if you *know* that you need this account. This may seem frivolous, but you do want to build a good rapport with this person.

Don't give up. Follow up on these calls. If someone won't meet with you at first, keep trying. After some time in business with a proven track record with other restaurants, approach him again. Suggest that a chef talk with a colleague who buys your herbs. Sooner or later, you will have him as an account. Persistence and patience pay!

4. Arrive early for your appointment. Allow plenty of time to park and find your way around. Bring a prepared product list, business cards, and a notebook to write in.

5. Dress like you mean business — casual, clean, and well groomed. Most chefs do not expect growers to be dressed in business suits, but a blazer presents a good image. Remember that this first impression is the only one you will give, and that it will last a long time.

Look the chef in the eye, smile, and offer a firm handshake. Introduce yourself using your first and last name and business name also. Remember that you are a confident salesperson as well as a grower. You have to sell yourself before you can sell your product. Be respectful of his position and appreciative of his time.

Thank him for taking the time to meet with you, and try to make this visit as short as possible. Give him a brief explanation of your qualifications as a grower and businessperson. He needs to know that you have the ability to follow through with what you offer.

6. Explain what you are doing or plan to do. Restaurants will be a big part of your business and you want to give the best service possible. Stress that you want to work together with him to supply all of his fresh herbs.

Tell him the benefits he would reap from doing business with you: Herbs will be delivered within 24 hours from picking and absolutely fresh; you offer diversity that can't be found elsewhere; you guarantee quality, competitive prices, bigger bunches (if you sell in bunches rather than by weight); you offer variable-size packaging, standing-order availability, and growing to order; and you provide the very best service.

7. If you brought along some samples, now is the time to show him. Bring at least one bunch each of the most popular herbs, or the ones that you know he uses. Make sure they are beautifully fresh. Keep them cool (except the basil), even if it means bringing a little cooler with you.

If you plan to sell in bunches, rather than by weight, make them the same size that you will be selling. Huge bunches at below the competition's price may breed confusion. He may wonder if you know what you're doing and what his orders would be like. You are here to build trust, not to overwhelm. Leave these herbs with him when you go. If you have no samples, let the chef know when you expect to have some that you can bring.

8. Ask questions. "What herbs do you use?" "What quantities?" "Whom do you get herbs from now? Are you happy with the quality?" "What would you like to use that you can't get now?" "How can I best serve you?" "Would you be interested in any specialized products such as fresh herb garnish or edible flowers?" Write down his responses. He will see that you are organized. You may think that you'll remember everything he said, but after meeting with five or six chefs, you may not.

If you are already growing and feel confident that you could supply him with what he needs, ask for his business — it's the only way to get it. If you are only on a fact-finding mission, ask him whether he would buy from you when you are set up and ready to supply him.

Tell the chef that you will keep in touch about your progress and let him know when you are ready. Give him a date when you anticipate being ready for business. Thank him for his time and help, offer a firm handshake, and leave.

Most chefs will be delighted with the prospect of getting really *fresh* herbs. They may show you what they have been getting, and some may even discuss prices they pay. Creative chefs usually are quite interested in trying new and different herbs. There is often a competitiveness among restaurants that can be to your advantage.

9. Send a short thank-you note following your meeting with a chef. This definitely helps build rapport and is important in cementing this very important relationship.

Selling to Supermarkets

Selling to supermarkets is sometimes known as direct wholesaling. You are, in effect, selling directly to the consumer without having to keep store hours. The price that you get for your product is less than if you sold directly to the consumer; prices generally fall in line closer to wholesale market prices. Selling to supermarkets requires that you assume some of the functions of a wholesale produce distributor.

You can expect last-minute and special orders in supermarket sales. You must keep up on trends and your local consumer base. In many cases it will be your job to educate the produce managers about herbs they should stock. Supermarket accounts are definitely more work than restaurants, but they provide more consistent income than any other type of account.

Carrying a Variety of Herbs

For large supermarkets, you must carry a large assortment of herbs. In busy markets you may have to supply as many as 30 different kinds. Because supermarkets usually prefer to buy just when they need produce, you must have a large inventory. They don't want to store any more than is necessary. This will increase your costs in labor, packaging, and delivery.

Many of these stores will sell a large volume of herbs each week. As time goes on, the customers will become aware of the fine quality and your orders will steadily increase. Most big stores will want delivery twice weekly. Many will set up standing orders rather than take the time to order anew each week.

Packaging costs are quite high with this type of sales. In small stores you could probably get by with packaging the herbs in plastic bags with a hand-written label. But upscale supermarkets and chain stores doing volume business will want more professional packaging. The herbs should be in a rigid tray so they can be stacked or displayed standing up in rows. Fresh-cut herbs in plastic bags are often bruised and crushed after one or two customers rummage through them looking for their favorites.

A professionally printed label is a must for bigger stores. There are ways to save money with your packaging and still have a good-looking display. (See chapter 13, Harvesting, Handling & Packaging.) Despite the added costs, selling to supermarkets can still be profitable. Just make sure that you are charging enough per pack to reflect your increased expenses.

Handling Market Ups and Downs

Supermarket sales also have ups and downs. The dips in sales do not seem to be as radical as they are in restaurant sales. Consumers who use fresh herbs are accustomed to cooking with them, no matter the time of year.

restaurant and supermarket approach styles

The difference in approaching a supermarket rather than a restaurant is in the packaging. When you meet with the market's produce manager, show him a package similar to what you will supply him with. This means that you must do a little more preparation prior to your initial contact. (See chapter 13 before deciding on what type of packaging you will do.)

You'd think that sales would dip in summer, when many customers grow their own herbs or buy them at farmer's markets. My experience, though, has been that supermarket sales increase between June and September! Sales may then dip after the first of the year for a month or so.

Most large supermarkets will expect you to supply them year-round, just as it would be with a wholesale distributor. Unless you have an abundance of greenhouse space, this can be a problem. Also, a disease problem or a crop failure could affect one or more herbs. Basil is particularly hard to grow during the winter, and it is the herb most in demand.

If your accounts are loyal, they will order the herbs you are out of from a produce distributor. This, however, does not make them happy, as it creates more work for them. An alternative would be for you to buy fresh-cut herbs and repackage them for your accounts. Many large commercial growers do that in the winter months. (Procedures for buying in herbs from suppliers are discussed in chapter 4, Doing Business.)

Approaching Supermarkets

The best supermarkets to start with are independent stores that are locally owned. Here the produce managers make the buying decisions. Some of the large chain stores do not allow their produce managers to buy directly from growers.

Many of the chains *will* allow the independent purchase of something as perishable as fresh herbs, but it will take a fair amount of work on your part to secure these accounts. Having a track record makes this job a little easier. It is much better to start with a smaller independent store until you become adept at handling this type of account.

Begin your search for accounts by visiting the produce sections of as many supermarkets as you can. Look closely for fresh herbs; sometimes they are hidden in out-of-the-way places. Usually they can be found in the sections where specialty items are located. When you find the herbs, write down the name, address, and phone numbers that are on the label. Note the quantity, quality, and type of packaging.

If the supplier is located close to your area, don't assume that you won't get this account. Many times the supplier does not deliver directly but works through a wholesale distributor. When this is the case, the herbs may have been stored in a warehouse anywhere from two to seven days before delivery, and this can decrease their shelf life. My largest supermarket account was buying this way. They were delighted to buy truly fresh herbs from me.

Ask someone who works in the produce department if they usually carry fresh herbs, who the produce manager is, if he does the produce buying, and when the best time is to reach him. If you don't ask these questions while in the store, ask them in a telephone call. Keep careful notes for each store you visit or call.

Preparing Packaging Samples

Before meeting with the produce manager, make two or three sample herb packages. Call or visit the nearest packaging supplier; it can be found in the telephone directory under food packaging, packaging materials, or paper suppliers. Ask for some samples along with a price list. Look for clear plastic packages that are 3 inches to 4 inches wide by 6 inches or 7 inches tall by 1 inch deep. Ask for some with locking lids (called clamshells), some with hinged lids, and some without lids. The latter is used with shrink wrap or inside a plastic bag. This type requires more work but is usually less expensive.

An alternative to plastic trays is a coated white cardboard bakery tray. A tray made by Veltone (#642) is 3.5 inches by 7.5 inches by 1 inch. Ask for a food-grade, gusseted bag sized similar to 4 inches by 2 inches by 12 inches. The bakery tray and the uncovered plastic tray both fit nicely into this bag. The bag can then be taped shut or closed with a twist tie. If you have trouble getting free samples, buy packaged herbs and reuse the trays. Bakeries may also have the cardboard trays.

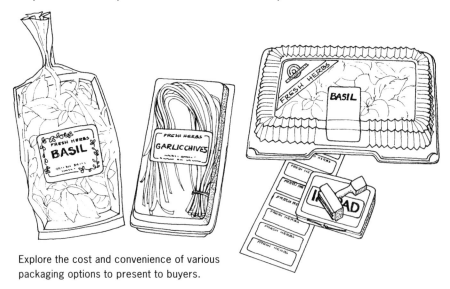

Explore the cost and convenience of various packaging options to present to buyers.

Next you'll need some makeshift labels. Buy a package of blank white labels 2" by 3" or a size that fits your packages. Use two colors of *waterproof* markers and design your label. Use red for the name of the herb to make it stand out.

Take the time to experiment with packaging your herbs. Arrange them in the package with the stem ends down, as some stores will stand the packages on end. Put slightly more in your package than your competitor packs. Take care to arrange the herbs to show them to their best advantage. They should be perfect, with no insect damage, soil, or yellow leaves.

Contacting Targeted Stores

When you feel comfortable with how your packages look, it is time to make contact with your targeted stores. Produce managers are accustomed to having growers approach them directly, especially during the summer, without writing a letter in advance. If you are more comfortable making the first contact by mail, do so. Use the same approach with produce managers as you would with chefs.

If the only stores available to you are large chains, you should definitely approach them. First meet with the produce manager and show him your samples. If he is interested in buying from you, ask him to tell the buyer at the main office about you so you can then call and make the arrangements with the main office buyer for an order.

The procedures to follow in securing these accounts will be explained by the main produce buyer; they vary with the type of chain. Be prepared to spend some time and money in long-distance phone calls. This takes some effort, but it can be done. If it is a large store doing good volume in produce, this extra work will be worth your while.

When you meet with the produce manager, tell him that you pick to order and deliver within 24 hours. This will give him extended shelf life and less waste. Offer him specialty packs that he cannot find elsewhere.

Explain that this is your market also and that you intend to promote the sale of these herbs through demonstrations, teaching, and so on. Offer to do standing orders that will make his job easier. Standing orders are made in advance for specific herb deliveries to be made regularly during the week. This saves the manager time because he doesn't have to take inventory and place an order each time he needs herbs. Standing orders are also a time-saver for you. Let him know that you will work closely with him to ensure that the most salable herbs are in his store.

Do not offer to sell on consignment or to take back unsold packages. Other produce growers and distributors do not do this, and making this offer may call into question your ability to be in business.

If you are ready to supply the store now, ask for the business. If you are not ready yet, give the manager a time frame as to when you expect to be able to supply him. If you have only a few herbs in quantity, offer to supplement his supply with those or to supply him in an emergency. Basil is a good herb to offer because it is the biggest selling herb in most areas at this time.

Most supermarkets refrigerate all of their herbs. Basil has a short shelf life and will turn brown when refrigerated; it should be kept at 45°F or higher. Be sure to tell the produce manager this when you meet with him or deliver his first order.

Selling to Wholesale Distributors

Servicing the needs of individual accounts can be very demanding. If you would prefer to spend more time growing and less time marketing, selling your herbs to distributors may be the approach for your business. You can sell all or most of your herbs to one buyer, thereby saving time and money, but you will receive less than premium prices for your product.

Wholesale distributors sell in volume to restaurants, grocery stores, and other establishments. Many of these places prefer to buy from distributors, as it is "one-stop shopping." The distributors deliver consistency, variety, quality, and dependable service. More often than not, they have a sales representative living and working in the area to ensure consistent service. In medium to large markets distributors are available to deliver daily.

There are many types of wholesale distributors, with just as many ways of buying produce and herbs. Some sell only produce (including herbs), some have all manner of food service items, and some of them are small specialty operations.

In the case of herbs, some distributors contract with growers before the season to buy all they can grow. Others will buy only on consignment. Distributors may buy from many growers during the season or from only one. Some will buy no herbs locally, preferring to import. Many distributors will buy in bulk and repackage the herbs using their own label.

Providing Specialized Service

First and foremost on the distributor's list of requirements for buying from you is quality. You absolutely must consistently deliver the freshest, highest quality possible. The distributor will settle for nothing less — nor should you.

The wholesale distributor must maintain a sizable inventory in order to meet the demand for product. This means that your herbs may be in the warehouse two days to a week before arriving at their final destination. Shelf life is of great importance to the distributor. Fresh herbs must be picked right before delivery to the distributor in order to extend their usefulness.

Postharvest handling is of prime importance to maintain quality. Herbs must be cooled immediately following harvest and remain cool through delivery to the distributor. This greatly increases the shelf life for all herbs except basil, which must not be stored at temperatures lower than 45°F. Distributors may have to be educated about basil. If you plan on selling large volumes of herbs to a distributor, carefully plan your storage and cooling areas to make sure you have enough space.

Most distributors will want you to deliver great volumes of herbs; it is costly for them to spend time trying to buy from too many growers. If it is a time- and money-saver for you to sell all of your product in one place, you then must strive to meet all of the needs of the distributor. This means growing, harvesting, packaging, and delivering a very large volume of herbs at one time.

Distributors want reliability. They need to know weeks in advance if you will be unable to supply them with a particular herb. They cannot afford outages. In addition, the distributor will want consistent packaging. Some will want herbs by weight — usually ½- and 1-pound weights. Certain distributors prefer bunches — usually bags containing 1 dozen or 3 bunches of herbs. Most will want a label with your business name, location, herb name, and quantity. Some herbs, such as curly parsley and cilantro, can be sold in cases containing 30 bunches.

You may also sell packaged herbs for resale to supermarkets. In this case you are using a smaller amount of herbs but the costs for packaging materials and labor are high. You may be required to use cardboard boxes that are standard for the industry. This is another expense to consider when looking at distributors as your primary market.

Delivery to the distributor can be expensive if you live a long way from the warehouse. The distributor will expect you to bear the cost of delivery to it. If you will be delivering a large volume of herbs, consider buying a refrigerated truck. Some distributors will make arrangements with you to meet with their delivery trucks when (and if) they are in the area closest to you. This way you can load directly into their truck and save time and money.

In most cases, the distributor expects you to set prices — he will negotiate with you if he feels your prices are too high. Wholesale prices for herbs do fluctuate based on national availability, but they still seem to stay more stable than other types of produce.

Distributors will require specific packaging of herbs.

The wholesaler may agree to pay more for your herbs because of the freshness and quality, especially if he has been buying from out of state. Wholesalers can pass on the higher prices to their customers, who are usually willing to pay more for the better quality.

Approaching Wholesale Distributors

There are many ways for you to sell to wholesale distributors — a few herbs once in a while, a big portion of your summer crop, or everything you can grow. You must decide how much of your business you want to devote to this market. Then you can begin looking for a compatible wholesaler.

One way to approach distributors is to make contact with the local produce sales representative for the wholesaler. You can find the name and

 setting prices for sale to distributors

You should know what the usual prices are in your area before approaching any wholesale distributors. Subscribe to the *USDA Fruit and Vegetable Market News* for this information. This service lists produce (and herb) prices paid by wholesalers in major cities across the country. Contact the headquarters for a list of reporting cities and subscribe to the service from the city nearest you (see Resources).

number of the sales rep by asking any of the restaurants or supermarkets that you have contact with.

Call the representative and introduce yourself. Ask him how his company buys from growers, the buyer's name and telephone number, and the best time to call. Be sure to build rapport with this person and leave your telephone number with him. Tell him that you could supply him with herbs locally in an emergency situation. Although it might seem that you would be in competition with him for accounts, he is actually a valuable ally. I sold many hundreds of dollars worth of herbs to sales representatives in their quests to meet the demands of last-minute orders from chefs!

Another way to find out which distributor is best for you is to call the companies. Ask for the name of the produce buyer for herbs and what the best time is to call her. Usually Tuesday or Wednesday is a good day on which to make contact.

Most produce buyers are very busy, so when you talk to them initially, make your questions short and to the point. "How do you buy herbs?" "Do you buy locally? How often? In what quantities?" "Would you buy herbs from me?" "Can I bring in some samples?" You should be knowledgeable about your products and honest about how much you could supply.

When a company seems right for you, ask for an appointment to meet the buyer, show some samples, and see her facility. This step is not mandatory but it is educational to see how your herbs would be handled.

When you meet with the produce buyer, there are other important questions to ask. "How much of what herbs would you buy?" "How often would you buy?" "How do you want the herbs packaged?" "Could I load into your trucks?" "What are your deadlines for notification of shortages?" "How and when would I get paid?"

If you decide to sell to distributors, remember that communication is vitally important. Most produce buyers have an understanding of the problems a grower may face. If you expect a shortage, let them know as soon as possible so they can order elsewhere. This simple step can mean the difference between having and not having a market for your herbs.

Selling to Brokers

Operating along the same lines as the wholesale distributor is the broker. These are people who handle the sale of produce without having possession of it — they sell to distributors and chain stores. Many times a broker will be looking for last minute orders that he needs to have shipped quickly to one of his accounts. This means that you must be prepared by setting up an account with a shipping company that has refrigerated trucks or, in the case of air freight, coolers (styrofoam works best) and ice packs. Brokers usually charge a commission of 3 percent to 10 percent of the sale. If you find a broker willing to work with fresh herbs, it may be worth further discussion to determine whether this would be a viable situation for you. Approaching brokers is much the same as with wholesale distributors.

Selling at Farmer's Markets

Selling your herbs at a busy Saturday farmer's market can be a most rewarding business activity. It brings not only immediate money but also increased public awareness of you, your expertise, and your product. Contact with the public can be delightful when you spend most of your time with plants. The business contacts you make, friends you gain, and the fellowship with other growers make all the work worth it.

Setting Up a Booth

You will make more money and have a more interesting booth with the addition of other items to sell. You should definitely focus on fresh-cut herbs, but do bring herb plants and herbal products. You want customers to think of you as "the herb person."

Grow and bring gourmet salad greens, mini-vegetables, fresh herb mixes with recipes, salad herb mixes, and a "pesto special" (6 to 12 bunches, or 1-pound bags, of basil with a recipe). See how quickly you gain a reputation as the "gourmet herb person."

These are but a few ideas you can use to make your booth interesting and colorful. Give out advice freely on cooking with fresh herbs, recipes, and growing herbs. You will gain customers, and have fun and learn at the same time.

Many vegetable growers will sell a few herbs at the market, usually basil and dill heads in season. If you grow in the greenhouse, you will have fresh herbs to sell when the other growers have only seedlings. Most of the growers don't know how to keep herbs fresh. Keep a backup supply of herbs in a cooler in your vehicle and sell from your display bas-

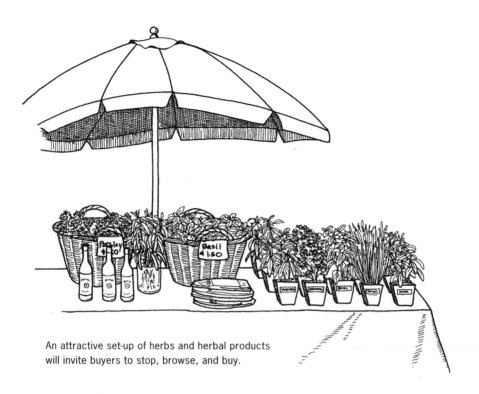

An attractive set-up of herbs and herbal products will invite buyers to stop, browse, and buy.

kets or containers. Put each bunch you sell in a thin plastic produce bag like the ones in the produce section at grocery stores. You can buy them in big rolls for a very reasonable price.

Display your herbs in attractive baskets or containers with a plastic tray in the bottom. Put 2 inches of water in the bottom and cover the top lightly with clear plastic. This helps the herbs to retain moisture. They will not wilt, and they will still be visible to customers. Keep them out of direct sunlight by using a colorful umbrella or canopy over your table.

Make an easy-to-read sign in large letters with a list of the herbs and their prices. Use a separate large basket for basil with a large sign on the front. You will want to bring a wide variety of fresh-cut herbs. When the customers discover your quality, variety, and quantity, you will find your business increasing.

The prices you charge per bunch or bag should reflect the quality and quantity. For example, if the supermarket charges $1.49 per package, you can get $2.00 per bunch because yours are larger and fresher. Rarely will a customer complain about the higher price. If you are selling herbs out of season, such as greenhouse basil in May, don't hesitate to charge more for your product. After all, it costs you more to grow it out of season.

Preparing for Farmer's Market Sale

If selling at a farmer's market interests you, do some research first. The chamber of commerce will tell you where and when the local markets are. Go to the market during the busiest time, usually between 8:30 and 11:30 on Saturday morning.

Stroll by each booth, taking note of who sells herbs. When you find a person already selling herbs, talk with her. Do not assume that she is there forever. You may wish to find another market or change your plans to fit this market. Go back to the market early when the growers are arriving and setting up. Talk to them. Ask their honest opinions of this way of selling. Their responses may surprise you!

Be forewarned that selling at a farmer's market is a great deal of work. They usually start early in the morning. Therefore, you must arrive very early and have your booth set up before the customers start filtering in. Some customers will always come *very* early anyway, just to beat the crowd to the best produce.

You will usually have to pick your herbs, plants, and other items the day before the market. Sometimes it takes all day to get ready (this after making Friday deliveries to your accounts) and to load up your vehicle. If you grow other produce to sell at the market, this also must be picked, washed, and packed the day before. I remember digging up and washing leeks and carrots by lantern light at 1:00 A.M. the night before the market!

Most farmer's markets require that you rent a stall for the entire season. This is good, as it ensures that you will be in the same spot each market day; the customers need to know where they can find you. Many farmer's markets have a waiting list of growers who want to rent a stall. Some markets will leave a few stalls unrented for the season for use by growers on a first-come, first-served basis.

Several farmer's markets have strict rules about what growers can sell. For instance, some allow to be sold only produce that is grown by the seller. Also be aware that most states have regulations governing the sale of products that are processed for human consumption, such as dried herbs, jams, and meats. Eggs may need to be candled and graded. The manager of the market will have the rules and regulations.

Approaching Farmer's Markets

Find out who the market manager is and talk to him. Request to be notified of the annual meeting. This usually happens during the winter. At this meeting officers are elected, business is taken care of, old friendships are renewed, and stalls are rented out. You usually must attend this meeting if you wish to rent a stall.

Farmer's markets are enjoying renewed popularity in this country. If there isn't a farmer's market in your area, think about starting one (although this is a complicated and time-consuming venture). For more information on starting a market, consult Resources.

Selling Retail at Your Farm

If you're selling primarily to restaurants, supermarkets, or distributors, you probably won't have the time to run a retail operation from your location. Retail requires that you make yourself available at any time to wait on customers, or you could hire an employee to serve customers. Not only is this costly, but also the people really want to talk to you, the grower. Your greenhouses must be set up to accommodate customers, rest rooms must be available, and display gardens are a necessity.

Allowing the public to roam freely through your growing houses is very unhealthful; they can bring disease and insects into your controlled environments. I once had a woman pull a small plant from her purse, toss it on the table in the midst of my fresh rosemary cuttings, and ask what was wrong with it. It was covered with powdery mildew!

If you really want to have a retail operation, consider offering other things for sale, such as herb plants and related herb products — herb wreaths and swags, recipes, spice charts, and decorative dried herb bunches. It is highly unlikely that you would make a profit selling only fresh-cut herbs. Your retail operation must be diverse and interesting. The expenditures involved in promotion and entertainment — such as herb festivals, classes, or formal herb gardens — for the customers are high.

If you value contact with the public, you can offer herb festivals or conduct workshops several times a year, during which you can maintain tight control over the activities of the public. Selling at a farmer's market also gives you the needed public contact.

riddle mill organic farm

Lake Wylie, South Carolina

This 36-acre certified organic herb farm has been since 1986 supplying restaurants and wholesalers with fresh-cut herbs and edible flowers. Tara Orlando grows more than 400 varieties of herbs for both culinary and medicinal uses. Riddle Mill Organic Farm has a retail store and display gardens, and conducts workshops. They provide guided tours for school groups, garden clubs, and bus tours, and are open to the public February through December. Mother's Day is celebrated with a festival and sale that attracts people from throughout the Southeast.

Located between zones 7 and 8, Riddle Mill has an outdoor harvest season from March through December. The soil is basically clay, augmented with composted manure and other organic matter. The herbs are grown in raised beds both outdoors and in the greenhouses. Some 5,700 square feet of greenhouse space is devoted to fresh-cut herbs and edible flowers. The houses are heated with liquid propane, and the soil mix in the greenhouses is the same as in the fields.

The best-selling herbs for Riddle Mill are basil and mint. All fresh-cut herbs are sold by weight. Tara says that 30 percent of the fresh-cut-herb sales are to wholesalers; 70 percent is to their 48 local restaurants and supermarket accounts, although they also ship out of state as far north as Nova Scotia and out west to the Ozarks. The local deliveries are made by Tara and one of nine part-time employees. In addition, potted herbs are sold wholesale and retail. They also market their products at the Charlotte, North Carolina, farmer's market.

According to Tara, the biggest challenge faced by Riddle Mill Organic Farm is that they need more hours in a day; they work an average of 10 to 14 hours daily year-round! The farm is closed to the public in January, but they continue servicing their restaurant and wholesale accounts. Her advice to aspiring growers is oh, so true: "If you desire it with conviction, go for it!"

Selling to Caterers

Selling herbs to caterers is very much like selling to restaurants. The main difference is that a caterer's business is often more sporadic. A caterer may have very large amounts of business in the summer months for weddings and graduation parties, for example. Unless they are located in a large convention area, though, winters can be quite slow, excluding the holidays.

There is a new breed of caterer or private chef on the market today. These are people working with individual families. In today's world of two-income households, more emphasis is placed on convenience and leisure activities. These new private chefs will go into a home and cook a week's worth of meals at a time. More often than not, they cook with fresh herbs. These chefs are working with select clientele and often with several families at a time.

You can find private chefs by looking in the telephone directory, newspaper classifieds listed under services, or by word of mouth. Some upscale supermarkets have their own chefs who do catering or cook for private clients. Do contact all the caterers you can find. Approach the larger companies first; they are more likely to cater upscale events. Also, many of the smaller companies may not use fresh herbs.

Other Potential Markets

Your main income will probably come from one or all of the big three: restaurants, supermarkets, and wholesale distributors. You could increase your profitability by adding some other outlets. Be careful, however, when expanding into other areas. When you spread yourself too thin, you can lose focus and become inefficient. Always channel your energy into your primary moneymaking activity.

Cooperatives

A cooperative, very simply, is a group of people who have joined together for the purpose of doing business. Herb cooperatives do exist and you should be able to find out if one is in your area as you do market research.

If there is not one locally and you feel this is an area you would like to explore, search the library or the Internet for information about starting one. This is a complex venture, but with the proper information, it can be done.

Health or Natural Food Stores

These stores usually sell to theˆ people interested in buying organically grown food and produce. They operate very much as do the larger supermarkets, but perhaps with less volume and packaging. Some of these stores will buy from you only if you are a certified organic grower. Many smaller stores will not demand this certification. If you grow organically, even only during the summer months outdoors, don't hesitate to contact these stores.

Food Processors

The large food processors usually use dried herbs because they hold their flavor during the long cooking process. Hormel Foods in Austin, Minnesota, buys dried herbs by the semi-truckload!

Many of the larger food processors have research and development departments staffed by chefs. At times they may wish to use fresh herbs, either in the development of new recipes or in catering meals for company executives. The publicity departments might use fresh herbs, too, for photographs. If you have any large food processors near you, call or write to let them know that you are there to serve their needs.

Look for small specialty-food processors. There are many locally based operations making everything from salsa and pizza to dressings, dips, and herb vinegars. These small processors can be a source of steady income for you.

Chefs' Associations

In most major cities or state regional areas, there is an organization of chefs. They usually meet monthly and often conduct tours or give educational seminars. Ask restaurant chefs for referrals to the local chef's society or association. Many of these associations allow people in the food service industry to join as associate members for a small fee. If they do not allow associate members to join, you could still offer to conduct a tour of your growing operation.

Joining this group gives you the opportunity to network with chefs and increase your business. You can also learn a great deal while having fun by attending the seminars and tours. If you offer a tour of your operation, show chefs how your herbs are grown and harvested. It would help them to better understand some problems you may encounter in supplying all of their needs. Your seminar will be well attended — most chefs love fresh herbs! I once invited the membership of the local chef's association out to my farm for a daylong seminar and tour. It was a learning experience for all of us. It not only turned into a very fun party but also brought a large increase in business for me!

Advertising Agencies

These agencies in larger metropolitan areas may need a small amount of fresh herbs to use in photographs or television commercials. While the volume you sell to them may be small, it would be great to see the product of your hard work show up in the media.

managing business details

Getting paid is the reason you are in business. You may just love growing herbs, but you can't give them away for long. You don't make money growing herbs — you make money selling them.

Invoices and Packing Slips

It would be wonderful to get paid when you deliver the herbs. This may work with some small businesses, but it rarely happens with most accounts. Most will issue you a check once a month or every two weeks.

The key to getting paid is your invoices. Your packing slip can serve as an invoice if you specify this with each account. It certainly simplifies bookkeeping. This means that you must give the packing slip/invoice to a responsible person with each delivery to make sure that it gets to the bookkeeper or person who issues the checks. This is the system that I used, and only twice in 10 years did an invoice get lost.

You may have to send your invoices directly to the bookkeeper. Find out if she prefers you to send all the invoices once a month or every time you make a delivery. The ideal situation is to send them the same day you make the delivery.

Your invoices can be computer generated, typed individually, even handwritten. You can buy invoice forms at office supply stores or make your own. If you buy forms, make sure they have enough lines to list all the different herbs you sell; otherwise, you may have to use two, which increases your costs. Purchased forms should be the self-copying type with two or three copies, depending on whether you must have separate packing slips and invoices. The invoice and packing slip will contain the same information, so you need only type or write it once. Use your business self-inking stamp for your letterhead, and be sure to stamp all copies.

The invoice/packing slip should contain the following information:
- Your business name, address, phone, and fax numbers
- The date and/or the date delivered
- Billing name and address
- Name of the person who placed the order, if applicable
- Purchase order number, if necessary
- Your invoice number (see Bookkeeping on page 51)
- Quantities and all herbs delivered listed individually
- Unit price and total price of each herb
- Total amount of the sale
- Terms of payment, if not previously agreed upon

Terms of Payment

Discuss the terms of payment before the first order. Cash flow is important to a small business. If you have to wait for long-overdue payments, you may not be able to pay your own bills. This can tarnish your credit rating as well as propel your business into a downward spiral. In my experience late payments were very rare, but you may encounter them.

Ask the person in charge of ordering *and* the bookkeeper when you will be paid. You can, if you choose, tell them that you will add interest if the invoice is not paid within 30 days. This is usually somewhere between 10 percent and 20 percent of the total invoice amount. Some vendors offer discounts if the invoice is paid within a certain period, perhaps 2 percent off the invoice amount on accounts paid within 10 days. These discounts are usually offered by very large companies; it is not necessary for you to offer one.

If you are confronted with unpaid invoices after 30 days, take immediate action. Send a statement by fax if possible to save mail time. This may alert the bookkeeping department of a lost invoice. Follow up with a polite phone call to troubleshoot rather than to express displeasure.

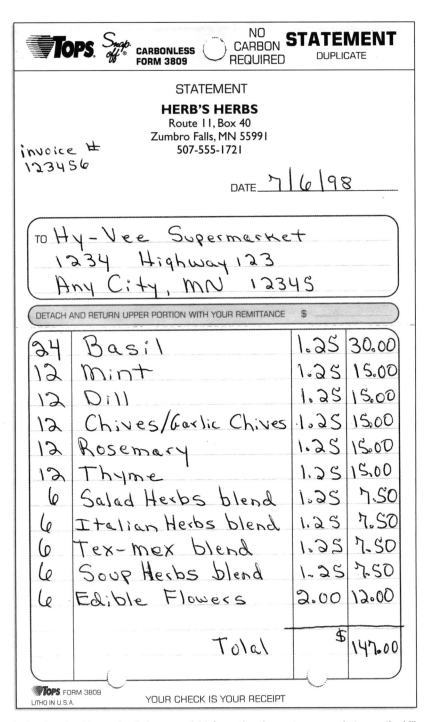

Tops. *Snap off* ® CARBONLESS FORM 3809 ○ NO CARBON REQUIRED **STATEMENT** DUPLICATE

STATEMENT

HERB'S HERBS
Route 11, Box 40
Zumbro Falls, MN 55991
507-555-1721

invoice # 123456

DATE 7/6/98

TO Hy-Vee Supermarket
1234 Highway 123
Any City, MN 12345

DETACH AND RETURN UPPER PORTION WITH YOUR REMITTANCE $

24	Basil	1.25	30.00
12	Mint	1.25	15.00
12	Dill	1.25	15.00
12	Chives/Garlic Chives	1.25	15.00
12	Rosemary	1.25	15.00
12	Thyme	1.25	15.00
6	Salad Herbs blend	1.25	7.50
6	Italian Herbs blend	1.25	7.50
6	Tex-mex blend	1.25	7.50
6	Soup Herbs blend	1.25	7.50
6	Edible Flowers	2.00	12.00
	Total	$	147.00

Tops FORM 3809
LITHO IN U.S.A.   YOUR CHECK IS YOUR RECEIPT

An invoice should contain all the essential information the customer needs to pay the bill.

Bookkeeping

Bookkeeping is a love-it-or-hate-it situation for the small grower. It is mandatory, however, for a successful business. It doesn't have to be a dreaded chore if you organize your system and take a few minutes each day to balance the books.

Bookkeeping by Hand

Your bookkeeping can be done by hand with the aid of a good system. There are several types of ledgers and charts available at office supply stores. Bertha Reppert, in her book *Growing Your Herb Business*, recommends a system called Ideal; it tracks all income and expenses in one ledger.

For those of us who don't have the time to do this on a daily basis, a filing cabinet will also work. It should have separate files for receipts for all expenses. I can attest to the fact that this is *not* the best way to keep books. It makes for a long and tedious session at tax time.

Computerized Bookkeeping

The use of a computer can organize your basic business functions by keeping track of expenses, income, payroll, inventory, and other services. Please note that you still must take the time to enter the data into the computer, just as you would take time keeping books by hand. One of the biggest advantages of computerized bookkeeping is the efficiency in assembling figures at tax time. There are several good software programs available; Quicken and QuickBooks are two of the most popular and easy to use.

A computer program can also do your taxes, although most small businesses hire a tax accountant for this purpose. Hiring a good business accountant is one of the best things you can do, even as a very small grower. Accountants give you guidance as to what is deductible, which often changes yearly, and on many other business details.

 business advice

My accountant, Janice Borgstrom-Durst, Borgstrom Accounting in Kasson, Minnesota, offers this bit of advice for those just starting out: "Get your business off the ground and worry about forming a corporation later, after it has proved to be successful."

Invoice Numbering System

There are some bookkeeping details that must be done on a regular basis. One of the most important is keeping track of invoices. Almost all of your accounts will require that you have a number on your invoice. This is how they keep track of payments, and you should too.

An invoice numbering system is quite simple and can be done by hand or by computer — the information is the same for both systems. The simplest method is to keep a notebook with a section for each of your accounts. Dividers with pockets are useful for holding the unpaid invoices. Each section should contain the account's name, address, phone number, and contact person, plus the names of other people you deal with at that place of business.

Assign that account a number prefix. For instance, "A" restaurant could be 10000, "B" restaurant 20000, and so on. Each time you fill an order for "A" restaurant, number the invoice sequentially; the first order is 10001, and the second would be 10002.

Before you deliver an order, write this information in your notebook with the date and the invoice number. Payments from your clients should include your invoice numbers on the check. Record the invoice numbers that are paid, the date, and their check number in your notebook.

When the invoices are paid, mark your copy with the same information. Staple them together, along with the check stub from the account, if there is more than one. Keep these paid invoices; they must be retained for tax purposes.

Mileage Log

Maintaining a mileage log is also a high priority. The number of miles that you drive to deliver your product or for any other reason involving your business is deductible at a rate of so many cents per mile. Keep a record of your mileage every time you drive for business purposes. Keeping this record daily is far better than trying to reconstruct it at the end of the year. Just keep a little notebook in your vehicle and write down your mileage every time you return home. This will become a habit after you have done it a few times. The business-related mileage deduction changes each year. This rate is published each year by the IRS in its tax return booklet. Your accountant also will have this information.

Checking Accounts and Credit Cards

A separate checking account for your business will build a track record with your bank, lend a sense of trust with your creditors, and have all your expenses paid from one place for tax purposes. A credit card in your business name can be useful for ordering items by mail. Seeds, plant labels, and other assorted items are not terribly expensive and are much easier to order using a credit card.

Defend your credit rating at all costs. It speaks for your integrity and ability to operate a business. It is the key to suppliers extending credit to you so you don't have to pay extra charges for cash on delivery (COD). A good credit rating is crucial for getting a bank loan for expansion, too.

quail mountain herbs

Watsonville, California

This large company is a wonderful example of what can be done if you have the drive, growing knowledge, and, most of all, the marketing skills. Quail Mountain Herbs is one of the largest herb growers and shippers in the country. Carol Schmidt, her husband, Kirk, and her brother Richard McCain started the business by growing herbs on one acre in California. Soon after, Carol's parents, John and Dolores McCain, joined the company. Now the five partners are growing on 200 acres and in over 1.4 million square feet of greenhouses nationwide and elsewhere. The production acreage, both outdoor and in greenhouses, continues to increase.

Quail Mountain Herbs grows hundreds of varieties of fresh-cut herbs and edible flowers as well as baby lettuces and specialty greens. They sell to international markets as well as locally. Herbs are also bought in and repackaged. Carol says the best-selling herbs are basil, chives, dill, and mint, which are sold both by the bunch and by weight. Should you ever need to buy in fresh-cut herbs, this should be the first place to call.

Marketing seems to be the key to the success of this company. Quail Mountain Herbs sells both wholesale and retail, and they believe in helping the buyers to sell the herbs. Even on this large scale, recipes are supplied with many of the herbs, both to the supermarkets and to chefs.

Carol says that they work too many hours year-round. Nevertheless, she loves what she is doing and offers this most important bit of marketing advice for those wanting to enter the herb business: "Sell it before you grow it."

Managing Employees

Let's just imagine this scenario. Suppose you've been busy running your business, handling it quite well by yourself. You consistently grow quality fresh herbs and the word spreads. You have become adept at marketing and more businesses seek your herbs. Soon you find yourself working 14 hours a day and you realize that you need some help if you are to get any sleep. The process of hiring an employee, even a part-time worker, introduces a whole series of complications into a situation that was supposed to make life easier for you.

First, you must apply for a federal employer identification number, which is necessary for paying employee taxes. Contact the closest Internal Revenue Service office for form SS-4. They will then send you all the necessary paperwork so you can begin paying taxes for these employees.

You will have to pay Social Security, state and federal taxes, unemployment taxes, and worker's compensation. These must be paid even if you hire a local teenager to do a few hours of seasonal work such as weeding or potting. Bookkeeping tasks associated with employees can be quite a burden, in both time and money.

Hiring Independent Contractors

The alternative to hiring employees is to use independent contractors, sometimes called subcontractors. You don't have to pay taxes for an independent contractor; he pays his own because he is self-employed. You can pay independent contractors up to $600 per year, after which you must supply them and the IRS with a 1099 form showing how much you have paid them. The use of these self-employed workers can save you money in taxes but you must be reasonably sure that they are independent contractors rather than employees.

The IRS has launched a program to tax these workers and to determine whether they are employees rather than subcontractors. If the IRS determines that contractors are actually employees, they can — and do — go after the grower for back taxes, including fines. The definition of an independent contractor, according to the IRS, is rather vague. Some of the key points the IRS will consider are: how much control you have over the activities of the worker, whether you have trained her, if she has her own business card and federal ID number, whether she works for other employers, and if she issues you an invoice for each job she does for you.

Using Employment Contractors

One way to acquire helpers, without all the paperwork and expenses of paying taxes for them, is to use the services of an employment contractor. These companies employ individuals, provide staffing for businesses, and pay the wages and taxes for these workers.

It may be difficult to find the right helpers through an agency, and even if you do, you might not be able to get them back when you need them, as they might be reassigned. You would pay more per hour than if you hired your own workers, but you are spared paying the taxes and all the paperwork that entails. I believe the savings from using employment contractors balances out the higher wages, especially considering the time saved from bookkeeping tasks.

Verifying Citizenship

There are other legalities that you should be aware of before hiring anyone. A recently enacted immigration reform law requires that all employers complete I-9 forms on all new employees hired, even if they are U.S. citizens. This is not just for large growers but for all businesses, large and small, in any industry. The purpose of this legislation is to try to limit illegal immigration through stronger border controls and increased enforcement of employer sanctions, among other measures.

The I-9 form is an employment eligibility verification. It states that a prospective employee is a U.S. citizen or has the necessary papers allowing him to work legally in this country. A copy of this form can be obtained from the Immigration and Naturalization Service (INS) or downloaded from its website. (See Resources.)

The produce, nursery, and processing industries are under scrutiny from the INS in all areas of the country. There have been surprise raids on companies even here in the Midwest!

Other Laws

If your business grows to 15 or more employees, you are subject to the Equal Employment Opportunity Commission (EEOC) enforcement of the Americans with Disabilities Act. This means, among many other things, that you must make accommodations for disabled workers. Contact the federal government, Department of Labor, for more information.

Worker Protection Standards are laws that you must abide by. See chapter 11, Controlling Insect Pests, for more details about these regulations.

Finding, Managing, and Training Employees

Growers may know very well how to care for their plants and run their business, but have little knowledge about how to recruit, train, and manage employees. Finding good employees can be difficult. Grower trade magazines are filled with comments about how hard it is to find, and keep, quality employees. It may not be such a struggle for you, especially if you need someone only a few hours a day, a couple of days a week, to help fill orders.

When interviewing prospective employees, try to match the applicant to the work you have through a thorough but casual conversation. You want someone who pays attention to detail rather than a person who is sloppy. An employee doesn't necessarily need to know much about herbs, but he should like to garden. Your helper shouldn't mind getting his hands dirty.

Once you have an employee, train him well. When he knows the job well, give him the freedom and authority to do the work. Allow him the responsibility of doing his job, and this will allow you to accomplish more.

Let the employee be creative, listen to what he has to say, and praise him often for doing something right. Appreciate him every time he works for you; a good employee can be your most valuable asset.

doing business

There is more to a successful business than just growing and selling herbs. There are many other aspects of doing business that must be dealt with to ensure you the greatest chance of doing well.

Creating a Product List

You need some way of showing the customer what you are selling. A product list or price sheet will show what herbs you have available for sale. The list tells potential buyers that they can purchase your herbs at the stated prices and on the indicated terms.

Your list should be professional looking and attractive — handwriting or expecting the customer to remember what you told him you grow just will not do. You can type this list yourself on your letterhead. It should contain your business and personal name, address, and phone and fax numbers. You can also add a little advertising or promotion to give it some flair: "Quality Fresh-Cut Herbs Available Year-Round!" or whatever you feel is exceptional about your business.

Update your product list whenever you change prices or add or subtract herbs. Mail your new product list to each of your accounts rather than handing it to them. Let the accounts know that you will be sending these, just as do the large wholesale distributors. Your list is a good marketing tool when you are looking for more business or just starting out. Send it or hand it out to all prospective accounts.

Herb's Herbs

Sandie Shores
Route 11, Box 40
Zumbro Falls, MN 55991
(507) 555-1721

Fresh-Cut Herbs Available Year-Round
Organically Grown Locally
Please order before noon the day prior to delivery!!!

herb	single bunch	dozen
Arugula	1.20	12.00
Basil	1.15	11.50
Basil, purple	1.15	11.00
Burnet, salad	1.25	12.00
Chervil	1.45	15.00
Chives	0.90	9.50
Chives, garlic	0.95	9.75
Cilantro	0.75	8.50
Dill Feathers	1.05	11.50
Lemon Thyme	1.25	12.00
Marjoram, Sweet	1.10	11.00
Mint, English	1.25	12.00
Oregano, Sicilian	1.15	11.00
Parsley, Italian	0.75	8.50
Rosemary	1.15	11.50
Rosemary Skewers	.25 each	
Sage	1.15	11.50
Sage, Pineapple	1.45	15.00
Savory, Winter	1.25	12.00
Sorrel	1.25	12.00
Tarragon	1.50	15.75
Thyme	1.10	11.00
Watercress	1.25	12.00
Fresh Herb Garnish		16 for $1.00

    Cut and packaged to your specifications.
    Washed and ready to use.

Garnishes are available in any of the aforementioned herbs, plus:
    Tricolor Sage (green, white, purple)
    Golden Sage (green and gold)
    Pineapple Mint (green and white, ruffled edges)
    Edible Flowers   .30 each

Packaged in moist florists foam in returnable containers:
    Pansies, multicolored
    Nasturtiums
    Miscellaneous herb flowers   .30 each

Make sure your product list is complete, up-to-date, and easy to read.

Noting Your Terms

Your product list should include any terms that you feel are necessary. My list said in bold letters, "Please order before noon the day prior to delivery!" All of my accounts knew that everything was picked to order and that I was very busy, and if they wanted the herbs (without an extra charge), they would have to comply with this. And they did.

Another term to consider is a minimum order. Decide how small an order you are willing to deliver. This might depend on how close this account is to other accounts, how much of a hassle it is to complete the delivery, and whether this is a long-term account. You may wish not to accept business that is not profitable enough.

I had one national name hotel account that would order one bunch of this and two bunches of that once a week. The restaurant was on the second floor and it often took me 10 minutes to get the service elevator and a half hour to complete the delivery. I stopped selling to this account because that half hour was better spent in the greenhouse. It wasn't worth the few dollars!

Varied Pricing

Your product sheet should list all the herbs you grow and their prices. You can break the prices down by weight or by bunch, any way that you sell them. Keep separate lists for supermarkets, restaurants, and distributors.

If you package herb garnish, include this information along with any special herbs that you grow just for garnish. Examples are tricolored sage, golden sage, and variegated pineapple mint. It is helpful with these to include a brief description of the item. State that garnish is cut and packaged to individual specifications and that ingredients are washed and ready to use. Edible flowers should also be listed, with their prices and colors.

Setting Your Prices

There are several methods of determining what to charge for your products. The most common involves figuring your costs of production, both variable and fixed, and adding your profit to that total. This requires complicated calculations, and more than likely you'll have to charge more than the market will bear for your herbs. Because of this, I don't recommend using this method to determine your pricing.

A popular pricing method is to charge prices similar to your competition's. Charging much less than the competition can cheapen the perceived value of your product. It also decreases the amount of your profit. People perceive a slightly higher priced item as a better-quality product than that of the competition. Not only is this the image you want your customers to have of your product, but it also must be the truth. Your product should be of better quality and fresher than anyone else's.

Competitive Pricing

As you begin your business and are trying to acquire an account base, you may have to set your prices just at, or slightly below, the competition's. Once you have established a track record of delivering a superior product, you can raise your prices to slightly above the competition. Don't set your prices too high, though, or you may price yourself right out of business.

So how do you find out what the competition is charging? First you have to know who your competition is. In many cases it is wholesale distributors. *The National Wholesale Herb Market News,* in their *Wholesale Herb Market News Report,* covers weekly wholesale bulk (pounds and bunches) herb prices to retailers in 15 cities. You can obtain this report from *The Fruit and Vegetable Market News* (see Resources). The report comes to you free, bimonthly, with your subscription to *The Herbal Connection,* a publication of the Herb Growing & Marketing Network (see Resources).

The prices listed in this report don't necessarily reflect what your accounts pay for fresh herbs. Perhaps you can find a sympathetic chef, produce manager, or purchasing agent who will share these prices with you. They are not supposed to do this, so be sure to express your appreciation when they do. You can always call the distributor and ask for a price list. Be honest when making this request. Some will send it, and some won't.

Supermarket packs can be priced by the individual tray, by the dozen, or by the case. Ask a prospective store account or distributor how they are priced in your area. In my region a set price is charged per tray. These range from $1 to $1.25 per tray. The supermarkets charge $1.69 to $1.99 for each tray.

Changing Prices

How often should you change your prices? Change them at least twice a year to reflect your added costs if you grow during the winter in the greenhouse. Keep a close eye on the national prices. If these rise or fall dramatically, even on just one herb, you can change your prices as well. There is no reason why you should not keep your prices in line with those charged by the big guys.

farmer's daughter herbs

Salt Lake City, Utah

In 1988, Dorothy Gifford started Farmer's Daughter Herbs. Her business is unusual in that she does not grow any herbs at all. She buys in all herbs and edible flowers and repackages them for her accounts. This business is a fine example of another way of being successful in the herb trade. Herbs are bought from local growers in season, including an apprenticeship program at Salt Lake City Community College. One grower supplies her with all the basil she needs. In the off season, she buys herbs from as many as five suppliers.

Dorothy has about 150 accounts divided equally between restaurants and supermarkets. Supermarket deliveries are accomplished in a rather unusual way. Her delivery person takes an assortment of packaged herbs to the stores and they choose which ones they want. Farmer's Daughter Herbs' most distant in-person delivery is in Provo, which is about 45 miles away. Herbs are shipped via UPS to 10 accounts in other states.

Dorothy has three employees and some occasional part-time people. One employee does all the local deliveries. Dorothy's daughter is getting involved in the business by teaching classes, and other family members grow some edible flowers for her in season.

Besides herbs and edible flowers, Farmer's Daughter Herbs also buys and resells arugula, spinach, mesclun, cilantro, and parsley in bulk. Dorothy packages and sells 20 varieties of herbs. Pansies are the most common edible flowers sold, but she also buys and sells bachelor's buttons, calendula, snapdragons, and whatever else she can find in season.

When asked what her biggest problems are and what advice she would give, Dorothy had much to say about grower practices. She feels that the herbs should be washed as soon as possible after harvest because she has many herbs come in dirty, wilted, and with insects. Growers need to learn how to properly harvest herbs and teach their employees as well.

Dorothy prefers field-grown herbs over those from the greenhouse. She feels that greenhouse-grown herbs are often leggy, soft, have poor color, and do not keep well after packaging. She advises growers to harvest very early in the day to maintain postharvest quality. When buying in herbs, be very careful whom you buy from and try to see the herbs before you buy. If this is not possible, be sure to tell them exactly what you expect to buy, namely quality and freshness.

Extra Charges

As previously mentioned, you may wish to add a service charge for last-minute or "emergency" orders. If you decide on this policy, first make these terms known to your accounts. Tell them personally and also write this on your product list.

One important key to maintaining account loyalty is your willingness to provide the ultimate in service. However, there is a cost for this service. It costs you to drop everything, pick an order, and deliver it. It might as well cost the account too. Rare is the account that would not be happy to pay the extra $10 or $20 to be rescued from a bad situation.

Buying Herbs for Resale

There may come a time when you are out of a certain herb, have a crop failure, or simply do not grow a particular herb. You can buy in a supply of herbs and repackage them for your accounts. Or you could just tell them that you don't have that and let them get it elsewhere, but, in essence, you would be telling your accounts to "go away."

If you are trying to be the sole supplier to an account, buying in herbs is one way to maintain your credibility. This is done by many growers, especially those having difficulty growing basil in winter.

Buying in herbs is not as expensive as you might think; it is certainly possible to make a profit operating this way. Dorothy Gifford operates her company, Farmer's Daughter Herbs, this way continually, although she does buy much from local growers in season.

Locating Distributors

There are several companies that supply herbs that you would need throughout the year (see Resources). You can also buy herbs from the wholesale distributor in your area. Wholesalers sell herbs by the pound. Most do not have a minimum order, but you will pay less in shipping charges when you order larger quantities. The average shipping charge is from $3 to $4 per pound. Most companies ship by next-day air. The herbs can be brought to your doorstep by a delivery company such as FedEx, or you could pick them up yourself at the airport. (The herbs are usually shipped in plastic foam coolers or in Styrofoam-lined boxes filled with ice packs to maintain their quality.)

If you do buy and repackage herbs, remember that under the law you then become a repacking house, and you must adhere to the regulations of the Good Manufacturing Practices law. (See Chapter 13 for information.) Remember, too, that if the herbs are grown out of this country, you must list the country of origin on your label.

Filling Orders

Most of your orders will come in by phone or fax. It goes without saying that you should have a telephone right in the greenhouse or office area, or you can carry a cell phone. Note that the high humidity levels in a greenhouse are not good for electronic equipment.

An answering machine is mandatory. Check the machine often. If an account has a question, a last-minute order, or cancellation, you don't want to miss it. Great service is one of the keys to a successful business. Be organized and keep notepaper and pens at all phone areas or always carry them with you. You can have special forms to write down orders, but plain notepaper works just as well.

Standing Orders

When you have had an account for a while, you may begin to see a pattern in the herbs that it orders. The business may be ordering nearly the same things for delivery on the same days of the week. This is the time to ask for a standing order. These are orders that you automatically deliver without a phone call or fax order. This will make your life much easier because you will be able to plan ahead. Standing orders are a great convenience for the accounts as well, and a good selling point. There will always be additions or cancellations, of course, but these won't be nearly as disruptive as continual phone calls.

Keep your standing orders in a notebook as well as on the computer. You want to have these lists even through a power outage or a computer breakdown.

Purchase Orders (POs)

Some larger corporate accounts may require that the person ordering gives you a purchase order number. This helps keep track of all purchases made by the company. Some will give you a different number each time they order; others give a "blanket" number that serves for all of their orders. There is not much extra work for you with this system except bookkeeping. You must include the purchase order number on your packing slip and/or invoice.

Making Deliveries

The delivery vehicle for the small business is usually a car or small truck. A refrigerated truck is ideal for transporting a large volume of herbs or for traveling long distances; however, this is usually reserved for the large grower.

Making deliveries, although sometimes tedious, can be a fun part of this business. It is satisfying to see how your hard work and quality are appreciated. It is also interesting to see the inner workings of a large kitchen or supermarket. Often the chefs would take me into the kitchen and show me some of the fancy gourmet things they created with my herbs and edible flowers. I would get some tasty samples and sometimes be given some great food and recipes to take home!

Controlling Temperature in Your Vehicle

It is important to maintain a constant temperature for the herbs; drastic temperature variations can cause rapid deterioration. Be sure to protect the herbs from the sun's rays during transport. When delivering in hot weather, use the air-conditioning in your vehicle to precool the car before loading the boxes of herbs. Portable coolers can be useful, but this often means you have to repack the herbs into boxes before going into the account. Growers improvise all sorts of larger portable coolers for their vehicles. One grower I know used styrofoam sheets glued together and cooled it with cloth-wrapped refreezable gel packs placed inside the container.

Special Handling

Each account is different in the way it wants its deliveries handled. Most will expect you to enter at the delivery or service entrance rather than the front door. Some will allow deliveries only at certain times, usually in the morning. Obviously you won't want to make deliveries to a restaurant during the noon rush.

When setting up an account, always find out where you should park when making a delivery. Parking areas or loading docks are usually full of big trucks. In some cities it is illegal for cars to park in these areas. You may need a permit — which can be obtained from the account — to park in certain places. A business sign on your vehicle may make it easier for you to use restricted parking areas. Removable magnetic signs are handy.

Most accounts will have someone check the order against the packing slip to make sure they match up, especially when you are just starting with them. This person may have to sign the packing slip and put away the order. Other companies will not require this after you have done business with them for some time. Certain accounts may want you to date the carton containing the herbs, especially those that are busy and use herbs constantly.

Many of my accounts, both restaurants and supermarkets, would have me put the herbs in the cooler and leave the invoice on a desk. This was most convenient for both of us. I didn't have to spend time searching for the right person, and they didn't have to take a person away from his or her job to put away the herbs.

Marketing

Marketing is a catchall phrase that, to many, means selling your product, but this is not accurate. Marketing means to focus on the consumer rather than the herbs you want to sell. It entails identifying and satisfying your customer's needs. Listen to your client rather than talking to her. Sales of your products will often follow successful marketing.

Advertising is not marketing either. Advertising is a means to inform the public and your clients about your products and services. It is a way to let them know that you are there and to help them think of you first when they need your products and services. If you are a grower who does not sell to the public, advertising, in most circumstances, is not needed. Word of mouth and your marketing efforts should be all that's necessary to acquire new business.

The steps involved in marketing are simple. Prospect for new business by gathering information about potential customers. Then, evaluate them by asking questions. (See sample questions and approaches in chapter 2, Finding Potential Markets.)

One way of identifying the needs of potential customers is to study your own sales records. Keep track of what herbs sell best, including those that were requested but not in stock. The market should dictate what you grow. If your field and greenhouse space is limited, get rid of the herbs that have not sold in six months or so and use the space for the ones that are in demand.

Be Creative

Your knowledge of your product and the marketplace will set you apart from the competition. Let your current and potential accounts know that you are an expert in your specialized field. Focus on the fact that you are local and not some far-off grower whom they'll never meet and who ships their herbs to a distributor.

There are many things you can do to establish yourself as an expert in your field. Keep track of what is hot. Read cooking and gardening magazines and tape cooking shows for viewing later. The public scrambles to buy herbs seen on television or read about, so be proactive rather than reactive.

Alert yourself to new trends. A trend often starts with local restaurants and can then follow into supermarket sales. Take notice of trends rather than fads. Fads can be costly because of the lead time for growing involved. You could take up valuable space, and by the time the crop is ready, the fad could be past. It may be difficult to distinguish between a fad and a trend. Generally, a fad comes on very quickly and leaves just as quickly. A trend is usually slower to evolve and its popularity lasts much longer.

If you don't use fresh herbs in cooking, learn to. You should be not only an expert grower but also a knowledgeable user of them. Create new sales by studying local ethnic and religious groups. Learn what and when their holidays are and what herbs they use in cooking those special meals. You can learn about other cultures' holidays and cooking techniques by reading about or taking classes in ethnic cooking. But, the best way is to make friends with people from different ethnic groups and ask them to teach you.

Educating Your Clients

Successful marketing often means creating a need in your clients. You can do this by educating the end user of your herbs. Teach classes on growing herbs or on cooking with them. Community education programs are a good alternative if you don't want to do this at your location. Sell herbs to others who teach classes.

Contact gourmet cooking clubs in your area. Join a group or ask them to send you their menus. In my market area there was a large gourmet cooking club consisting of many couples who loved to cook and eat. They sent out a monthly menu to all members and they would get together in small groups, on a certain day, to cook these meals. They all shopped in the same store and sent their menu, in advance, to the store (and to me) to make sure that there were enough ingredients in stock to supply all the participants. This was a nice boost in sales each month for me!

Place recipe cards near the display of herbs at your supermarket accounts. Or write a column for your local newspaper about cooking with herbs. See if you can do a series of short shows on cooking with or growing herbs at a local radio station. Send out newsletters to all your existing and prospective accounts. Include news about cooking trends, recipes, new varieties, and information about special growing conditions or concerns.

Do food demonstrations at your supermarket accounts. Plan them for their busiest time, usually on a weekend. Ask the store manager to advertise your appearance as a special event. You'd be surprised by how many people this will draw into the store. Be prepared to answer a lot of questions.

Make yourself and the work table attractive. Have available sample herbs for people to smell and taste. Show how to chop the herbs. Bring your own cutting board and knives. Make something that doesn't require cooking, such as a dip, cream cheese ball, or pesto. Have crackers or vegetable sticks available for tasting your creation. (The store should supply these.) Give out recipes.

Be sure to supply the store with extra packages of herbs, especially of the ones you will use in your demonstrations. You may have to give the store a small discount on this order. If you use a combination of herbs for a recipe, make up special packages of these and include the recipe.

You can give free samples and recipe ideas to the executive chefs at your accounts, but do this with caution. There may be a few chefs who view this as a sign of disrespect. After all, a chef is a professional and here you are, seemingly knowing more about using herbs than he does! In this case, just some free samples for someone to "experiment" with will suffice.

Set up a visitation day at your operation for chefs and/or produce managers. Invite key people from prospective accounts. Call it a little party, seminar, or just a tour. It is good for these people to know how the herbs are grown. Offer to teach a seminar about growing and handling herbs for produce managers and employees. This can also be done for wholesale distributors.

o'toole's herb farm

Madison, Florida

Betty "B" O'Toole and her husband, Jim, have been in the herb business since 1989. Betty describes their business as 40 percent wholesale plants, 20 percent fresh-cut herbs, and 40 percent retail. The retail portion of the business is growing. Although their most popular herbs are basil, garlic, chives, mint, parsley, rosemary, oregano, and tarragon, the company also grows shiitake mushrooms. All the fresh-cut herbs, which are sold by weight, go directly to restaurants and supermarkets within a 60-mile radius. They do not buy in herbs to repackage for sale. O'Toole's Herb Farm has six to eight accounts and Betty does the delivering — even though they have one full-time and four part-time employees.

There are display gardens open to the public, herb festivals, and workshops. The O'Tooles have several older farmhouses on the property. One house contains the office and a gallery of local and regional art, another house has been converted to a retail shop, and another is a guest house.

They have 1.1 acres of outdoor growing fields, some of which are in raised beds. In their area, zone 8, they grow outdoors year-round by picking different herbs in different seasons. The soil in this location is sandy loam, and the couple augment it with composted manure and green manure crops. They are certified organic growers, so any insect or disease problems are handled in an environmentally friendly way using only safe products and beneficial insects.

The two greenhouses, both 22 by 96 feet, are heated when necessary with liquid propane. They don't grow fresh-cut herbs in the greenhouses but do take the trimmings to sell from the large volume of potted herb plants they grow. The potted plants are in a mix of pine bark, peat, perlite, vermiculite, and chicken manure.

Rather than insects or disease difficulties, Betty says that her most common problem is fatigue! They work 10 to 12 hours a day in the summer and 9 or 10 hours a day in the winter, 6½ days a week, year-round. After six years, Betty reports that the business finally became profitable. Her opinion on growing fresh cut herbs? "This is not a get-rich-quick scheme, but the *most* rewarding occupation!"

Diversifying Your Business

No matter how hard you try, you may not be able to get more business from your existing accounts. Perhaps you have a limited number of possible accounts within your delivery area and you already work with them. So what can you do to increase your business?

You could look to the next largest metropolitan area for more business, but this presents a problem with deliveries. While you could make the deliveries yourself, you can also hire someone to do them or use a delivery or courier service. Your orders would have to be large enough to make the extra expense cost-effective, however.

You could diversify your business by broadening your product range. This is a good hedge against the serious risk of having most of your business concentrated in one or two accounts. Your largest account should be only a small percentage of your total business. Stores and restaurants frequently change hands or suffer business failures and close down. What would happen if you lost one or both of those accounts?

Do some research and test the market before spending a lot of time and money on a new venture. The trick is to find another niche and do it well. The possibilities are almost endless. The following are just a few suggestions to spur your imagination.

Crafts. Bundle the discarded parts of the herbs left over from packaging for use on the grill. Dry or package them for potpourri. Dry little bundles for decorative sales or sell them to someone who does crafts. Try growing catnip and creating cat toys. Grow everlasting flowers to sell or craft them yourself into wreaths and swags.

Mail order. Sell fresh-cut herbs or herb plants by mail order through your own catalog or through a distributor. Open up your farm to include sales to the public, or have herb festivals.

Decorative plants. Concentrate on growing herb plants and sell them wholesale to nurseries and/or to the public. Make herb topiaries or bonsai.

Produce. Grow other items — medicinal herbs, salad greens, vegetables, Asian or other ethnic vegetables, baby and gourmet vegetables.

Herbal products. You could construct a separate "commercial" kitchen or rent one to make herbal products such as dried herbs and blends, vinegars, jellies, teas, and cosmetics.

Catering. If you excel at cooking with herbs, why not start your own catering business or open your own tearoom? Develop your own herbal recipes and write a book.

Expanding Your Business

If you have a good business in your area and there is an increasing demand for your herbs, will you be ready for it? New business may always be welcome, but you can grow too fast if you are not prepared for it. Your business can quickly grow beyond your ability to finance it or even manage it if you have not planned for expansion.

It is best to continue operation at your maximum financial potential before you decide to expand. Occasionally, when you want a new piece of equipment or a new greenhouse, or you have too much overhead, you have to expand in order to pay for it. As always, the demand for your product should pull you along rather than you pushing it and trying to flood the market.

Planning Ahead

Business expansion has its price. As your sales increase, so do your costs. You'll need more money for seeds, packaging supplies, employees, and all other aspects of doing business. Your workload will increase dramatically with each boom in business.

Many times growth happens gradually and the pace of work consumes too much time to allow the business owner to plan. Soon the business is running ineffectively and begins to lose money even with the increase in business. It can be a whirlwind, wild ride. This growth can be a real stroke for your ego or a disaster in disguise.

Lots of us go into this business because of our love for growing and working with herbs. We want to find a way to make money while supporting our herb habit or want to make a living from the land. These were some of my reasons for starting Herb's Herbs.

But it wasn't too many years after starting out that I found I was too busy to work with the herbs much. It became a luxury for me to be in the greenhouse. I had little time to spend with family, friends, or just to be alone. My time was spent delegating the work to employees, making deliveries, marketing, and tending to the myriad details necessary for running a smooth operation. I began to lose sight of why I started the business in the first place. Luckily, I had planned ahead for this growth; that made it easier to sell the business and farm when the time came.

This can happen to you too. Decide before your business grows whether you really want to get bigger. Take the time to develop a mission statement for you and your business before you face the challenges of expansion. Be honest with yourself and decide what your goals and objectives are. What kind of image do you want to project? What are your personal and business philosophies? If a larger business is included in your plans for the future, design for that expansion now. Decide what steps are necessary to complete the expansion and implement them carefully when the time comes.

Learning How to Say No

There is nothing wrong with saying no to new business if you decide that you don't want to grow. Operating a small, one-person business can give you just as much pride and satisfaction as running a large multi-employee operation. You can always change your decision to remain small.

I once got a call from the produce buyer of a distributor that was the supplier for a very large national chain of supermarkets. Someone had seen my product in one of my stores and, duly impressed, asked if I would be interested in supplying the distributor with all its fresh-cut herbs. I briefly entertained the idea until I found out how many truckloads they wanted each week! For one very small moment I considered the vast expansion I would have to undergo to accomplish this task. Then I remembered that I was already working 14 hours a day. Most times in life, less is more!

 processing herbs in a commercial kitchen

Most states have regulations requiring that certain conditions be met in a kitchen before food can be processed. Meeting the standards usually requires the use of expensive equipment, and entails frequent inspections by the State Department of Health or the United States Department of Agriculture. Contact your State Department of Health for more information.

Managing Your Accounts
and Keeping Them Loyal

Each type of account, although they operate differently, wants the same thing from a supplier: quality, consistency, and service. The types of accounts differ only in how you supply them with these things. Try to understand their business and their needs.

Give your accounts good value. This is defined as a combination of quality, service, and price. Understand that it is the responsibility of the buyers at your accounts to get the best product at the best price. If you can give them this, along with the best service and consistency, your accounts will be loyal to you.

Quality, Consistency, and Service

You must have an underlying passion for quality and a commitment to excellence. Sell only the very best herbs that you can. Don't be afraid to discard imperfect herbs; you can use them in many other ways. Make sure your herbs arrive fresh and in top condition. Follow up on storage at the receiving end to make sure they are kept fresh.

Treat each customer as if he is the most important account to your business. Farmer Jones Farms (The Chef's Garden) in Huron, Ohio, has a creative way of doing this. They take a good-quality color photograph of a visiting chef with one of the owners in the field. This photograph is placed in a frame along with a nicely worded statement about how the chef and the restaurant use the best-quality produce available. This hangs in the lobby of the restaurant and makes the chef feel very special and appreciated.

All people at your accounts should be treated with respect. Be personable and friendly when making deliveries. Of course you're busy, but take the time to meet with the manager, chef, or buyer just to chat and follow up on sales. Show interest in them as people, not just as buyers of your produce.

Keep open the lines of communication. Let them know that you want to know any problems or concerns they might have about your herbs. Be flexible and understanding of business fluctuations. All restaurants and supermarkets have slow times, and so will you.

Guarantee your herbs. If there is a problem with an herb, replace it. Buy it in if you must to satisfy your account. Sometimes you'll have large quantities of a certain variety of an unsold herb — give some away to your accounts. It can't hurt and it could lead to more sales. Go the extra distance with service. This is something you can excel at, and it will create a win-win situation for you and your accounts.

Respect your commitment to deliver when you say you will. This helps establish a relationship of trust. When a delay is unavoidable, let your accounts know as early as possible.

Maintaining Supermarket Accounts

Although the supermarket is your account, the end user of your herbs is the person to whom you gear your product. It will pay you to understand the consumer, both in sales and in account loyalty. Many customers are in two-job families, and they lead a very busy existence. They want rewarding, fulfilling, enjoyable lives but they also want convenience. You can help supply customers with these things by educating them and providing them with ways to use your herbs quickly and easily.

Appearance is important to supermarket accounts, not only of your herbs but also of the packaging and their display. Offer to help them with their display or to straighten up herbs (but only if you can spare the time; they might come to expect it every time you deliver herbs). Suggest that they place complementary and tie-in products together, such as basil next to tomatoes with a recipe for fresh tomato sauce. This can lead to impulse buying.

You might suggest adding the supermarket name to your label. Brand recognition can help sales. This also helps to personalize the store's relationship with its customers.

Supermarkets prefer to have small deliveries made frequently rather than taking up their display or storage space. This will increase your costs and time spent delivering, but it is a service you must offer willingly and with enthusiasm.

Maintaining Restaurant Accounts

Chefs are the trendsetters in the cooking world. They dictate, more or less, what ends up in the retail marketplace. They appreciate fine quality and like knowing where their herbs and produce come from. They often have a reputation for being temperamental "prima donnas." While this can be true, it has been rare in my experience. If you treat all chefs with respect and learn to keep your own ego in check, you should have no problems, even with any "difficult" ones you may encounter.

Chefs can be a big help to you and your business when you cultivate a relationship with them. Not only can you be their herb supplier, but you could custom-grow other produce for them as well. Custom-growing requires detailed communication between the chef and you. You must know what he expects concerning variety, size, volume, and timing of your produce. You could meet with the chef and go over seed catalogs and plan the crops together. The chef must learn what your problems can be and the variables that may cause them.

Communication is essential when working with any restaurant account. Both chefs and growers are working under pressure and don't always understand each other's problems and concerns. Some chefs are not familiar with how herbs are grown, the amount of work it takes, or the lead time necessary to have an herb ready for market. Sometimes, the grower may not know what gourmet cooking is, how a large volume of meals are prepared, how recipes are developed, or how to garnish a plate. Dine in these restaurants and let them know that you do. Ask to see how they use your herbs. Let them know you are interested in how they prepare foods. Find out what problems they encounter. This might even lead to a free meal now and then!

Consistent flavor of the herbs is crucial to restaurants. If you are considering a change to another herb cultivar, grow a few trial plants before changing your whole crop. Let the chef know you why are considering a change and let him try the new herb for "his approval." You don't want to risk losing his business because he doesn't like the flavor of the new herb.

Appearance of the herbs is also important, especially those herbs used for garnish and any edible flowers you sell. Most herbs are chopped up and used for flavoring foods, but that does not mean that you can sell blemished herbs. Always sell only top quality.

Give restaurants plenty of notice if you foresee a shortage of an herb. This will give them time to order it from another supplier. Offering them a special discounted price on an herb that you have in abundance will help both of you. They can use it in special dishes and you can profit on something that might have gone to waste.

Work closely with those restaurants that change menus often or offer daily or weekly specials. Ask the chef to let you know as soon as he plans changes or for a schedule of specials.

Be considerate of restaurant schedules. Don't try to visit with the chef or kitchen staff during peak mealtime hours. Ask the chef when the best times are for you to talk with her. Few growers realize how very busy chefs are.

Your awareness of these things, along with your very best service and quality, should earn you the loyalty of your restaurant accounts. However, loyalty aside, you still may experience some changes or losses in your business with restaurant accounts.

It is customary for chefs to move to a new restaurant every few years. This can be a large loss for you if the new chef does not use herbs or uses ones that you do not grow in volume. But it can also be very good for you — the new chef may use more herbs than did the previous one.

Advance notice of a change in chef will do you little good unless you have the opportunity to meet with her before she takes over. A new chef will usually change the menu but will often wait a bit before doing so. Try to meet with the new chef as soon as possible to determine how you can best serve her needs.

Increasing Your Chances of Success

A fresh-cut-herb business is a wonderfully rewarding way to make a living. But do remember that this business is a great deal of work. It requires long hours and a commitment to quality. It is a *business* and must be treated that way. You must work toward making a profit.

Take the time to enjoy your family as well as your work. If you allow your business to consume all of your time and energy, you may grow to dislike your work. Remember why you started working with herbs in the first place.

This is a fast-evolving field that requires you to keep up with changes and new products. This can increase your chances for success and enable you to enjoy your work more. Read trade publications, share information with other growers, attend educational conferences, and never, never stop learning.

building and maintaining a greenhouse

selecting a greenhouse

A greenhouse gives you the opportunity to control the weather. Frost, rain, storms, and winter are not a problem for the greenhouse-grown plant. You can control the environment and provide exactly what the plants need for optimum growth, no longer at the mercy of Mother Nature.

Most herb growers will eventually want a greenhouse. A greenhouse can be used to start plants early for transplanting outdoors, to extend the growing season, or to grow herbs year-round. Although you may start herbs on a sunny windowsill or even under fluorescent lights in a basement, this can require an elaborate setup for a large volume. And as plants grow, they will need increased space and light. Without proper light the

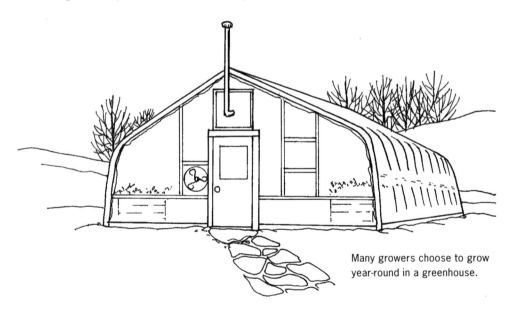

Many growers choose to grow year-round in a greenhouse.

plants become leggy and the growth soft. Better-quality plants result by using a greenhouse to prepare them for transplanting outdoors.

Why Grow in a Greenhouse?

Growers in all but the very southernmost regions (zone 8) may be limited to selling annual herbs only in season without a greenhouse. Most perennials are slow growing and can take six weeks or more after coming out of dormancy to reach cutting stage or to provide a good yield. If they are grown through the winter in a greenhouse, however, this waiting time is eliminated. A greenhouse can also be used in late winter to force perennials out of dormancy early so they will be ready to harvest when many of the annual herbs are.

Many commercial growers of fresh-cut herbs use the greenhouse only to extend the growing season. Crops are planted in the greenhouse in very early spring. Herbs are harvested indoors until heating the greenhouse becomes necessary. The plants are allowed to freeze or go dormant in the winter. In the winter, these growers buy in herbs and repackage them for their accounts.

Some growers use the greenhouse to grow herbs all year. This type of year-round structure requires heating, ventilation equipment, and artificial lighting. An operation such as this can be expensive, depending on your location and weather patterns. As a year-round grower in southern Minnesota, I can testify to the added expense and difficulty of growing this way. It is, however, a glorious sensation after trudging through snow and cold wind to step into a warm, humid, green environment. Instant summer!

If you want to grow year-round, do some research to find out how much the equipment will cost. Get estimates of the average utility costs

 winter herb supplies

It may be a better option to buy in herbs from another source to repackage and sell to your accounts in order to supply them year-round rather than try to grow them all yourself. Do some research to determine which would be the more cost-effective for you — greenhouse growing or buying in. (See chapter 4, Doing Business, for information on buying in herbs.)

from your power and fuel companies. The heating contractor or greenhouse supplier can give you estimates of how many British thermal units (Btu) it will take to keep the greenhouse at the desired temperature.

You will also need to use artificial lighting and ventilation fans. Find out how many amps each light, fan, and any other equipment draws. Consult with your electric company for estimates on operating costs.

Types of Greenhouses

There are many types of greenhouses available today. When determining which one is best for your situation, consider the future. Look at your business plan and where you want to be in three, five, and ten years. The following are some things to consider when planning your first greenhouse. (See Resources for a list of greenhouse manufacturers.)

The Attached Greenhouse

This greenhouse is attached to a structure, usually a house or work building. The greenhouse should be positioned on the south wall. This type of greenhouse is usually cheaper to build and maintain because its total size is about half that of a free-standing greenhouse. It is usually designed and built by the owner, although one can be purchased. Space is limited, but it may be enough to start plants for the outdoor growing

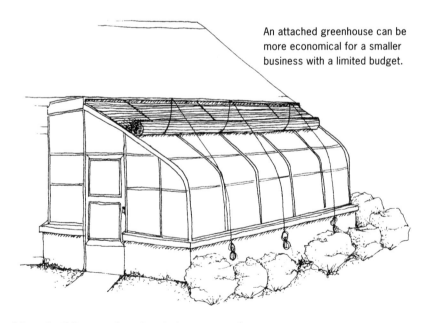

An attached greenhouse can be more economical for a smaller business with a limited budget.

season. For those on a limited budget, this is a way to get started on greenhouse growing without much expense.

There are "lean-to" greenhouses made from steel hoops; sunrooms; and other types of attached greenhouses available from manufacturers. It may be possible to send some heat from this type of greenhouse into the adjacent structure during sunny winter days. It is easier to heat an attached greenhouse because it receives some heat from the adjacent structure.

To be most efficient, the greenhouse length should be twice its width. The covering for the greenhouse, called glazing, is usually glass or rigid plastic. The roof should be slanted for maximum light transmission; otherwise, the plants will lean toward the sun. Pay extra attention to ventilation and doorways.

The Pit Greenhouse

The pit type of greenhouse usually has the north wall built into a hillside, or the whole floor can be a few feet below ground level. This greenhouse is easier to heat and is often used as a seed germination house because the structure holds its heat longer and can be kept warmer than aboveground types.

Construction costs are higher with this greenhouse due to belowground excavation, and the costs of bringing power and water to the site. The walls below ground level should be of concrete or masonry blocks.

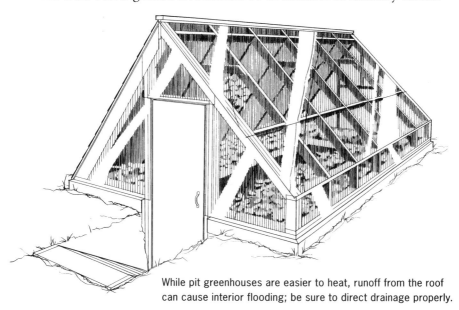

While pit greenhouses are easier to heat, runoff from the roof can cause interior flooding; be sure to direct drainage properly.

Stairways into the house must be made of materials other than wood; wood stairs can rot fairly quickly when they are in direct contact with moist soil, and are also very slippery when wet. Take note of drainage so runoff from the roof does not go directly into the greenhouse. Flooding of the floor is the most common problem with the pit greenhouse.

The Freestanding Greenhouse

This type of greenhouse is best for the commercial grower. It allows more room for growing, offers more sunlight, and can be placed where you want it. There is room for expansion, adjacent work buildings, and a work area away from living quarters. Freestanding structures are more costly than the attached greenhouse because electric and water services must be brought to the area.

There are many types of prefabricated freestanding greenhouses on the market. The materials used in construction depend on the type of covering (glazing) used.

Framework. Wood-frame greenhouses are the most versatile and can be covered with almost any type of glazing. When exposed to the high humidity levels common to greenhouses, wood will rot in time. Wood must be painted or waterproofed frequently, making for high maintenance costs.

Pressure-treated wood should not be used for greenhouse construction. Chromium, copper, and arsenic (CCA) are the most common chemicals used to pressure-treat wood. These chemicals have been shown to leach from wood when exposed to water or soil. During times of high humidity in the greenhouse, condensation may drip off the frame and onto plants. Reportedly, arsenic-treated wood can be used if it is washed (scrubbed) at least six times to leach out the chemicals prior to construction. However, this chemical is toxic to some plants, and you must exercise caution.

Foundation. Foundation types vary with the kind of greenhouse and individual manufacturer. Some foundations are concrete, concrete blocks, or in the case of the double poly hoop type greenhouses, pipes buried in the ground. Many types of greenhouses use a 2-inch by 12-inch wood base around the foundation. A long-lasting type of wood such as redwood or cedar should be used for this application. Other wood may be used for the base, but it should be completely waterproofed before use.

Glazing. Greenhouses with galvanized steel or aluminum frames are most often covered with a rigid type of covering, usually glass or polycarbonate. These can be slanted or straight-sided. They usually come in panels that are put together on-site. The glazing material is set into the frames of the panels and held in place with watertight caulking.

Hoop Houses

Steel-pipe arch or hoop houses are constructed of 1⅝-inch or 2-inch galvanized steel pipes (sometimes called hoops). Smaller pipes, called purlins, run the length of the greenhouse inside to add stability. Three or five purlins can be used. Metal wind braces — which are placed high up inside the greenhouse across the width and attached to the steel hoops on both sides — may be used also, depending on the wind in your location and the size of the greenhouse. Building codes in some areas may require the use of wind braces.

Hoops are placed 4 feet or 6 feet apart depending on snow loads (the average amount of snowfall in your area). Most often the hoops arrive on-site in two pieces. A shorter pipe connects the two pipes in the center, thus forming the hoop. The purlins are fitted together and attached to the hoops with brackets.

The foundation is usually 4-foot-long pipes (called sockets) that are buried in the ground, usually 3 feet deep or more. The hoops slip over these foundation pipes and are bolted into place. The entire structure is bolted together and attached to the baseboard. Glazing is usually two layers of poly film, but polycarbonate sheet glazing can also be used. End walls can be glazing or wood.

Hoop-type greenhouses are available in a variety of styles from many different manufacturers. Widths can be from 12 feet to 36 feet. Lengths are usually in multiples of 4 or 6 feet. Styles range from the circular hoop (or quonset) to straight-sided with peaked roof, from slope-sided with peaked roof to straight-sided with circular roof, among many more. (See examples on page 84).

Gothic or peaked-roof styles reduce condensation drip inside the greenhouse better than the quonsets. The moisture tends to run down the slope of the roof rather than dripping from the top. The water does catch and drip from the purlins, but it is easier to deal with from a specific area rather than from all areas of the ceiling. Some greenhouse manufacturers offer purlins with gutters to collect condensation runoff.

Hoop-Style Greenhouse Models

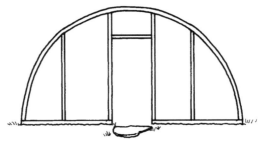

Circular hoop or quonset

Straight-sided with peaked roof

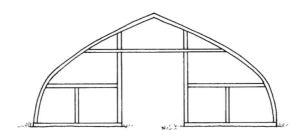

Slope-sided with peaked roof

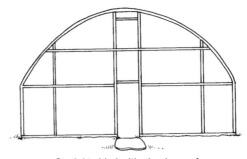

Straight-sided with circular roof

Gutter Connected Greenhouses

These are freestanding units that are placed parallel to each other. They are connected with galvanized steel or aluminum gutters to allow for drainage. This arrangement permits hallways to be built on one or both ends so that the greenhouses can all be accessed from inside the hall.

Gutter-connected greenhouses can be free span (sometimes called clear span) or individual divided bays. The free span has no interior walls, making for one large, open space. Some manufacturers offer poly curtains that can be closed to divide a free-span area, if desired. These greenhouses are available in many of the roof and sidewall styles described on page 83.

Some manufacturers offer individual greenhouses that can be gutter-connected to another greenhouse at a later time. Should your plans call for more than one greenhouse eventually, do consider building this type initially.

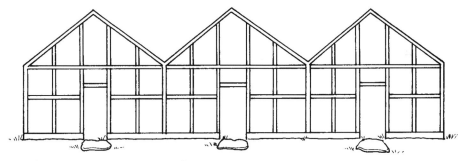

Gutter-connected greenhouses offer a convenient way to expand your growing space.

Roof-Vented Greenhouses

All greenhouses require some type of ventilating system. Fans draw fresh air through the greenhouse, but the cooling effect is far greater with roof vents because hot air rises. Fans can and do break down, and they are

 gothic-style greenhouses

Gothic or peaked-roof styles are better for areas that receive abundant snowfall. These structures are stronger and the slope of the roof encourages the snow to melt and slide off. Hoop- and quonset-style greenhouses are usually used for overwintering nursery stock, as sales houses, or as large cold frames.

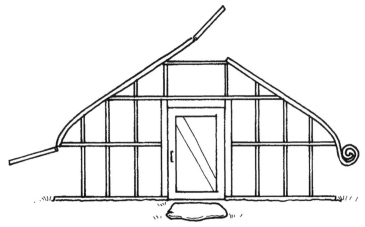
A roof-vented greenhouse takes advantage of natural ventilation.

expensive to run. Anyone who has ever worked in a greenhouse during sunny, hot, humid days would value the natural ventilation of roof vents.

Roof venting is available in many styles, including the wing vent, ridge vent, and rolltop or retractable roofs. They can be operated manually or automatically. All or part of the roof opens to allow hot, humid air to escape. When combined with side curtains that either drop open or are rolled up, extra natural ventilation occurs. In many cases, roof vents can be retrofitted to most types of existing greenhouses.

Rolltop systems roll up all or part of the poly that covers the roof. This allows natural ventilation and maximum sunlight to reach the plants. It is an excellent way to harden off plants. Growers not using the greenhouse in winter can retract the roof then. Rolling up the roof during the summer protects it from harmful degradation by the sun's ultraviolet rays, thus making the roof last longer.

Greenhouse Glazing

Choosing the material with which to cover your greenhouse is a most important decision to make. New materials are being developed frequently, so learn about the latest technology before making your choice. The glazing can be a factor in productivity, as the various coverings allow different levels of light through to the plants inside. If you plan to grow during the winter, it is important for plants to receive the maximum amount of sunlight available.

"There is a 1 percent increase in plant growth during the winter for every 1 percent increase in light transmission of your glazing material," says John Bartok Jr., Extension agricultural engineer at the University of Connecticut and an author as well. If you are growing during the winter in a northern location, the glazing on your greenhouse could mean the difference between success and failure.

The glazing material also has a great effect on heating costs. Most growers will heat their greenhouses, either in spring and fall or all through the winter. Obviously, it is wise to choose a covering that will save money.

Glass

The conventional glass greenhouse is the most expensive to construct. It requires a strong, rigid frame and is difficult to work with. Light transmission is high with glass. The newer types of tempered safety glass have a long life span. Older glass and single layers of glass can, with age, become brittle and crack. Glass panels, when cracked or broken, are often difficult for the grower to replace.

Manufacturers of glass greenhouses claim that glass is the most impact-resistant of all glazing. However, I have heard some horror stories of growers hiding under benches during a hailstorm while glass fragments flew through the air!

Rigid Structured Sheets

There are many types of plastic sheets on the market today, with even more being developed. Included in this category are polyvinyl chloride (PVC), plexiglass, fiberglass, acrylic, polycarbonate, and Lexan. Each of these types of sheets has a different level of light transmission. All of the double-layered rigid sheets retain more heat than does glass.

Structured sheets are light and easy to work with. Many of the sheet types fasten directly into the greenhouse frame designed specifically for them. Some types can be screwed onto the frame.

Structured sheets of all types degrade in time and must be replaced. Some actually turn yellow and greatly reduce light transmission. Many kinds of sheets have ridges, which tend to collect dust and dirt, so you must wash them periodically. Check with individual manufacturers as to the specifications of their products. Ask for names of growers nearby using their product. Find out whether they are happy with their choice in coverings.

Double Polyethylene Film

Polyethylene (poly) film seems to be the covering of choice among growers. Poly films are light and thus require a less rigid support. Another benefit of poly coverings is high heat retention. They are most often used in double layers with the hoop-type frames, which makes for a less expensive greenhouse. Glass and structured sheet greenhouses are at least twice the cost of double poly hoop structures. Looking for a cost-effective way to grow indoors? The double poly greenhouse should be your first choice.

Manufacturers are constantly upgrading their films. Poly films are now available that contain additives that absorb and retain more heat. Many have antifog properties to reduce dripping from condensation. Many films have additives that diffuse sunlight, thus creating less shading on plants. Most films contain UV stabilizers to inhibit destruction by the sun. A tri-layer film manufactured by FVG America, Inc. is said to block the UV wavelength that allows the fungus *Botrytis cinerea* to spread.

Most greenhouse films available today should last three or four years. This is an ongoing cost to consider when choosing a covering. Poly films do sometimes tear or get holes poked in them. (Thinking it was great fun, my cat walked up the side of my greenhouse one day.) You must always have a roll of "poly patch" tape on hand to repair damage.

Double layers of poly are inflated with a small air inflation fan. This fan is attached to the side of the greenhouse and blows air between the two layers, much like a balloon. This keeps the poly tight and able to withstand rain, hail, and wind. It also provides insulation against the elements. If the end walls are also double poly–covered, a jumper tube — a plastic tube that allows the air to flow from the sidewall to the endwall — can be installed from the length of the greenhouse to the end wall to provide inflation there.

Poly films are usually packaged in long rolls. Some types are unrolled all the way along the length of the greenhouse on the ground and then pulled over the top. Other films come in fan-fold rolls and are unrolled over the top. Either way, a crew is needed to cover the frame. (This should always be done on a day without wind.) There are also several different types of systems available that attach to a tractor, which will mechanically

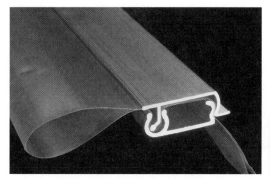
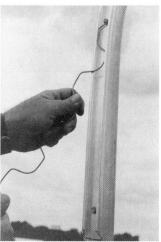

Polyethylene film can be attached to the greenhouse walls with a locking aluminum frame (above) or a spring-type metal wire that fits inside the base rail (right).

lift and unroll the poly on top of the greenhouse. People are still needed, however, to attach the poly to the baseboard.

Most manufacturers have standard-size rolls of 100, 130, and 150 feet. Of course, they will always cut poly to your specifications, but your greenhouse will be less expensive if you size it to match these roll lengths. Allow 2 to 4 feet extra for attaching it to each end wall.

Poly attachment devices. There are many different types of poly locking devices available for attaching the poly to the base and end walls. Most are two-piece aluminum rails. The base rail is attached to the baseboard of the greenhouse, usually using self-tapping screws. This must be done before the greenhouse is covered.

Another type of locking device is a wider aluminum rail that is attached before the poly covering. The poly is held in place by a spring type of metal wire that fits inside the base rail. This type is said to cause less damage to the poly.

Films can also be attached to the wood baseboard by wrapping the poly several times around latex-painted wood laths. The lath is then nailed or stapled to the wood base.

Batten tape is a vinyl stapling tape. It is used over both layers of poly and stapled to the wood baseboard. Some manufacturers do not recommend the use of batten tape, however, because it can tear the poly and pull loose. Both the wood-lath and batten-tape methods are inexpensive and perform reasonably well.

Greenhouse Equipment

Selecting the proper equipment for heating, ventilating, and lighting the greenhouse is crucial for the profitable operation of your business. Utility costs can be expensive, depending on your location and weather patterns. Indeed, many who grow all winter in the greenhouse say that the extra income from summer merely subsidizes the added costs during winter. By using the latest in energy-saving technology, you can realize more profit.

Heating

The even distribution and circulation of warm air is the most important part of providing heat to the plants. Most of the heating in a greenhouse is from solar gain. On the average, however, heating costs are the largest expenditure for the commercial grower operating a greenhouse year-round.

The heating system you will use depends on your location. Growers in the South may need only a heater to take the chill out of the air when temperatures approach freezing. In the North, however, a heating system that produces at least a 60-degree heat rise is necessary. The proper size and amount of Btu a heating system produces are of utmost importance. There is quite an array of heating systems available. Hot-water and steam boilers, and forced-air and radiant heaters are a few of the choices. Fuel can be natural gas, liquid propane (LP) gas, fuel oil, wood, or other combustibles.

Hot-water and steam boilers are designed to be used with piping to distribute the heat. The heat can be placed where it is needed — under benches, around the perimeter, beneath raised beds, or buried in the floor. These systems work well with radiation fin-type heaters. Hot-water or steam boilers can be used with any type of fuel.

Forced-air furnaces usually have lower initial installation costs. They are available in many sizes and models. They use oil, natural gas, LP gas, and some brands use methane gas. Some forced-air heaters are available with the Fan-Jet (sometimes called convection tubes) accessory. This is a kind of blower-and-fan assembly that directs the warm air into a large-diameter clear plastic tube running the length of the greenhouse. It can be hung from the ceiling or close to the floor. Holes are punched in the tube to allow even distribution of air throughout the greenhouse. These tubes do degrade from the sunlight and must be replaced every year or two.

Many growers use a forced-air furnace suspended from the ceiling. The warm air is allowed to blow freely into the greenhouse. Baffles, which are metal fins, direct the air to the sides and floor.

Wood-burning furnaces and boilers can also be used. Situate the burner outside the greenhouse, either outdoors or in an adjacent building. This prevents wayward smoke from building up on the glazing and foliage. Placement of the furnace and chimney should be downwind to prevent any cinders or hot ashes from landing on the greenhouse glazing. Thermostats are available that shut off the blower when the fire stops producing heat. After all, you don't want cold air blowing into the greenhouse.

Wood should not be used as your primary heating system, especially in very cold climates, unless you have a backup system. You may not always want to go to the greenhouse to reload the furnace when it is 20° below zero at 3 A.M.! A nice feature of blowing wood-heated air into the greenhouse is that it is a very dry heat. This is much appreciated on cold, cloudy winter days in the humid greenhouse.

The type of fuel you use depends on your location. In most rural areas, natural gas is not available. Your choices in conventional fuels for primary heating are then oil or liquid propane.

Fuel oil is a little more efficient than propane because it has more Btu per gallon. It comes in different grades and viscosities, but it does seem to burn dirtier. Propane burns clean and the equipment doesn't require much maintenance.

Both fuels require outdoor, aboveground storage tanks. Some fuel suppliers offer a lifetime or yearly lease on storage tanks, or you may buy your own. Fuel suppliers often have programs whereby you can pre-purchase or prepay fuel at a lower cost before the heating season begins.

Each system — and its application to your greenhouse — is different. Do some research to compare heat-loss calculations to determine which system is best for your situation. Greenhouse and heating-equipment manufacturers can give you specific information to help you in this decision-making process.

Cooling Systems

In some parts of the country, cooling is more important in the greenhouse than is heating. But unless you live in an area where the temperatures regularly soar above 100°F, you probably will not need a cooling system.

Most herbs love the heat and, if adequately watered, will grow rapidly in high temperatures. Even mint, an herb that likes cooler conditions, did well in full sun planted on the north side in my greenhouses. Most summer days the temperature was well above 100°F. The stress of high temperatures seems to cause the herbs to produce more volatile oils; thus, they are more flavorful. If you live where cooling is needed, there are several systems available.

Fan and pad systems are the most common method of cooling greenhouses. They consist of exhaust fans at one end of the greenhouse that pull air through evaporative cooling pads at the other end. This causes temperatures cooler at the pad end than at the exhaust end. A good air circulation system within the greenhouse helps to even out the temperature.

Positive-pressure cooling systems use the same evaporative cooling pads as the fan and pad systems, but the cool air is pushed into the greenhouse rather than being pulled through. Most of these systems force the air through the cooling pads and into a convection tube (similar to those used for heating), where it is then pushed by motorized pumps. This creates pressure inside the greenhouse and the hot air is forced out through roof vents. Positive-pressure systems generally end up being more costly than the fan and pad systems because of the increased electricity needed.

A mist system is a method of evaporative cooling that uses fog or mist inside the greenhouse. As this mist evaporates, heat is removed from the air. Misting systems can be purchased as complete systems or rigged up by the grower using "fog" nozzles (available from suppliers). It can also be as simple as attaching a fog nozzle onto a watering wand and spraying mist around the greenhouse manually. Don't use these systems unless the water is extremely clean and without calcium and salt. These minerals quickly clog the nozzles and leave deposits on leaf surfaces.

Another method of cooling with mist or water is to spray the outside of the greenhouse. This will quickly cool the greenhouse, but of course on hot sunny days it will quickly heat up again. External misting must be done frequently to be effective under these circumstances.

Shading material, or shade cloth, is painted on the greenhouse glazing and is often used by those growing ornamental and bedding plants. Some herb growers, especially those in the South, use shade cloth with good results. If you consider using this material, remember that most herbs grow best in full sun.

Roof vents and retractable roofs are the most effective and economical way to cool a greenhouse because hot air — even moist hot air — rises. For further information about these systems, see the section on roof-vented greenhouses on pages 85-86.

Ventilation Fans

Good air quality is critical to produce healthy plants in the greenhouse. This means replacing stale humid air with fresh dry air. You can accomplish this air exchange with ventilation fans.

Every greenhouse should have ventilation fans, even those with rollup sides and roof vents. Ventilation must still occur during the winter, early spring, and late fall when temperatures outdoors are too low for you to roll up the sides. Solar gain is high during these months and greenhouse temperatures can rise quickly on sunny days.

Most fans are installed on the end wall away from the prevailing wind direction. In greenhouses longer than 150 feet, install the fan in the center with fresh-air inlets on each end wall. With end wall installation, the fresh-air inlets should be on the end wall opposite the fan, so fresh air is pulled through the greenhouse.

The fresh-air inlets should be twice as large as the fan opening. It is important to provide a large enough area for fresh air to enter the greenhouse. The fan motor may stall or burn out if there is a limited amount of air to move. This situation can also create a vacuum inside the greenhouse.

A ventilation fan is critical to have in every greenhouse, even those with alternative venting systems.

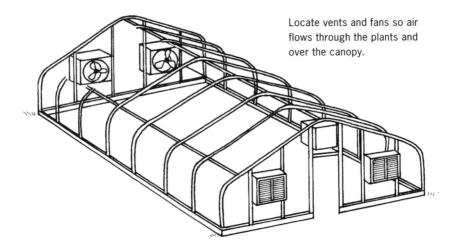

Locate vents and fans so air flows through the plants and over the canopy.

Fan location. The ideal fan setup is two variable-speed or two-speed fans with motorized shutters providing fresh-air intake. Having two fans provides backup in case one fan breaks down. Motorized shutters are activated when the fans come on. The entire system operates automatically when connected to a thermostat. Ventilation needs will change with the seasons. Variable-speed or two-speed fans provide the flexibility you need for all conditions.

Locate fans so that air flows through the plants and over the canopy. The fan

Motorized aluminum shutters open automatically when the fan is turned on.

should have aluminum shutters on the outside. These are not motorized but open and shut as the fan operates. The fan should be covered with hardware cloth or screening inside the greenhouse to guard against injury and prevent lightweight items from being sucked into the fan.

The ventilation fan should be mandatory even for those on a limited budget, but in a pinch you could do without the motorized shutters and thermostat. Fresh-air inlets could simply be windows and doors (with screens) that you open manually. The biggest problem with a manually operated ventilation system is that someone must be there to turn it on and off.

Some growers have installed attic ventilation fans in their greenhouses with success; others have had these fans fail quickly. Not all of the fans will operate for long in the vertical position; many need more rigid support. They are not nearly as expensive as greenhouse fans, and care should be taken when purchasing fans made for the home. Cover them with plastic when not in use to protect them from high humidity levels.

Early-spring and late-fall weather conditions can change rapidly, and greenhouse temperatures will rise very quickly if the sun comes out and ventilation is not provided. This can be lethal to plants, especially seedlings, that are not adequately watered. And high temperatures are stressful to plants, even heat-loving herbs.

To provide adequate summer ventilation, the fan should move one volume of air per minute to a height of 8 feet by 8 cubic feet per minute (cfm). The formula to determine the fan capacity is to multiply the greenhouse width by its length by 8. This figure is expressed as cfm. This figure will help you determine what size fan to purchase. Many greenhouse-supply companies can help you with your ventilation needs. A good salesperson will make the calculations and recommend the proper-size system for your greenhouse.

Fan types. Providing the greenhouse with fresh air and proper temperatures is essential for growing quality plants. Good air circulation is necessary to ensure that the plants can benefit from these conditions. The best way to provide air circulation is the horizontal air flow fan (HAF). These are usually placed above head height in the greenhouse. They operate constantly and keep the air moving in a circular motion. The air also mixes in an up-and-down motion. This constant movement of air provides many benefits to the plants and grower.

A horizontal air flow (HAF) fan provides the best circulation.

By moving the air throughout the greenhouse, the temperatures are more uniform. Cool and warm spots are virtually eliminated, and you'll realize great savings on fuel costs. Because the air is moving constantly, condensation is reduced, thereby making a less humid environment. Leaf surfaces stay drier, thus reducing the chances for plant diseases to take hold. In the summer HAF fans provide a cooling effect on plant surfaces.

One horizontal air flow fan is usually placed 10 feet from the end wall facing the length of the greenhouse. The next fan is placed 10 feet from the end wall in the diagonal corner. This keeps the air moving in a circular motion. Two fans, depending on their capacity, should be enough for a 60-foot-long (or smaller) greenhouse. Check with the supplier for specifications for your setup.

If your budget does not yet allow the purchase of HAF fans, try home-type box fans. These, like attic fans, are not made for constant use or the high humidity levels found in most greenhouses, but the important thing is to keep air moving. In my first greenhouse I placed a series of window fans throughout. They kept the air moving, but they failed on a regular basis. Small oscillating fans can be used to keep air moving in "dead air" spots.

Small, "squirrel cage" fans are used to inflate double layers of poly glazing. They are available in a variety of sizes, depending on the needs of your greenhouse. The blower outlet is attached through the first layer of poly. They usually mount onto the side of a hoop with brackets. The mounting bracket kit is not always included with the fan. Some are prewired with a three-conductor plug; others must be "hard-wired" into the electrical service.

These fans are designed to operate constantly but they do eventually fail. They are relatively inexpensive, so purchase two fans and keep one as a backup. If the poly glazing deflates on a very windy day, it could only take a short time for it to be destroyed. Inflation fans are much less costly than new poly glazing.

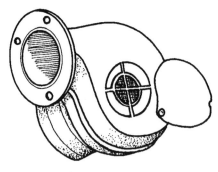

Poly-inflation fans, which come in a variety of sizes, are used to inflate double layers of poly glazing.

Thermostats

The type of thermostat you use depends on your equipment. If ventilation fans are automated with motorized shutters, you'll need a two-stage thermostat. This controls both heating and ventilation equipment and allows the greenhouse environment to be monitored automatically.

Synchronize the two-stage thermostat temperature settings so that there is a two- to five-degree difference between heating and ventilation. The furnace should not run at the same time as the ventilation fan; this could result in exhaust fumes from the furnace being sucked into the greenhouse. These exhaust fumes can do severe damage to plants.

You can achieve better temperature control in the greenhouse with an electronic thermostat. The difference in price between the mechanical and electronic thermostats is marginal. However, electronic thermostats have been shown to be more accurate, thereby saving the grower money in fuel. Protect electronic thermostats from moisture.

Thermostats in the greenhouse must be placed properly to be effective. Place them out of direct sunlight, and not on an outside wall. The most efficient system is to place a thermostat in a weather-guard box and hang it in a central location away from the direct warm air of the heating system. The height of the thermostat should be just slightly above the canopy of the tallest plants. Too high, and there can be as much as a 10-degree difference between the plants and the thermostat.

Computerized operating systems are available that control the operation of all mechanical equipment. These are ideal for growers with a big, multiple-greenhouse operation. They are, of course, expensive, but also cost effective for the large enterprise.

Root-Zone Heating

Heating the soil where plants are growing has many benefits. You can lower the air temperature in the greenhouse and save money in fuel costs. Yields are increased and quality is better, especially if you grow the herbs during the winter.

If the temperature at the root zone is slightly higher than in the air, this has the effect of evaporating moisture constantly from the soil. This is quite beneficial, especially during winter, when the humidity levels are high because the greenhouse is closed tightly against the cold.

There are several methods to provide soil heat. Most can be used either buried in the ground or in raised beds. Some systems, such as the

Basil is a heat-loving herb and the most difficult to grow in winter. This herb really benefits from heated soil. In my experience, most of the other herbs grew well during winter without heated soil.

brand Gro-Mat, can also be used under gravel, mulch, buried in concrete, or on and under benches.

One method uses PVC tubing or metal pipes buried in the soil. Steam or circulating hot water provides the heat. A system such as this usually requires an additional heating plant and can be costly to install. Once the pipes are installed, the grower is locked in to using those beds for the herbs that need soil heat if all the beds are not soil heated.

Electric heating cables allow flexibility in placement, as you can move them easily. Only heating cable specifically made to be buried in the soil should be used. It is available in many different lengths. If lots of cable is going to be used, purchase it in large rolls and install the plugs yourself. Electric cable is available with or without attached thermostats. Thermostats are available that allow four cables to be plugged in. These have a copper temperature sensor that is buried in the soil of one bed, a temperature control dial, and an indicator light.

Use an inexpensive soil thermometer to monitor the soil temperature in beds without thermostats. If heating cables will be used, the grower must install electric outlets in the greenhouse where needed.

Many growers have designed their own system for soil heating. As you plan your greenhouse, consider future soil heating needs you may have.

Lighting

Every greenhouse should have some lighting. You many want to work at night or just enter the greenhouse to enjoy your plants. These lights can be incandescent, fluorescent, or floodlights.

In North America (especially in northern areas), expect very low levels of sunlight from mid-October to mid-February. The low angle of the sun and shorter day length cause plant growth to stop. Perennial herbs go into dormancy during this period and will not grow. Annual herbs also will not grow well during this time. In the greenhouse you are creating an artificial environment for the plants. The plants should react to the conditions as though it were June, not December.

If herbs are to be grown during the winter, supplemental lighting will be needed to extend the daylight hours and increase the amount of foot-candles — a measurement of light intensity; most herbs need at least 1,000 foot-candles to produce good growth — reaching the leaf surfaces. You will see a better yield, shorter growing time, and better-quality crop with supplemental lighting.

Fluorescent lighting, which provides a cool light, has some green-house applications. It is mostly used in germination areas. Place the lights very close to the plants and move them upward as the plants grow. They are not used when high light levels are needed.

Incandescent lighting is mostly considered a supplement to other lighting systems. It does provide light in the far red spectrum, which aids plant growth and branching.

High-intensity discharge lights are most often used by growers because they are highly efficient and have a long life. They are high-wattage lights, however, and require ballast to operate. The ballast may be attached to the light or placed in a remote location. The lights usually weigh between 20 and 40 pounds. A metal reflector placed above the light is used to determine the light pattern. These lights are available in 110 volts or 240 volts. The 110-volt lights can be ordered with plugs or with-out, to be hard-wired into your existing wiring. Bulbs are available in wattages from 150 to 1,000; 400 to 430 watts are the most popular. The bulbs usually cost $50 and up, and have an average life of 24,000 hours.

High-pressure sodium lights, one of several types of high-intensity discharge lights, are the choice of most growers. They are the most efficient at providing the most foot-candles per wattage. These lights have an orange or yellow glow when they are operating. You may be familiar with them — they are often used in street lighting. High-pressure sodium lights produce heat as well as light, an advantage during the winter.

High-pressure sodium lights are highly efficient.

The reflector above the bulb of a high-pressure sodium light directs the light downward to the plants. It must be shiny and kept clean. Some lights can be purchased with coverings over the bulbs to protect them from water damage. These reduce the amount of light reaching the plants and aren't cost-effective. It is better to use extra care when watering and working near the lights.

The easiest way to operate the lights is with a timer. The plants need uniformity and you may not always be able to turn the lights on and off at just the right moment. The timer should be a professional, weatherproof, heavy-duty model. It should have at least two on and off settings. The timer should be installed and wired at the same time as the lights. Consult your electrician as to the proper size of timer for your situation.

The amount of supplemental lighting you'll want is determined by many factors: The crop you are growing, location, the months during which you plan to use the lights, cost of electricity, the type of beds, and the size of the greenhouse are all items to be considered in choosing a lighting system.

Lighting system manufacturers use computer programs that can choose the right system and placement of lights for your specific application.(See Resources for a list of manufacturers.)

Regulating Carbon Dioxide (CO_2) Levels

Plants need carbon dioxide for photosynthesis and to grow well. Carbon dioxide is released into the air by decomposing organic matter, the burning of fossil fuels, and our very own respirations. Soils high in compost and organic matter produce generous amounts of carbon dioxide.

The average level of CO_2 in our atmosphere is 300 to 340 parts per million (ppm). Research has shown that levels as high as 1,000 to 1,500 ppm greatly benefit crop production. At these levels, plant growth can increase by 25 percent and production time can decrease by as much as 50 percent with some crops, including greenhouse vegetables.

Too much carbon dioxide in the air can have detrimental effects, such as lower yields and foliar death. Optimum levels are different for each crop; what is an optimum CO_2 level for one crop may be toxic to another. For example, 1,500 ppm is toxic to cucumbers and 2,200 ppm is toxic to tomatoes. Yet without adequate levels of CO_2, plant growth slows. Growth can completely stop when levels drop to 100 ppm. Carbon dioxide levels outdoors and in well-ventilated greenhouses are usually more than adequate for good plant growth.

Problems arise when greenhouses are closed tightly during cold weather to conserve fuel. Then, as plants use the CO_2, it is not unusual for the levels to drop to 100 ppm.

As you can see, maintaining adequate levels of carbon dioxide during the winter is necessary for good production of fresh-cut herbs. The optimum level for individual herbs is not known, but it may be necessary to supply additional CO_2 in the winter growing house.

If the temperatures are right, the lighting is adequate, and your plants are still not growing well, low CO_2 levels may be the problem. The ambient air level, or circulating levels of CO_2, might be adequate in the winter greenhouse, but not at the leaf surface. If the air is stagnant, the leaves draw the CO_2 from the area nearest them. This creates an envelope of carbon dioxide-depleted air at the leaf surface. You can correct this by providing good air circulation.

Increasing Winter CO_2 Levels

There are several ways we can actually increase the levels of CO_2 during the winter. The first way is to ventilate the greenhouse with outside air as much as possible, but this is difficult, of course, when the outside temperatures are below freezing.

The burning of fossil fuels, such as wood, liquid propane, and natural gas, increases the CO_2 in the air. If the furnace is located inside the greenhouse, the combustion of these fuels should increase the carbon dioxide level to some degree, even in those situations where most gases exit through the chimney.

The addition of compost or decaying organic matter can do much to maintain adequate levels of CO_2 in the greenhouse. Nevertheless, it isn't a good idea to have an active compost pile inside. A better choice is to renew the annual beds with lots of mature compost and organic matter before planting the crop to be grown over winter.

Commercial growers not wanting to risk these "hit-or-miss" methods may want to purchase a carbon dioxide enrichment system (see Resources). There are two types to choose from. The first, and simpler, is the pressurized liquid tank system, which vaporizes the gas into the air. This system requires a separate device to measure the CO_2 level in the air. The other is the carbon dioxide generator, which involves *complete* combustion of fuels. Some units provide automatic monitoring of the CO_2 level in the air. These are not terribly costly — $3,000 to $5,000 — but they do increase the fuel costs for operating the winter greenhouse.

Plants do not use carbon dioxide in the dark or when foot-candles are fewer than 500. If artificial supplementation of CO_2 is used, it should start (if it is an automatic system) one hour before sunrise and stop one hour before sunset. If you employ supplemental lighting, the level of carbon dioxide should be increased during the time the lights are operating.

construction know-how

There are many factors to consider when building a greenhouse. The first priority is to reexamine your goals and needs for a greenhouse before you begin construction. If you fully understand the environment you are going to need (now and in the future), it will be easier to choose the components for the greenhouse.

Work closely with the manufacturers of the various equipment systems so the greenhouse is fully integrated. If all the equipment does not blend together properly, you may end up spending more money to correct the situation. Be very specific when getting price quotes and ordering equipment.

Many manufacturers offer total greenhouse packages. These packages include basically everything you'll need to build a complete greenhouse system except for lighting and soil heating. For the first-time greenhouse builder, this may be the best way to purchase. It eliminates the possibility of omitting any important components. However, hand-selecting various equipment often results in better equipment at less cost.

There are many types of rigid greenhouse systems available today, and each one requires a different method of construction. Some of these greenhouse types may be built by the grower, but many require an experienced contractor. The steel-pipe arch or hoop-type greenhouse can be easily built by the grower with a little help from friends. This is the most popular kind of greenhouse erected today, and it will be our focus here.

Choosing the Greenhouse Site

Choosing the location for the greenhouse is the most important decision you'll make. Consider any future expansion you may wish to do. Even if your plans never call for another greenhouse, do choose a site that would accommodate more than one. Allow at least twice as much room as you think you'll need.

Pick a site that receives unobstructed sunlight, especially from the south. It should be free of hills and ridges, although a hill on the north side could be valuable protection from the cold north winds of winter. Consider the track of the sun during winter. The greenhouse should sit higher than the surrounding area to allow for drainage. You should know where the water will run in the event of large amounts of rainfall or spring floods.

If your location gets snow, make sure to leave enough space around the greenhouse for its removal. Snow slides off the greenhouse and large banks of snow can build up around the base. If not removed, they can press into the sides or block the sunlight.

Choosing a site where outbuildings are already located can be a money- and step-saver. You should have a work and storage area close to the greenhouse. If you can build the greenhouse adjacent to an existing structure, do so, even if it requires some remodeling. Consider the site in relationship to utilities. Bringing water and electricity to the site is expensive, and most power and plumbing contractors charge by the foot.

The driveway to the site should be wide enough to accommodate big delivery trucks. Make sure there are no low-hanging wires or other obstructions. Also, be sure to allow enough room for customer or employee parking.

Orientation of the greenhouse is important. Louis Kren, staff writer for *Greenhouse Grower* magazine, says: "If you're located above 40 degrees latitude, single greenhouses would be best positioned so the ridges run east to west. Below 40 degrees latitude, north to south is the ideal orientation. Ridge and furrow [gutter-connected] greenhouses should always be oriented north to south."

Obtaining a Building Permit

Make sure that your local township board is receptive to the idea of a greenhouse business before you make a purchase. You may have to assure them that this business is agriculture and not manufacturing. Find out if any long-range changes in zoning regulations are planned; changes that occur in the future could allow housing subdivisions to be built close to your greenhouses. In some farming and rural areas there may be no building codes in place, but building codes in a county or township are changing rapidly.

Contact the building and housing inspector as soon as you have a plan for your greenhouse. Find out what building codes are in place for wind and snow loads, property setback, and construction requirements. "Load" refers to the average high wind and total snow fall in a season for your area; the structure must be able to withstand these. Property setbacks require that any structure be built a certain number of feet away from a public right-of-way. Building codes differ from state to state and from county to county. Greenhouse building codes may be nonexistent in your area: This may mean educating local officials about the structure and its uses. If no codes are in place, a board may take the time to adopt a code before issuing you a permit. Allow yourself plenty of time in advance of your planned construction schedule to deal with these possibilities.

Code Definitions

There are three building codes, called model codes, universally used for greenhouses. These codes contain definitions of production and retail greenhouses. Your local inspectors may have adopted one of these. The building inspector will examine your plans and the site before issuing a building permit.

If you are building a poly-covered greenhouse, you may be able to classify it as a temporary greenhouse. This can be a benefit for tax purposes or in the cost of the building permit, because permanent structures are taxed higher. This classification may be hard to obtain, as there seem to be no definite policies in the codes as to what constitutes a temporary greenhouse. The determining factors have to do with manner of heating and type of foundation. If the greenhouse is heated for year-round use, it won't be considered temporary. Likewise, if the foundation pipes are anchored in concrete or if the floor is concrete, the temporary classification wouldn't be appropriate. A greenhouse cannot be classified as temporary if it will be open to the public.

The best way to deal with these issues is to be informed. Contact county and town officials as soon as you find a property you would like to build on. If you already own a farm, contact officials when you arrive at a plan to build a greenhouse. Work on the project with local building officials from conception to completion in order to avoid any costly surprises.

Site Preparation

The pad (land) that the greenhouse sits on should be solid and level. It should be at least 1 foot above the surrounding land to allow for drainage. It may be necessary to hire a contractor with a bulldozer to do this work.

Evaluate where water would run in the event of heavy rain, spring runoff, or a flash flood. If there is any danger of flooding, you may need a waterway to divert the flow away from the area. If this becomes necessary, be sure the contractor you hire is knowledgeable in conservation work. Cost sharing may be available to you with the Department of Soil and Water Conservation.

Locating Utility Lines

You will need to know the location of buried cables and gas lines before construction can begin. Contact the telephone company, electricity supplier, and gas company. They will send someone to the site to locate any underground wires or lines. This service is usually free.

Creating the Greenhouse Floor

The timing of underground installation of water and electrical lines depends on the type of floor to be installed. With concrete flooring, the underground lines must be in place before the floor base is prepared. Sand is usually used as a base for a concrete floor and must be leveled before the concrete is poured. Floor drainage pipes should also be installed at this time.

If the greenhouse floor is to be of gravel, the underground lines can be installed before or after the greenhouse pad is prepared. It is desirable to have at least a few inches to 1 foot of gravel on the floor, depending on the soil type. This allows for water to drain through the gravel, and then floor drainpipes are not necessary. It is easier to place the foundation sockets (pipes) and baseboard before the gravel is leveled.

Be specific when meeting with the excavation contractor about your needs and the timing of other construction. Always get estimates in writing.

Adding Water and Electrical Service

Both water and electrical service lines to the greenhouse should be underground. The depth at which these are buried is governed by local regulations and codes. These two lines should not be buried together in the same trench, although that sometimes is done.

The disturbed ground in a trench moves as it settles, which may occur years after the work was completed. This movement can cause a power line to stretch and break or a water line to leak. Digging to make the repair can be quite difficult if the lines are together or if the electrical wire is 3 feet above the water line in the same trench. The danger, of course, is accidental damage to the power line and possible electrocution of the contractor making the repairs. The building inspector, electrical contractor, and plumbing contractor should all know what the ordinances are in your area.

Water Lines

Where winters are cold, the water line should be buried below the frost line. This could be as deep as 6 feet in far northern parts of the country. The water line should come up inside the greenhouse and be connected to a hydrant. A corner or end-wall location for the hydrant is best. In a very long greenhouse you may want a hydrant at each end or one in the middle.

The hydrant should be of very high quality, as it will be used a great deal. It should also be a brand currently on the market to ensure that repair parts are readily available. A leaking hydrant is a nuisance and costly to repair, and it wastes time, money, and a valuable resource.

Electrical Supply

If you live on the property where your greenhouse is located, provide separate power service to the greenhouse with its own meter. This is useful for tax purposes, in order to divide your home and business expenses.

Your electric company will usually bring power only to the point of the meter. There may be costs involved, especially if utility poles need to be installed. Some costs may be waived depending on how much power you plan to use. For instance, if you will grow during winter and be using supplemental lighting, you may satisfy the dollar amount required by the electrical company to waive some construction costs. Contact the construction department of your electricity provider as soon as you decide on a greenhouse site.

Look at every piece of electrical equipment that you will be using. Most equipment will list the amperage that it uses. Look into the future. Even if you do not plan on growing during the winter, find out what kind of amperage lights and furnaces use. Add these numbers to determine the total amperage. Then double it! This will give you the amount of electrical service you should have.

Most electricians will tell you that a 100-amp service breaker box is all that you'll need. I strongly suggest that you install 200-amp service. This is only a little more expensive to put in at the time of construction as compared to upgrading later. Install weatherproof outlets on both ends of the greenhouse. Again, install twice as many as you think you'll need. Should you want to run a small fan in a corner or set up a germination area, it is much easier just to plug it in rather than using extension cords — and extension cords are a safety hazard in the moist greenhouse environment.

If there is any possibility that you might be using electric soil-heating cable, plan ahead for this. Install weatherproof outlets close to each raised bed. The plug-in cords on these cables are usually quite short.

In some localities the home/business owner may be allowed to do the inside wiring himself. This work will be inspected by the state electrical inspector several times during the work and again upon completion. If the wiring is not done correctly to code, it may be more expensive to fix it than hiring an electrician would have been in the beginning.

Minimum standards for wiring installations are provided by the National Electrical Code (NEC) and probably by state and local governments as well.

 plan ahead!

You must have enough power available in the greenhouse to supply all of your equipment. Not having a large enough electrical service is one of the biggest mistakes made by new greenhouse growers. It is terribly frustrating to have to unplug something before you can plug in something else. The burden placed on an undersized electrical system also can cause machines to lose efficiency and burn out. Don't let this happen to you.

Some local codes may require you to install ground-fault circuit interrupters, which can cause problems in a greenhouse because they are sensitive to moisture. Some codes dictate that wire used in the greenhouse be the more expensive type that can withstand moisture and be placed in a PVC conduit. It is best to have a licensed electrician install new systems and to inspect them regularly.

Laying Out the Greenhouse

The method described below of laying out the foundation and building of the greenhouse is the one we used to build my greenhouses.

Setting Your Greenhouse

Use a compass to set the ridge direction of the greenhouse, either north to south or east to west. The compass needle points to magnetic north, not the North Pole. The difference between the two is called the angle of declination. This varies from region to region and is shown on most topographical maps. If you want to be exact in siting your greenhouse, obtain a map and reset the compass to true north. I have never been that exact when laying out my greenhouses, yet they have always worked just fine.

Locate the first corner and pound in a stake. Then measure the length of the greenhouse and pound in a stake at the opposite end. Now measure the width of the greenhouse on each end and pound in stakes. Measure it diagonally, and then measure it again to be sure

 starting tools

Before beginning the greenhouse layout, gather together:
- a 25-foot, or longer, tape measure
- string or rope longer than the greenhouse length, with each end tied to a sturdy post (or materials for your own means of leveling)
- a heavy hammer
- a level
- enough stakes or posts for each foundation socket
- instructions from the greenhouse manufacturer

it is correct. Make sure the greenhouse will be exactly where you want it and that it has square corners. There are other methods for laying out a building; don't be afraid to ask for help from contractors or knowledgeable friends.

Starting the Foundation

The type of foundation required depends on your location and the type of greenhouse. The foundation for steel, aluminum, or wood-frame greenhouses should be poured concrete or concrete blocks. In northern parts of the country the foundation should extend below the frost line. Your contractor or building inspector will know the codes for your area. Unless you are experienced with this type of construction, the foundation should be built by the contractor.

Steel-pipe arch greenhouses usually have a pipe foundation. The sockets (pipes) for the foundation are usually 3 to 4 feet long and can be set by the grower. Some manufacturers recommend that concrete footings be poured and the sockets encased in concrete or piers. Many greenhouse growers (including myself) have used a simpler method with very good results.

After pounding stakes to mark the four corners of your greenhouse, measure where the foundation posts will be placed and pound in a stake to mark each spot. Most hoops are placed 4 feet apart. Remember that there will be a hoop on each end wall, so be sure to count how many hoops there will be down the length.

Dig holes at each marked spot. You can do this by hand with a post-hole digger or by using a gas-powered auger or post-hole digger. The holes should be at least a foot wider and deeper than the sockets.

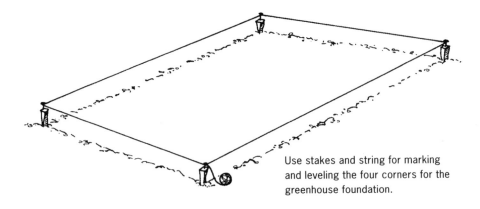

Use stakes and string for marking and leveling the four corners for the greenhouse foundation.

The tops of the pipes should be at least 6 inches above the ground. String a line from one end to the other at exactly the same height, or use whatever system you are comfortable with, to make sure that all the posts are set at the same height. The pipes should be level and straight in the ground.

The top of these pipes usually has a hole drilled through it for bolting it to the hoops. Make sure the hole is on the top and that it is facing the right direction. Look at the instructions again. If there is any doubt, assemble one complete hoop and determine which way the holes should line up. This voice of experience says that this is one mistake you do not want to make!

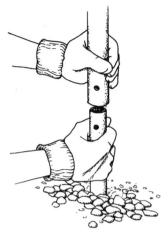

Make sure the hole in the foundation post lines up with the direction of the hoop.

Concrete (a gravel-mix type is best) purchased in 60-pound bags should be mixed in small batches if you are doing this job yourself. Fill the bottom of the hole with concrete and place the socket in the concrete. Put some rocks in the concrete to add volume and stability. Add more concrete until the hole is filled at least halfway. Allow several days for the concrete to dry before attaching the hoops.

Assembling the Greenhouse

It's good to have a small crew to help you in this phase of assembly. The hoops are tall and somewhat heavy so it's helpful to have a person to hold each end of the hoop, and another person to help guide the hoop ends into the sockets.

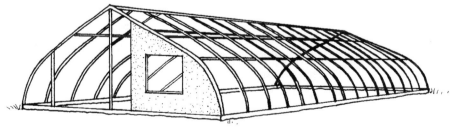

Assembling a gothic or peaked-roof greenhouse may require two people to line up all the pipes at the peak.

Before continuing on with assembly and placing the hoops, gather
the following:
- two step ladders (6 and 10 feet tall)
- power drills
- sockets and wrenches to fit the bolts and nuts supplied
  by the manufacturer
- a long-handled, heavy pipe wrench
- a heavy hammer

Bolting the Hoops

Gothic or peaked-roof greenhouses usually have two sections of
hoops that are to be joined at the peak by a shorter-angled pipe or brackets. Assemble the hoops on the ground and then place them in the foundation pipes if the center height is tall. It is sometimes more difficult this
way and requires at least two people, one on each end of the hoop.
Shorter height hoops can be placed in the sockets first and joined at the
peak using a ladder.

Place the hoop end into (over, in some cases) the foundation socket,
line up the holes, and bolt together. This sounds simple, but it is not
always so. The holes may not line up perfectly, the socket may need to be
turned a bit, the hoop may not slide easily over the pipe. Some manufacturers are more precise than others in drilling these holes. The hoops do
have some flexibility and this helps in maneuvering the hoops to fit.
When bolting the pipes together, be sure
to have the long end of the bolt facing
inward, away from the poly glazing.

Fastening the Purlins

Most greenhouses have three or five
purlins that run the length of the greenhouse. These smaller pipes either interlock
or are bolted together. They are fastened
on the *inside* of the hoops with brackets
supplied by the manufacturer.

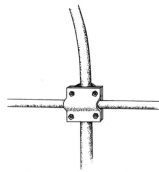

Purlin brackets are fastened on the
inside of the hoops.

Before putting up the purlins, lay them out on the ground as if they were joined together to ensure they are complete. The directions should tell you where to place the purlins. In most cases, one purlin should be directly down the center of the roof. The others are spaced evenly down the roof to a point about halfway down the sidewall.

In most cases it is easier to attach the purlins to the hoops in sections. They are too flexible and long to attach the entire length at one time unless there are many hands to help hold them. Again, always point screw or bolt ends toward the inside.

Attaching the Baseboards

A wood baseboard should be attached around the perimeter of the greenhouse. The baseboard is used so there is something to attach the poly and sockets to. This board should be at least 2 inches by 6 inches. Larger boards up to 2 inches by 12 inches are preferable, so that a portion can be buried in the ground. The boards should be redwood, cedar, or painted with a waterproofing agent.

Place the baseboard on the *outside* of the hoops. It is then attached to each socket using a U-shaped bracket, usually supplied by the manufacturer. You can make your own out of heavy metal pipe strapping. Tightly fasten the board to the sockets, making sure that screws or bolts do not come all the way through the wood; this could cause damage to the poly glazing.

Constructing End Walls

The end walls of the greenhouse are usually built by the grower using locally purchased 2 by 4s. They can be covered with wood, double poly, or rigid double-wall plastic. End-wall brackets attach the hoop to the wood frame of the end wall. These brackets are available from many suppliers, or you can use heavy metal pipe strapping.

Each end wall should have a door, preferably of aluminum. Wood doors are acceptable as long as they are painted well. They tend to warp, crack, or swell, however, due to the greenhouse's high humidity. The doors should be hinged to open away from the prevailing wind.

The end wall opposite the ventilation fan should have an opening for motorized shutters and a double-sized door. The large doorway allows you to move equipment and supplies into the greenhouse with ease. The double doors can be open in warm weather to increase airflow into the greenhouse. A screen door large enough to cover the open double doors will keep out insects, birds, and animals. Hang this screen door from a sliding track so that it rolls away from the door opening.

Solid wood end walls limit the amount of light entering the greenhouse, especially during the winter. They are warmer during winter and provide cooling shade in summer, and sometimes have windows that can be opened. This allows small amounts of fresh air for winter ventilation and furnace combustion. Even end walls to be covered by double poly should be built with 2 by 4s. Wood should be used to frame in the doors, windows, and all around the end hoop. This provides a base to which you can attach the poly for the roof and end walls.

Regular maintenance is important with both types of end walls. Wood must be painted frequently to prevent damage from high humidity. The double poly on the end walls must be replaced each time the roof poly is replaced. Rigid-plastic end walls last longer but are more expensive initially.

Covering the Greenhouse

Several steps must be completed before you cover the greenhouse. Have the air inflation fan ready to install. Be familiar with the directions for installation and choose the spot where it will be. This is something that should be done as soon as the poly is fastened all the way around.

If you will be installing a system to roll up the side walls, some types must be installed before the greenhouse is covered; refer to the manufacturer's instructions. If you plan to roll the poly up by hand, the board that you will attach it to must be installed and painted before covering the greenhouse. Have on hand a roll of poly patching tape in case of accidental tears.

Installing Poly Locking Devices

Your system and materials for attaching the poly to the base should be ready to use. Install the bottom rail on the outside baseboard along the full length of the greenhouse on both sides. It should be a few inches off the ground and level. (See chapter 5, Selecting a Greenhouse, for more information on poly locking devices.)

Painting

Before covering the greenhouse, coat all wood (even the baseboard, if you like) with two layers of white latex paint. Never use oil-based paint in a greenhouse; it can damage the poly. Pay special attention to painting the wood on the end walls inside and outside the structure.

Coat the metal hoops with two layers of white latex paint where they will come in contact with the poly. This is especially important in hot climates. As the sun heats the metal, it

Metal hoops should be painted with white latex paint where they will be in contact with the poly.

may cause the poly to dry out, and also can cause premature film failure. In extremely hot climates, cover the bows with foam insulating tape on the outside.

Puttin' on the Poly

This may seem like a daunting job for first-timers, especially with a tall greenhouse. But it really isn't as hard as it looks. It's a wonderful feeling when it's done to stand and look at your very first greenhouse! Again, it is necessary to have some help for this job, especially with long greenhouses. A tall step ladder is a boon as well.

Install the poly on a day without wind or rain. It only takes a small gust to blow the poly around, which would make the job extremely difficult and could damage this covering. If moisture is allowed to collect

 covering quonset greenhouses

Quonset-type greenhouse frames are easier to cover by pulling on attached ropes than are peaked-roof greenhouse frames — the poly will sometimes catch on the peak and must be loosened from inside. The second layer of poly is pulled on the same way, and will slide easily over the first. Make sure that the first layer is held in place while pulling over the second layer.

between the two layers of poly, it will take a very long time for it to dry. It will also diminish the amount of light the plants receive. When covering the greenhouse on a hot day, do not stretch the poly too tight. As the weather cools, the poly could contract and weaken.

Most poly will come with installation instructions from the manufacturer. Poly that has an anti-condensation coating must be installed with the coating inside the greenhouse. Other types of coated or treated poly must be installed in the correct position or they will not work properly.

Some rolls of poly are installed by unrolling them on the ground down the length of the greenhouse. You then pull the poly over the top. Make sure the ground is clean and dry. The simplest way to pull the poly over the greenhouse is to tie long ropes to the film every 12 feet or so. Throw the ropes over the top of the hoops. If you have a person to pull each rope, this job can be completed quickly.

Some poly comes fan-folded on a long roll. This type of installation requires a few strong people and two tall step ladders or a mechanical device to hoist and unroll the poly. Both rolls are hoisted to the roof and placed over the top purlin. Most of these rolls have an indentation in the middle for just this purpose. Make sure that as they are unrolled, they are in the correct position, inside and outside.

The first roll is hoisted to the top and unrolled just enough so that there is enough to attach to the end wall. The second roll is hoisted up and placed just behind the first roll where it has been partially unrolled. Both rolls are then unrolled, one right after the other. The rolls should be supported by a person on each side on ladders. As the poly is unrolled, the fan folds can be pulled down by people standing outside the greenhouse frame.

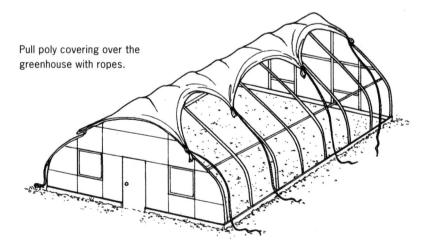

Pull poly covering over the greenhouse with ropes.

Attaching the Poly

Stretch and smooth the poly as much as possible to avoid wrinkles. Water can collect in the wrinkles, causing the growth of algae and weakening of the material. The inside layer should be stretched tight. The outside layer can be a little looser, but without wrinkles, to accommodate inflation.

After the poly is positioned exactly where you want it, begin to attach it. There are several approaches: Start at one end and work down one side; go back and forth between sides and work down the length; start in the middle on one side and work both ways down the length; or fasten it to the end walls at the peak first. There are no set rules; your method depends on many things, such as how many wrinkles must be smoothed out, how long the greenhouse is, how many helpers you have, and weather conditions.

As both layers of poly are stretched, hold them tight over the base rail. Snap the top rail securely into place. With most types, the top rail hooks over the top of the base rail and the bottom is snapped shut.

Some locking devices fit very snugly and may be hard to close. If this is the case, place a block of wood over the rail and smack it sharply with a hammer. That should seal the top rail in place without damaging it. Make sure the poly is in the correct position before locking it in place. Many of these locking devices can cut the poly. You'll have to repair any cuts or tears with patching tape, which may make the material too thick to fit into the locking device.

Poly locking devices can be used to secure the poly to the end walls. This is usually the last step. Wood lath can also be used to hold the poly onto the end walls.

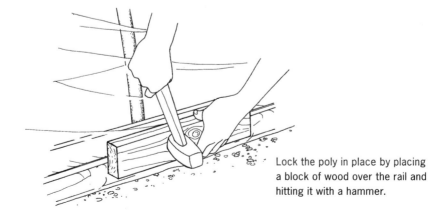

Lock the poly in place by placing a block of wood over the rail and hitting it with a hammer.

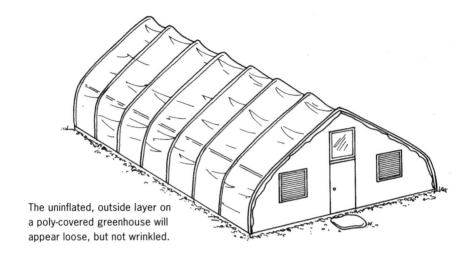

The uninflated, outside layer on a poly-covered greenhouse will appear loose, but not wrinkled.

After the poly is locked in place, paint all the way around the outside rail along with 2 inches of the poly above the rail. This helps to keep the poly cool and prevents stretching. Use a whitewash approved for use on poly. This is especially important on the south side, where the sun strikes.

Double Poly Inflation

Now that the poly is completely attached to the baseboard and end walls, it's time to install the air inflation fan. Try to do this as soon as possible. If a wind should come up before the poly is fully inflated, the film may flap or ripple, possibly resulting in damage. Large greenhouses take a while to become fully inflated.

Use outside air to inflate the two layers of poly — this air is less humid. Moisture allowed to collect between the layers of poly can decrease the amount of sunlight that reaches the plants. Outside air will also have less exposure to chemicals, which can cause poly damage.

A jumper tube can be used to transport outside air to the fan. An inexpensive clothes dryer venting kit works just as well with a little tinkering. Be sure that the outside air vent is screened to prevent insects and debris from entering the fan.

After the fan is installed, turn it on and open it all the way. Your inflation fan should have an air volume control and directions on how to use it. In most cases this control is a flat metal piece that slides across the air intake on the fan. Open this metal plate all the way to push the maximum amount of air between the poly layers.

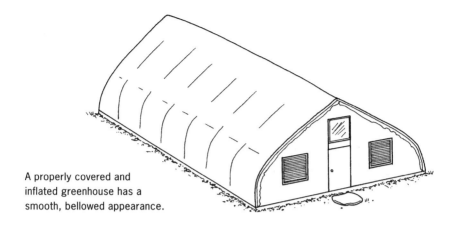

A properly covered and inflated greenhouse has a smooth, bellowed appearance.

Some types of air inflation fans have a variable speed switch. For initial inflation, set the fan speed on high to increase the pressure. Do not leave the greenhouse unattended during this initial inflation, although it can take several hours for the initial inflation to be completed. If the poly becomes overinflated, tears could result, or the poly may pull loose from the baseboard or end walls. We made this mistake during the initial inflation of my first greenhouse. There was so much pressure between the layers of poly that it pulled the baseboard right out of the ground!

As the poly layers begin to fill out, turn down the speed or partially close off the fan's air-flow intake. The outside roof should have a smooth, billowed appearance. The inside layer should dip slightly below the hoops. The film surface should be taut and smooth. When this is achieved, turn down the fan to slow speed or nearly close the air intake. Monitor the inflation closely until you are comfortable that it is inflated properly and that it will stay that way. Examine the greenhouse thoroughly for any signs of leaking. The poly should be airtight.

Many poly manufacturers recommend the use of air or water manometers to measure the correct air pressure between the poly layers. These can be installed permanently in the roof of the greenhouse for ease in checking the pressure. The manufacturer of the manometer will tell you what the correct pressure reading is for your greenhouse.

Many growers use the "hand pressure" and visual methods for determining the correct inflation level. When you push on it gently with your hand, the poly will give a little but still feel firm and tight. With experience, a visual check can tell you whether the greenhouse is inflated properly.

Outside air temperature has an important connection to the pressure between the poly layers because hot air expands and cold air contracts. On hot sunny days, it may be necessary to lower the air pressure because overinflation can cause the poly to stretch and weaken. Wrinkles may then form when the poly cools. On cold windy days you may have to increase the air pressure to keep the film taut.

Setting Up Work and Storage Buildings

Locate the work and storage facilities as close as possible to the greenhouse. A building adjacent to the greenhouse is best for the packaging area. The ground between the two structures should be level, without steps. The walkway between the greenhouse and the packaging building should be enclosed for protection of both you and your herbs during inclement weather.

Packaging Area

The packaging building should be heated for working during cold-weather months. You can package herbs in the greenhouse in winter when the sun is low. However, a workbench area large enough for several people to work around is a waste of valuable greenhouse space, especially during the winter. There may also be health department regulations in your state regarding produce-packaging areas.

If the building has a southern exposure, shade windows in summer to keep the building cool. Provide good ventilation and fans to keep the air moving.

The larger the building, the better. You will need a cooler, or several refrigerators, to store the packaged herbs. You'll need an area for packaging supplies, a place for hand tools, a rest area, rest rooms, and possibly an office. Workbenches 8 or 10 feet long can accommodate several people at one time. Benches at two heights, for sitting and standing, are an added convenience.

The floor should be of concrete or tile for ease in cleaning. A gentle slope toward a drain allows you to hose down the floor. It is surprising how dirty the floor can get after a full day of picking and packaging herbs. If this building will also be used as a potting shed, keep the packaging and potting areas separate. The packaging area must be kept absolutely clean.

There should be running water for washing herbs and hands. A deep sink, such as a laundry sink, is helpful for "swishing" bunches of herbs. The sink should drain into the sewer or septic system. If only fresh water is used in this sink, it may be possible to drain it outdoors and use the "gray" water for irrigation. Check your local codes before making this decision.

Storage Building

A building close to the greenhouse will be needed to store many items. Remember that most plastic bags containing supplies should be stored out of direct sunlight because they will degrade when exposed for long periods of time. Equipment for outdoor growing also needs a place to be protected from the weather. Tractor, tiller, cultivator, rakes, and shovels all can be housed in this building.

Bags of perlite and bales and bags of peat mix and soil are quite heavy — these are items you may not wish to carry a long distance. If you plan to sell herb plants, there will be boxes of new pots, hanging baskets, flats, and plant labels. Many growers reuse pots and flats after washing and sterilizing them. Of course, these supplies have a way of taking up a lot of space until they are washed and organized.

Greenhouse Watering Systems

The Clean Water Act, passed in the early 1970s, focuses on controlling sources of water pollution. The Environmental Protection Agency (EPA) claims that the leading source of river water pollution is agriculture, which includes commercial growers and greenhouses.

Many states now have, and many more are in the process of adopting, regulations that require agricultural operations to collect and recycle irrigation runoff and rainwater from greenhouses. Even if your state does not have regulations now, have a plan in place before beginning greenhouse construction because tighter controls will probably be in your future. These regulations may not affect your operation if you grow only in soil-filled raised beds or have only a small amount of potted plants. Call the EPA in your state to find out the status of any regulations.

New technology is being developed for the collection, filtration, and recycling of water from greenhouses. Because of new developments in this field and the fact that none of this may apply to raised-bed cultivation, only a brief overview of the existing systems will be discussed.

Choosing a Water-Collection System

The questions to be considered before deciding on a water-collection system are: What are the costs? What can you afford to spend? What do you hope to achieve with a system? How much runoff will you generate? How much room do you have available? What is your climate?

There are several types of systems used today by growers of potted plants and hydroponic vegetables. The basic components of a water-recycling system are a means to collect the water, a storage container, a filtration system, and pumps to recirculate the water.

Water collection is usually achieved by placing troughs under the benches containing potted plants. When the plants are watered from above, the water then leaches into the troughs below. Ebb-and-flow benches are flooded with water and the plants take up moisture from below through holes in the bottom of their pots. The excess water is then returned to the storage container or troughs.

Water-holding containers come in two basic types. Tanks made of man-made materials can be stored above- or belowground, indoors or outside, and in lined or unlined ponds. Check water quality at least once a month. Fertilizer levels should be monitored more frequently. High levels of nitrogen can be diluted with fresh water.

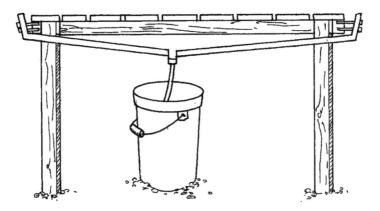

A simple, homemade water collection system utilizes hoses and a bucket to channel excess water.

Because of their heavy weight, tanks should sit on the ground rather than on any type of legs. Keep belowground tanks full or they may collapse due to pressure from the surrounding ground. In cold-winter climates, tanks stored outdoors must be heated to prevent freezing. Open tanks stored in greenhouses can raise the humidity, which is not desirable during cold weather, when greenhouses are closed up tight.

Ponds can be useful for the large grower and those wishing to collect rainwater. They are, however, not without problems. Unlike an enclosed tank, ponds collect all sorts of debris and algae, and filtering is required before pondwater can be used on plants. Safety is another major concern with ponds: They could require fencing, warning signs, and restricting public access. Heavy rains may cause flooding of the pond, but you can control flooding with a series of ponds to catch rainwater before it enters the irrigation pond.

Filtering the Water

The water from all systems must be filtered and disinfected before it is used again on plants. Disease pathogens or fungi may be present in the water, which could cause serious plant damage. There are several technologies available to disinfect water, but they are expensive — and many are less than completely effective. Research is continuing to find a better system to disinfect recycled water so that it is not hazardous to plants. All of these systems require careful planning.

 pond information

Interested in installing a pond? Design help and possible cost-sharing may be available. Contact the closest State Soil and Water Conservation District office for further information. They work in partnership with, and receive technical assistance from, the U.S. Department of Agriculture Soil Conservation Department.

interiors

The interior layout depends on how you will grow herbs and the size of the greenhouse. You can grow in containers, raised beds, beds in the ground, or a combination of these. Ease of use should be the first consideration; efficient use of space, the second.

Benches and Beds

Most growers will want some bench area for potted stock plants and flats of seedlings. Place these on the north or ends of the greenhouse where they won't shade the growing beds. Space-saving hanging baskets can be

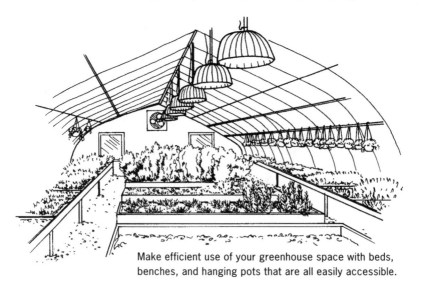

Make efficient use of your greenhouse space with beds, benches, and hanging pots that are all easily accessible.

used for low-growing plants such as thymes, pansies, and nasturtiums. Hang the baskets from the purlins or from long pipes suspended from the hoops, just above head height to prevent bumping into them. These, too, are best hung on the north side to prevent shading of the growing beds. If you hang them directly on the purlins, place some sort of cushioning between the pot hanger and the poly to protect it from damage.

The beds or benches should be accessible from at least three sides. If they are wider than 4 feet, you may have trouble reaching the plants on the inside. Allow enough space between the beds or benches for easy passage and turning around, usually 2 to 3 feet. The main aisle (or aisles) should be wider to allow people and equipment to get around.

Designing the Interior Watering System

There are many elaborate watering systems. Drip irrigation, misting systems, fertilizer injectors, and high hose systems may have a place in your operation. In most cases, the hose and hand-held watering wands will suffice. These allow you the flexibility of watering only those plants or beds that need it.

Hoses and Wands

You may want to have two hoses, one for each side of the greenhouse. Dragging hose all around the greenhouse is one of the minor irritations for most growers! The hoses should be long enough to reach the far end of the greenhouse. Buy top-quality, flexible rubber hose; this won't kink or bend at the connection site. It will stand up to intense heat and to being dragged over gravel or concrete repeatedly.

The two lengths of hose can be attached to a "Y" connector with a valve to direct the water flow down either hose. This can be placed on the hydrant or on a short hose, which saves having to switch hoses often and allows you to direct water to either hose without waste.

Each hose should have a shut-off valve placed permanently on the end so you can turn the watering wands off with ease. All connectors and shut-off valves should be brass or metal of the highest quality you can find. There are some "quick connector" kits available, but these do not stand up to heavy greenhouse use. Always have extra hose washers on hand.

You should have at least one watering wand. The 3-foot length enables you to cover more area. There are several types of nozzles. The best for all-around use is the "water breaker" type; these break the water into many small streams. A nozzle that produces mist or fog can be helpful for misting cuttings or cooling the air and plants during intense heat.

A water-breaker nozzle is most versatile for watering.

Planting Options

There are several ways of using your greenhouse space to grow herbs. In many cases, a combination of methods can be used because of the variety of herbs that will be grown. For instance, stock plants can be grown in large containers, while some tender perennial herbs with large root systems (such as lemongrass) can be planted in the ground.

Ground Beds

For those on a limited budget, ground beds are an inexpensive way to begin. This is certainly similar to gardening outdoors, but there are many disadvantages to growing this way in the greenhouse. My first greenhouse was all in-ground beds and I can attest to the problems. Much time was spent stooping, bending, and squatting to weed and harvest. Can your back and knees take this?

It is easy to damage plants by dragging hoses, wayward steps, or tools. It's also more difficult to control moisture and fertilizer levels in ground beds. If the soil around the perimeter of the greenhouse is dry, the moisture from indoor beds will be wicked away quite quickly. The opposite happens when the outside soil is wet. This, coupled with the humid environment, makes it difficult for the beds to dry.

Ground-dwelling insects, and even snakes, have access to these beds. It is harder to control the spread and sprawl of herbs such as mint, thyme, oregano, and sorrel. Weeds that spread by underground stolon (runners) can find their way into your beds from outside.

Ground beds do have a place in the greenhouse — they are great for growing tall annual herbs such as dill. A better situation for this herb is to make a ground bed only 1 foot high. This way you can control the soil mix and moisture.

Container Growing

Herbs can be grown in pots, 1-gallon size or larger, and placed on benches. If they are on the ground, put plastic under the pots to prevent the roots from finding their way into the ground. Container growing allows better moisture and fertilizer control, but the plants may have to be watered twice a day during hot weather.

While herbs for fresh-cut harvesting can be grown in pots, they simply do not produce enough yield this way, and leaf size is usually smaller. I have heard and read countless stories from growers (including myself) who have found this to be the case.

Container-grown herbs do have a place in the fresh-cut-herb greenhouse. Pots are good for stock perennial herbs. These herbs can be brought out of dormancy in the house before the greenhouse is started in late winter for those who do not grow year-round. They can also be used to try unusual and new varieties without taking up much space. Put these pots outside when warm weather arrives.

Some herbs are best grown in pots. There may be some varieties that you want to have smaller leaves for garnish sales, such as tricolor sage, golden sage, and variegated pineapple mint.

Raised Beds

There are many advantages to growing herbs in raised beds. Raised beds allow you to vary the soil mix for different herbs. Moisture and fertilizer levels can be controlled. You can install soil-heating equipment (initially) without having to dig. Harvesting and weeding are more comfortable, and the herbs just grow better!

Building raised beds in the greenhouse is a simple matter if you make them before the area is full of plants. You can construct them of just about any material with the exception of pressure-treated wood. Most growers make wood frames from 2 by 4s. The sides of the beds are covered with wood boards attached with long heavy wood screws. Used wood boards are inexpensive but the high humidity will rot the boards in less time than new wood. The boards can then pull loose from the frame, causing soil to spill out and plant roots to be exposed.

In one of my greenhouses, the 4-foot by 10-inch raised beds are framed with 2 by 4s. The sides are made from plexiglass sign facings recycled from a small sign-making outfit. The company charged me nothing for these, as it is costly to dispose of them. The plexiglass was placed inside the wood

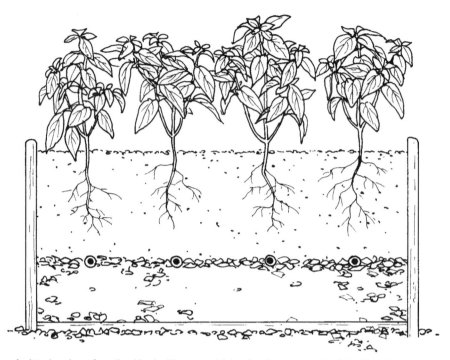

An interior view of a raised bed with a gravel lining for drainage and soil heating pipes installed before planting (above). Raised beds allow you to tailor the soil mixture to a particular herb's needs (below).

Three design options for arranging greenhouse raised beds

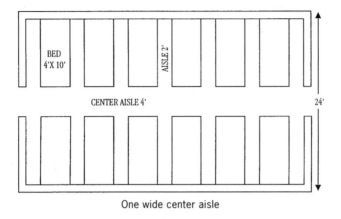

One wide center aisle

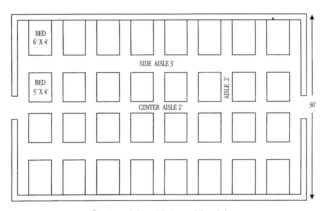

Center aisle with two side aisles

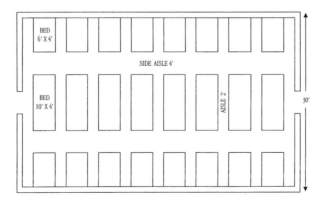

Two wide side aisles

frame, leaving an inch on each corner exposed for aeration. This has worked very well in the beds and will never rot or degrade.

Line the bottom of the beds with gravel for drainage. A few inches to 1 foot of gravel should be adequate. Soil-heating pipes or PVC can be placed below, above, or within the gravel before you add the soil mix. Fill the rest of the bed with soil mix to a couple of inches from the top. This is a tedious job initially and requires lots of material. You can put all of the components into the bed and mix it with hand tools. A soil-mixing machine or cement mixer works well.

The size of the beds should depend on your greenhouse size, design, and your personal preference. Each area of the bed should be accessible without stretching. If the beds are longer than 4 feet, reinforce them in the middle to prevent bowing. The beds should be at least 2 feet high; this allows for a foot of gravel for drainage and a foot of soil mix. Be forewarned, however: The 2-foot-high beds make for many a bruised knee! A better height is 3 feet.

Remember to allow enough space between the beds for easy passage and bending over. Flats of plants can be placed between the beds if they are spaced 20 inches apart. This is a fairly narrow area for most people to work in.

Installing Heating Cable

Don't lay electric soil-heating cable on top of the gravel in raised beds. Most of these cables are made of thin coated wire that could break with the pressure of heavy soil pressing it against the rocks. There should be at least 2 inches of soil mix on top of the gravel before laying in the cable.

Don't let the cable wires touch each other. The easiest way to install cable is to attach it to vinyl-coated, woven-wire fencing. This fencing is purchased in rolls and you can cut it to fit the inside of the raised bed. Attach the wire before placing it in the bed.

Plug in the wire before working with it. This will

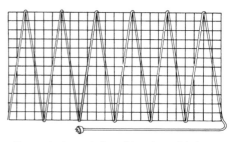

The easiest way to install heating cable in a raised bed is to attach it to vinyl-coated fencing first.

soften it and make it easier to adjust. The wire can be attached to the fencing with small pieces of vinyl-coated "twistie wire." Leave some slack in the wire when you attach it. Allow plenty of room on the plug end of the cable to reach the outlet.

When the cable is completely attached to the fencing, lay it in the bed with the fencing facing *up*. This will protect the wire against damage from tools when you work in the beds.

Soil Mix

Rich, heavy soil is not necessary for growing herbs. But good drainage is absolutely required to achieve high-quality herbs in the greenhouse. Most herbs don't like "wet feet" and will grow poorly in heavy, wet soil. A soil mix that is too heavy holds too much water and doesn't allow enough oxygen to reach the plant roots.

The soil mix for greenhouse raised beds should be light enough to dry out completely within four or five days in the high-humidity conditions during the cool months. It also must be heavy enough for the roots to support large plants without pulling out of the soil. Good soil drainage is achieved by incorporating amendments into the soil. There are several different components to consider when planning your soil mix.

Topsoil

Topsoil provides nourishment and stability to the mix. With the addition of soil, the nutrient-holding capacity increases. A growing mix that includes soil has more buffering capabilities and is more forgiving of nutritional, water quality, and pH problems.

Steam-sterilize this soil to avoid introducing disease pathogens, insects, and weed seeds into the greenhouse. Soil can be sterilized by steaming it at 160° to 200°F for 1 hour. Many authorities recommend steam sterilizing soil for 30 minutes at 180°F. By sterilizing for an hour, you will compensate for the moisture content in the soil, which takes longer to heat up. Soil sterilizers are available from greenhouse-supply companies.

When you think about the quantity of soil necessary to fill your beds, you'll see that sterilizing the soil yourself would take a very long time. Consider buying already sterilized topsoil in bags instead.

Some growers will not sterilize the soil for the initial filling of greenhouse beds. I didn't when filling the beds in my last greenhouse because of the large volume needed (two dump-truck loads) and time constraints. Luck prevailed, as it may for you, and I did not have any serious problems with disease or insects. I wouldn't chance it again, however.

Oxalis (often called wood sorrel and looks like clover) seed, which is a true pest in greenhouse beds, usually is not killed at 200°F. It is very difficult to eradicate once it is firmly established. Steam sterilizing kills many bad organisms, but many "good guys" are destroyed also. A variety of beneficial nematodes and bacteria are now available commercially to introduce into the soil mix.

Peat Moss

Peat moss conditions the soil by loosening it with added texture while at the same time binding the components together when wet. It holds water and gives the soil structure. It is almost universally used by growers, in both beds and potted plants.

Peat moss mixes are available in many different configurations. Most manufacturers mix peat moss with perlite, vermiculite, wetting agents, and an assortment of other ingredients. Wetting agents are added because peat does not absorb water easily at first. It is also available without amendments. Because the nutrients in peat moss are released very slowly, many manufacturers add a "starter" fertilizer to their mixes. It is available without this fertilizer for those who wish to grow strictly organically.

Peat mixes are usually sold in compressed bales. This is a space saver, but it's also a time waster. In order to use the peat from compressed bales, you must break it up; this means taking the time and energy to scrape the peat off the bale and manually break it up into little pieces. Peat mixes are also available that are loose and ready to use.

Always wear a dust mask when working with peat. Small particles can float into the air and into your lungs.

Perlite

This is crushed volcanic rock — the little while pellets you see many times in potting-soil mixes for the home gardener. Perlite adds volume, provides aeration, and contributes to good drainage. This should be a major ingredient in your soil mix.

Perlite comes packaged in large bags and is surprisingly light. The coarse grade is the best to use. As perlite is so light in texture, some of it does break down into a powder when handled roughly. You can break a perlite pellet into a powder with two fingers. This powder is a hazard to your lungs and eyes. Whenever you open a bag, wear a dust mask and wet down the entire contents of the bag with water before taking out a single pellet. This will control the dust.

Sand and Fine Gravel

These are used by some growers in their soil mix because they provide good drainage. I have found that using sand without perlite in the mix compacts the soil. Nonetheless, sand has several uses in the greenhouse.

Sand can be purchased by truckload or in bags from your local building-supply store. It often comes labeled as play sand. If purchased in bags, the sand is often moist and heavy. Either way you buy it, sand should be sterilized before you put it in your beds.

Fine gravel also can have a place in your soil mix, especially if your mix will contain lots of soil. Gravel helps to prevent the mix from compacting and provides good drainage. The gravel should be no larger than "pea rock," so named because it is the size of fresh peas.

Other Soil Amendments

Other materials can be used in your soil mix as well. Compost is a valuable addition to the soil mix because it offers more nutrients, and sometimes disease suppression. If you wish to have a sterilized mix in your beds, though, compost generally cannot be used because it is not sterile.

Some growers add pine bark, rice hulls, shredded coconut hulls, or vermiculite to the soil mix. Vermiculite has a tendency to hold water, so it should not be used in volume in greenhouse raised beds.

Topsoil	Peat moss	Perlite	Sand
30%	30%	30%	10%
25%	40%	25%	10%
25%	25%	25%	25%
30%	30%	10%	30%

recommended soil mix ratios

Soil Mix Ratios

The ideal ratio of ingredients in a soil mix will vary with the climate, greenhouse conditions, and the herbs to be planted. See Part IV: Successfully Growing More Than 20 Herbs and Flowers and the cultural recommendations chart on pages 136–137 for recommendations for each herb's preference as to soil type.

If you are growing in a hot, dry climate, you may want a heavier mix that holds more moisture. In northern areas, a lighter soil mix is more desirable. It will allow the beds to dry out more quickly, especially if you plan to grow year-round in the greenhouse.

In southern Minnesota, where I grew year-round in the greenhouses, I used a very light soil mix in most of the raised beds. In the chart on page 133, my soil mix is the first. The rest of the chart lists ratios used by other herb growers from around the country.

operation and maintenance

The objective of greenhouse growing is to create an environment as close as possible to the natural outdoor conditions most beneficial for plant growth. This is accomplished by controlling the systems in the greenhouse carefully. Most greenhouses do not "run" themselves, especially those operated year-round. Even if all the main equipment (heating, ventilation, and lighting) is automated, you still must constantly monitor the environment.

Managing Microclimates

Each greenhouse has several microclimates. Some areas are colder, hotter, sunnier, or more shaded than others. Conditions and temperatures can vary greatly from one area to another even on the same side of the greenhouse.

Next to the glazing on the south side of the greenhouse will be the hottest during a sunny day but the coolest at night. The north side of the greenhouse is cooler and receives less light. You can determine where these areas are by placing minimum/maximum thermometers in different parts of the greenhouse, which will record the coolest and warmest temperatures reached in a 24-hour period.

These microclimates can be used to your advantage because not all herbs favor the same growing conditions. Some grow taller than others; some like cooler conditions. Basil likes sun and heat; mint prefers cooler conditions. See the cultural recommendations chart on pages 136–137 for growing conditions for most of the culinary herbs.

cultural recommendations for culinary herbs

Herb	Life Span	Soil Type	Light
Arugula	Annual	Loam	Part/full sun
Basil	Tender annual	Rich loam	Full sun
Bay	Tender perennial	Loam	Part/full sun
Chervil	Annual	Moist, rich	Part shade
Chives	Perennial	Moist, rich	Part/full sun
Cilantro	Annual	Rich loam	Full sun
Dill	Annual	Rich loam	Full sun
Epazote	Annual	Light loam	Full sun
Fennel	Annual/Perennial	Rich loam	Full sun
Lemon balm	Perennial	Moist, rich	Part/full sun
Lemongrass	Tender perennial	Moist, rich	Full sun
Lovage	Perennial	Moist, rich	Part/full sun
Marjoram, sweet	Tender perennial	Light, rich	Full sun
Mint	Perennial	Moist, rich	Part/full sun
Oregano	Perennial	Light, dry	Full sun
Parsley	Biennial	Moist, rich	Part/full sun
Rosemary	Tender perennial	Alkaline, light	Full sun
Sage	Perennial	Alkaline, light	Full sun
Salad burnet	Perennial	Light loam	Full sun
Savory, summer	Annual	Light, rich	Full sun
Savory, winter	Perennial	Light	Full sun
Sorrel	Perennial	Moist, rich	Part/full sun
Tarragon	Perennial	Light, rich	Full sun
Thyme	Perennial	Light	Part/full sun

Providing Ventilation

Good ventilation and air flow are two of the most important elements in producing healthy greenhouse-grown herbs. Ventilation refers to the exhausting of greenhouse air while at the same time replacing it with fresh outdoor air.

Fungal diseases can kill plants, and these diseases thrive in humid, stagnant air. Some, such as downy and powdery mildew, render the

Herb	Spacing (inches)	Mature Height (including flowers)	Hardy to (degrees Fahrenheit)	Temperature Preference
Arugula	4	to 12"	25	Cool
Basil	18	to 36"	40	Warm
Bay	pot or 20'	to 25'	15	Warm
Chervil	9	24"	20	Cool
Chives	12	18"	-40	Cool
Cilantro	clusters, 12	to 36"	25	Cool
Dill	12	to 48"	29	Warm
Epazote	6	to 48"	32	Warm/hot
Fennel	18	60"	-10	Cool
Lemon balm	24	24"	-20	Cool
Lemongrass	18	to 36"	40	Warm
Lovage	36	72"	-35	Warm
Marjoram, sweet	12	to 18"	32	Warm
Mint	18	to 36"	-20	Cool
Oregano	18	24"	-20	Warm
Parsley	12	24"	15	Cool
Rosemary	36	to 48"	20	Warm
Sage	36	to 36"	-20	Warm
Salad burnet	18	24"	-40	Cool
Savory, summer	12	24"	33	Warm
Savory, winter	12	8"	-10	Warm
Sorrel	12	24"	-20	Cool
Tarragon	24	36"	-20	Warm
Thyme	18	to 12"	-20	Cool

foliage unsalable. At the present time there are very few compounds to fight fungal diseases labeled for use on herbs. The best way to deal with these diseases is to prevent them in the first place. By removing humid, stagnant air, you eliminate conditions that most fungal diseases like.

When operating any ventilation fan, the air should be drawn through the greenhouse from openings on the opposite end. Close tightly any doors or windows that are located on the same side as the ventilation fan or the air will "short-circuit"; instead of being drawn through the greenhouse, the air will be drawn from the open window or door on the same side as the fan. The fresh air will exit through the fan and the rest of the greenhouse is left with still air.

Make sure that the shutters in front of the fan are opened fully. If not, the fan won't work efficiently and may even stall. The area directly in front of the fan should be unobstructed. This is not a good area to place any plants or people!

Cool-Weather Ventilation

Humidity is more of a problem during cool weather because greenhouses are closed up to conserve energy and fuel costs. Poly-covered houses are especially airtight and require more ventilation during cool weather.

Opening vents and turning on the heat will dry the atmosphere quickly. However, if the ventilation fan (especially at high speeds) and furnace operate simultaneously, the furnace exhaust fumes may be sucked into the greenhouse. Take care to avoid this, as it can damage plants.

Operate fans at a very low speed during the coldest months. If your fans do not have variable speed controls, place a house-type box fan in a window facing outward. You will be surprised at how much humid air is removed from the greenhouse this way. Remember to allow a small amount of cool, dry air to enter the greenhouse to lower the humidity level.

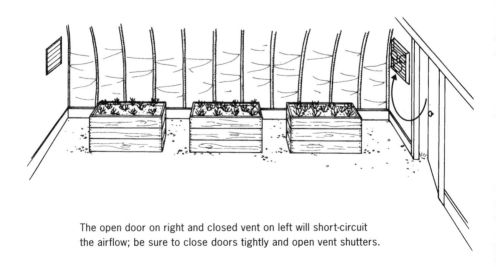

The open door on right and closed vent on left will short-circuit the airflow; be sure to close doors tightly and open vent shutters.

working around temperature changes

The people working in the greenhouse may need more cooling than the herbs! During the hottest months you may wish to arrange your schedule so that watering and harvesting are done early in the morning or in the evening. Harvest before the leaf surfaces become overly warm; this prolongs the shelf life of the cuttings.

Cool-weather ventilation can be a problem because you also must bring fresh, cold, outside air into the greenhouse. If this cold air is mixed with warm air as it enters the greenhouse, it should not damage plants. This is easily done with the convection or fan-jet tube heating system. (See chapter 5, Selecting a Greenhouse, for more information.) You may have to improvise ways to prevent plant damage by placing baffles to direct the incoming air away from plants.

Humidity levels increase dramatically overnight, when the greenhouse is closed up tightly. During this time the plant leaves are wet and fungal diseases can grow quickly. One strategy to counteract this situation is to ventilate the greenhouse completely just before sunset. After ventilating, turn up the heat a bit to warm the air quickly. Many growers will object to this practice because of the added costs in fuel. You must balance the extra fuel use with the possible costs associated from losing many plants to disease.

There is no set formula for cold-weather ventilation. Each situation and greenhouse is different. Experiment until you find the best way to provide your plants with fresh, dry air in the season when it is needed most.

Warm-Weather Ventilation

During warm weather, high humidity in the greenhouse is generally not a problem; ventilation will remove warm air. Many growers leave screened greenhouse doors, windows, vents, and side curtains open night and day to prevent the humidity from rising beyond what it is outdoors. Ventilation fans can be left on low speed overnight if needed.

Daytime ventilation is a necessity during warm weather. If your system is not automated, it is imperative that someone attend the greenhouse to open windows or vents and turn on the fans. Temperatures in the greenhouse can rise to well over 100°F in only a few minutes of sunshine without ventilation.

operation and maintenance **139**

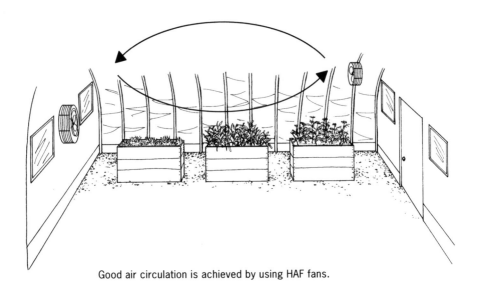

Good air circulation is achieved by using HAF fans.

Increasing Air Circulation

Good air circulation in the greenhouse is an absolute necessity. Do not confuse this with ventilation. Leaf wetness and condensation are the real problems that cause plant diseases to grow. By providing good air circulation, the leaves stay dryer and fungal diseases have a harder time getting started. There must be enough room among plants for the air to circulate, so avoid the temptation to overcrowd them.

There are several things that can be done to ensure good air circulation in the greenhouse. The first thing is to keep air moving at all times. This can be accomplished with fans of any type. Horizontal air flow (HAF), as described in chapter 5, Selecting a Greenhouse, is the best system because these fans move the entire mass of air in the greenhouse constantly.

Again, a variety of fans can be used in place of HAF fans if you place them around the greenhouse. When situating these fans, don't forget the corners where you might have a few potted plants. A small oscillating fan works well in these stagnant-air areas.

Fan Maintenance

Keep fan blades as clean as possible for efficient operation. Be sure to lubricate the fan motors, if needed, on schedule. Many of the newer fan models have sealed bearings and do not require lubrication.

Cover the fan motors with plastic, sealed tightly, when not in use. This will protect them from rusting due to high humidity. Ventilation fan failure during the heat of summer could be a disaster! Have a few portable fans available as backup in case a fan breaks down.

Controlling Temperature

Duplication of ideal outdoor growing conditions should be the goal when controlling the temperature in the greenhouse. Outdoor temperatures fluctuate, and they will in the greenhouse, too.

Heating

Most herbs are heat-loving plants; their growth slows when the temperature goes too low. I have found that all the herbs grow well at a daytime minimum temperature of 65°F. If the sun is shining, the temperature may rise well above 70°, even on the coldest winter days. On very cold, cloudy winter days, it is acceptable to lower the temperature a few degrees, just as conditions would be outdoors.

In the greenhouse, comparable to the outdoors, there should be a temperature difference between day and night. Night temperatures should be at least 5 to 10 degrees cooler than those during the day. The heat should be turned down at sunset. If you are using supplemental lighting, lower the

heater maintenance

It is imperative to have your entire heating system checked by a professional annually before the heating season begins. Fuel efficiency will increase after furnaces are cleaned and adjusted. Heaters that are not working properly can cause air pollution in the greenhouse, which can severely damage plants.

When outdoor temperatures are consistently below freezing, the fresh-air inlets used for furnace combustion should be checked daily. These openings can become iced over due to excessive humidity inside.

If you have fuel storage tanks, be sure to check fuel levels frequently — daily during very cold weather. It would be a shame to lose your entire herb crop because you ran out of fuel!

temperature when you turn off the lights for the night. Herbs grow well with nighttime temperatures between 55° and 62°F. It is possible to lower nighttime temperatures to 45°F, but plant growth may be stunted.

If you are not using automated thermostats, the temperature should be turned up as soon as you enter the greenhouse in the morning. Get into the habit of doing this, since adjusting the thermostats easily can be forgotten if it is already sunny and warm in the greenhouse when you first go in. If the clouds move in and it is cold outside, the temperature can drop rapidly in the greenhouse.

Always provide fresh air for furnace combustion. Burners use oxygen, and if fresh air is not provided, they could use all the oxygen in the greenhouse, especially on very cold nights. This results in incomplete combustion that can release damaging gases into the greenhouse.

The fresh-air opening should be at least 1 square inch per 2,000 Btu of your furnace size. Check with the furnace manufacturer for its recommendations.

Soil Heating

Root-zone heating during the winter is almost mandatory for sensitive herbs, especially basil. Not only does it provide better growing conditions, but it also helps control moisture in the soil and in the plants. Basil is susceptible to soil-borne fungal diseases. These diseases thrive and grow in cool wet soil. When the air is very humid, as it often is during the winter months, soil does not dry out rapidly. This creates perfect conditions for fungi to spread.

Heating the root zone to a temperature a few degrees higher than the air temperature forces the moisture out of the soil and plants through evaporation. This allows you to have better control of moisture and fertilizer. Soil temperatures should be between 68° and 72°F. Basil, however, grows better with the warmer soil temperature of 72° to 74°F.

Root-zone temperatures that are too high can have adverse effects on plants by causing too much evaporation from within the plants or burning the roots. This is stressful to plants and can cause serious damage. The temperature of the soil should be monitored; simply place soil thermometers in the soil and check the reading several times daily.

In addition, today there are various electronic temperature probes on the market. Some can be connected to a computer and some are connected to a system that records the temperature and, subsequently, adjusts it.

Thermostats that have temperature settings should also be checked frequently to make sure they are working properly. Some growers simply stick a finger into the soil to determine the temperature and moisture level. While some people have good luck with this technique, it isn't very accurate. Mistakes could lead to serious plant damage.

Root-zone heating should begin in the fall when the day length shortens and the angle of the sun is lower. This is when the greenhouse is closed up more and the humidity inside rises. Discontinue the heating in late spring as the temperatures rise. The higher angle of the sun will warm the soil during the day and the beds will usually hold the heat overnight.

Cooling

If you live in an area of the country where cooling is needed, see the section on cooling systems in Greenhouse Equipment on page 90. In most locations, greenhouse cooling during the summer is best achieved by ventilation to exhaust the heat buildup.

On very hot days you can cool the herbs by misting or fogging the leaf surfaces: Just attach a fogging nozzle to the watering wand. Be sure to stop misting early enough in the afternoon so that leaf surfaces are dry before nightfall.

Supplemental Lighting

Growers who grow herbs during the winter should use supplemental lighting to extend the day length. This fools the plants into believing it is summer even when it's cold outside. Even while using supplemental lighting, you will see a definite increase in plant growth when the natural daylight lengthens. Many growers celebrate winter solstice with exuberance!

For those in the northern regions of the country, provide lighting from October through mid-March. Growers in the southern states may be able to shorten this period. There are no set rules: Let the growth and condition of your plants be your guide. Experiment until you find what works best for your situation.

There are several schools of thought on when to supply supplemental lighting to your crop. One practice calls for lighting in the middle of the night, usually for 4 hours. This method breaks up a long night into two short ones, and is often used by growers of ornamental plants to induce flowering. Obviously you would want to have your lights on timers to implement this technique. An automated thermostat is useful to turn up the heat during this time.

Another method is to use the lights for 16 hours a day during the winter. This results in better plant growth, but large utility bills. Some growers use the lights all day only on very cloudy days.

The most common method is to extend daylight by lighting in the hours after sunset and before sunrise. My system was set so that the lights would come on a half hour before sunset and go off when the plants had received 16 hours of light, including natural daylight. Day length changes constantly, so change the timer settings every week or two. In early spring, the lighting time can be gradually reduced to only an hour or two.

Authorities recommend that lighting be based on the amount of solar radiation received each day. This requires the use of expensive instruments and numerous calculations. Should you wish to explore this further, contact your local horticultural Extension agent.

Light Maintenance

Take care when working around lights, especially if the bulbs are not covered with a protective lens. Do not splash water on the lights, watch where the ends of shovels and rakes are, and keep the light cords from hanging down. Although these bulbs have a long life span, it is a good idea to have one or two extra bulbs on hand.

Some insects are attracted to the lights and you must wipe these off periodically. Clean the bulbs and reflectors often with a dry soft cloth. Don't touch the bulb with your bare fingers; the oil from your skin could weaken the glass.

Hints for Successful Greenhouse Growing

There are lots of little details to learn about greenhouses, their care, and growing plants within them. Presented below are some hints on a wide range of topics that I hope will help you, your greenhouse, and your business thrive.

Care of Poly Glazing

Some vinyl and other plastics contain chemicals that can cause poly to deteriorate. Do not allow any of these to come in contact with the glazing. Do not spray any chemicals directly onto the poly, as these can also cause the poly to deteriorate. Check the inflation of double-poly greenhouses daily. Over- or under-inflation can cause the poly to weaken.

Always keep on hand a roll of patching tape made especially for poly. Store this tape in a cool dry place close to the greenhouse. Inspect the glazing frequently for tears and holes. If a hole needs repair, clean the area around the tear and wipe it dry before patching. The patch should be at least twice as large as the hole to ensure a long-lasting repair.

Poly films must be washed periodically. Algae and insects accumulate on the film, especially after a long humid winter. Use diluted dishwashing soap and a soft rag and sponge (a sponge mop works well) to gently wash the poly, then rinse it with plain water. Do this cleanup early on a warm morning when the greenhouse will be ventilated all day.

Animals wreak havoc on greenhouse poly. Cats with claws can cause considerable damage in areas you may not be able to reach to repair. Male dogs and cats will often mark their territory repeatedly, which can include your greenhouse. Their urine can "melt" the poly glazing.

Disposal of used poly after replacement can be a dilemma. There are companies that recycle used poly, but it must be kept clean and then packaged in a special way. Look for ads for these recyclers in grower trade magazines, or check with your greenhouse supplier.

Used poly can be cut up and used for tunnels or cold frames for an early start in field growing. (See information on tunnels and cold frames in chapter 10 on page 199.) Local farmers may buy used glazing to cover machinery or hay.

Stormy-Weather Greenhouse Care

Storms of all kinds strike fear in the hearts of greenhouse growers everywhere. But there are steps you can take to minimize damage when storms rage.

Probably the greatest concern is the loss of electrical power to the greenhouse. In cold weather, power outage means the greenhouse won't be heated, which could result in freezing temperatures with much crop loss. In hot weather, loss of power can cause greenhouse overheating, also with possible crop damage. In rural locations, where the water is pumped from the well by an electric pump, loss of power could mean plants die of water deprivation. When power goes, there's also the gradual loss of inflation between the double layers of poly glazing. Even the smallest wind can cause damage to the poly if it is not kept taut.

A gas-powered electric generator could mean the difference between profit and loss for your business. Generators are relatively inexpensive when compared to the amount of money that could be lost due to a power failure.

The generator wattage you'll need depends on how much power you require to run the equipment necessary to prevent plant damage. If you live off the property and don't have an alarm system, consider a generator that starts automatically during a power outage. These are more expensive than the manually operated types, but may be well worth the price.

When a winter storm with heavy snow is forecast, make sure that your fuel storage tanks are full. It could be some time before fuel trucks reach you after a heavy storm. Turn up the heat in the greenhouse to 70°F a few hours before the storm begins so that the snow will easily melt off the glazing. Brace up any weak points on the glazing before you expect heavy snowfall.

Heavy rain and hailstorms are usually not much of a problem in the double poly–covered greenhouse, unless the hail is very large. You may want to increase the pressure slightly between the poly layers to keep it extra taut; this helps the rain and hail to bounce off. Large hail can cause

 be prepared

The key to avoiding storm damage to the greenhouse is prevention. Constantly be aware of weather conditions. Watch or listen to forecasts daily. In the greenhouse, keep a portable weather radio tuned to the local National Oceanic and Atmospheric Agency (NOAA). An electric (with battery backup) weather radio for your home is nice to have. Purchase the kind that has an alarm that sounds when a weather watch or warning is issued by NOAA.

considerable damage to glass greenhouses, especially those covered with older glass, which is more brittle. You may not want to be in the greenhouse under these conditions because rain and hail make a lot of noise in the double-poly house.

Strong winds can cause severe damage to greenhouses of all types. The wind load varies, depending on the direction of the wind, wind breaks, and openings in the greenhouse. All greenhouses should be able to withstand 80 mph gusts if they are built to code. Make sure the area around the greenhouse is free of debris that could blow around and through the poly glazing. Check nearby trees for dead limbs that could fall onto the greenhouse. Check metal chimney pipes to make sure they are properly secured.

All doors, windows, louvers, and vents should be closed tightly. If wind is allowed to enter the greenhouse, it creates more wind inside. With double-poly houses, strong winds create a lifting effect and the greenhouse actually rises and sways with the gusts.

Slightly increase the air pressure between the poly layers to keep them taut. Make sure that the poly is fastened securely to the baseboards and end walls. Consider reinforcing the attachment with wood lath below the point where it is currently fastened. In very strong winds you may want to create a vacuum in the double-poly house. Turn the exhaust fan on low with all the doors, windows, and vents closed. This causes the poly to be sucked tightly against the hoops. If the greenhouse does not have wind (diagonal) bracing inside, you can brace it by bolting metal strapping tape from the ridge to the foundation.

In some situations, such as hurricanes and tornadoes, damage cannot be avoided. Preparing for a storm should also involve plans for what you will do during the storm and after it's over. The most important step immediately following a natural disaster is to make sure there is no threat of danger. Watch out for leaking fuel lines, live electrical wires, broken glass, and weakened structures. After that, do everything you can to protect your plants and business from further damage and losses. The grower who plans ahead will have a better chance of returning to business successfully.

Landscaping around the Greenhouse

The area around the greenhouse should be gravel rather than plants or grass. Many of us who are short of growing space do plant around the outside of the greenhouse. It is certainly more attractive than gravel.

However, be aware that this makes a happy home for insects, which can then easily find a way into your greenhouse. Never allow weeds to grow around the greenhouse.

If grass is growing around the greenhouse, take care when mowing it. Mowing in the wrong direction can cause clippings and debris to be thrown against the glazing and possibly cause damage. Mowing will also disturb insects, and they then will migrate into the greenhouse.

Herbicides

Never use herbicides, especially those containing 2,4-D, in or around the greenhouse. The liquid itself does not have to come in contact with the plants for damage to occur. The vapors themselves can cause severe damage, or even death, to plants — and to humans.

Do not reuse sprayers that have contained herbicides. I know a grower who used a hose-end sprayer containing an herbicide to kill some weeds on her property. She then rinsed out the sprayer and filled it with fertilizer to feed her ornamental seedlings in the greenhouse. Her entire crop of flowers died!

Cleanliness

It is important to keep the greenhouse clean and orderly. Debris on the floor, twisted hoses in the aisles, tipped-over buckets, and the like are accidents waiting to happen.

Don't drop weeds and plant debris on the floor; they may harbor insects and disease and attract "critters." Keep buckets or pails in the greenhouse for this use only. Empty these daily away from your greenhouse.

Clippers, scissors, knives, and hand tools should be disinfected often. Allow them to soak for 30 minutes in a solution of 1 part laundry bleach to 9 parts water. Rinse and dry them carefully. Store these tools out of the greenhouse in a dry location.

If you reuse pots and flats, always disinfect them before putting anything else in them. This same bleach solution works well. Bleach is not "labeled" for greenhouse use, although many growers do use it. There are, however, disinfectants available specifically for greenhouse use.

Be a frequent hand washer. Diseases can be spread from plant to plant by your hands and tools. If you use tobacco in any form, wash your hands before touching plants. Tobacco mosaic virus is spread from tobacco to your plants. Never smoke in the greenhouse, of course.

Critters

Mice, shrews, voles, snakes, and assorted other animals can take up residence in the greenhouse. It may even become a mini-ecosystem, with the snakes there to eat the mice. But you really don't want these creatures in your greenhouse. Besides the "scare factor" of critters darting around, they spread disease and do considerable damage. Mice and other animals burrow and nest in the soil in the beds. Mice will also eat very young seedlings. Anger and frustration are only two of the feelings you'll have upon entering the greenhouse in the morning to find all the tops eaten from five flats of basil seedlings!

If you have mice in the greenhouse and important plants just germinating, don't leave the flats unprotected. Make a barrier with hardware cloth — this will keep the mice out and let air and light in. Hardware cloth is not cloth at all but large mesh (¼- to ½-inch) metal screening. Cut and bend a section into a box shape so the flat is completely contained within it. Be sure it is tall enough so the seedlings can grow for a few weeks without growing through the mesh.

Repeating live traps placed around the greenhouse will help to keep the mouse population under control. Snakes, unfortunately, must be caught by hand and deposited far away. Some growers of ornamentals put cats on mouse patrol in the greenhouse. However, cats are not allowed in greenhouses where food crops are grown.

Watering

When watering raised beds, be sure to soak them thoroughly. Often it is necessary to water each bed many times to be sure the water reaches the roots. Water beds and pots only when they are dry. Overwatering encourages disease.

 herb tip: rosemary

Rosemary becomes quite tall and bushy with age. This can be a problem at watering time. The water takes a while to trickle down to the soil, and the foliage takes just as long to dry. Soaker hoses are a good alternative for these thicker, older herbs.

Water gently and evenly throughout the bed using a small volume of water. A hard, forceful spray of water will compress and compact the surface of the soil. Mud then splashes onto the herbs, making it necessary to wash them after harvest. Young seedlings and tender herbs may be soaked or lying on the soil, even after a gentle watering. Gently lift the herbs and shake the water from them. This will help them to dry more quickly and keep them cleaner.

Always water in the morning, especially during the cooler months, when the greenhouse is closed. This allows the foliage to dry before nightfall. You will not have to water so often during cool weather. Water temperature, especially during the cool months, should be at least as warm as room temperature. Very cold water "shocks" the plants and causes delayed growth or foliar damage.

During warm weather, of course, you will have to water more frequently. The plants will appreciate a cool, not cold, shower during the extremely hot days in the greenhouse, and so might you. Always time these showers so the foliage can dry completely before nightfall.

Soaker hoses are an easy way to water beds with thick or tall perennial herbs. They also allow you to do other things while the beds are being watered. See chapter 23, Rosemary, to learn how soaker hoses can be used most efficiently.

Always keep the ends of the hose and wands off the floor. They could pick up organisms from the ground and spread them to the plants as you water.

Selecting Clothing

Take care when choosing colors to wear when working in and around the greenhouse. Insects are attracted to certain colors and can hitch a ride on your clothes from outdoors into the greenhouse and from one greenhouse to another. Yellow is the worst color to wear — most insects are attracted to it (what shall we blondes do?). Thrips are attracted to blue; flea beetles love white.

Most importantly, always choose clothing that is comfortable and allows for movement.

Using Sounds to Help Your Plants Grow

There is more to growing plants than sunlight hours, temperature, water, and soil fertility. Mother Nature is complete. Pay close attention to a plant's natural outdoor growing environment. What do you hear? Birds!

If you really want to duplicate this complete environment in the greenhouse, especially during winter, play birdsong tapes. I have been doing this for years and the herbs really do respond to the sounds of birds! And so do people!

Tapes are available with a voice announcing the type of bird singing or without human voices. Try both types if you like, but do play these tapes. You can keep a radio turned on in the greenhouse while not playing tapes. Classical music often has the same pitch and tones as birdsong.

Radios and tape players, as well as telephones, fax machines, and answering machines, can be adversely affected by the high humidity often found in greenhouses. Remove them at night, or at least cover them tightly with plastic.

Playing tape recordings of actual birdsongs in the greenhouse helps create a natural growing environment for your herbs.

Nurturing

The good feelings you send to your herbs have a direct effect on their ability to thrive. We all know people who have "green thumbs" and some who don't. Not all people have the ability to grow plants despite their good intentions. Some people just don't know how to nourish plants or don't have the appropriate nurturing skills.

It is not some magical "vibration" that we "green thumbs" have. It is just that we love our plants and have the ability to communicate that to them. When you work with these herbs, do so with love in your heart, no matter how busy you are, and your herbs will thrive.

growing
and nurturing
your plants

starting seeds and plant propagation

There are several ways to produce new plants. Growing plants by seed is called sexual reproduction because the seeds are the result of a union between male and female plant components.

Asexual propagation, on the other hand, is accomplished by vegetative means — through layering, division of the plant or roots, rhizomes or bulbs, and most commonly by rooting cuttings. Tissue culture, or micro-propagation, is another method of asexual propagation. Simply put, each cell in a plant has the ability to reproduce an entire plant exactly like the parent plant. Tissue culture is done by growing plant cells in glass dishes using an artificial medium and aseptic conditions. This method of propagation has many benefits, but it is also quite complicated. More information can be found in books about tissue culture propagation.

You can buy started plants, sometimes called plugs or liners, but it is far better to propagate your own. There may be times when, for one reason or another, you must buy started plants. In this case, buy them locally so you can see that you are getting healthy, pest-free plants of the variety you want. Shipping is very hard on young plants, even when done with the best of methods.

Keeping Records

Good record keeping is essential for the successful grower. It will help you evaluate the performance of herb varieties, what cultural practices work for what herbs, and what doesn't work for you in your situation. This will also be an invaluable tool to help you with crop scheduling.

Keep a notebook in the greenhouse or work area and jot down notes daily. Store this book in a plastic bag to protect it from humidity and dirt; it won't do much good if the ink has run all over the pages from accidental water splashes.

Recording the Growth of Seeds

When starting plants by seed, keep records of the lot number (found on the seed packet) and supplier of the seeds, date planted, temperatures provided, medium used for covering the seed, light or darkness, and the date of germination. The variety, supplier, lot number, and date should also be marked on the planting container.

As the seedlings grow, make notes of percentage of germination, seedling quality, any disease problems, and any fertilizer used and when it was applied. As the plants continue to grow, record where and when they were transplanted (including outdoor average nighttime temperatures), if they were ready when needed, quantity and quality of harvests, disease or insect problems and postharvest holding qualities (the ability of the cut herb to retain its freshness).

Recording the Growth of Cuttings

The same rules apply when starting plants by cuttings. Keep notes of variety, date stuck (placed in rooting media), number of cuttings, what stock plants are used, length of cuttings, rooting medium, rooting hormone and strength used, temperature of bottom heat, misting timetable, when and how many struck roots, and when they were ready for transplanting.

If these herbs, from both seeds and cuttings, will be sold as potted plants, also make note whether they were of acceptable size and quality at time of sale. This information will be of great help in scheduling your seeding or propagation dates in the years to come.

Enter this data into the computer when you have a moment. It can be a great time-saver to let the computer sort out your schedules for planting, along with cultural information needed for each herb. This is especially helpful when you grow a large variety of herbs for sale as potted plants.

germination and growth record

Herb	supplier and lot #	date planted	temp provided
light or dark	covering media	germ. date	% germ.
quality	fertilizer used	date transplanted	where
on schedule?	date of 1st harvest	comments	

cutting and growth record

Herb	stock	date/number stuck	cutting length
media	rooting hormone	temp.	misting timetable
date rooted	% rooted	date transplanted	date of 1st harvest
		comments	

You may want to set up your records on individual cards or in a computer database system for easy reference of how each herb has performed from year to year.

Crop Scheduling

The first step is to assemble the germination and cultural requirements for each seed that you will plant. You'll find this information in part IV: Successfully Growing More Than 20 Herbs and Flowers and in the cultural recommendations charts on pages 136–137. Some commercial-grower catalogs and seed packets also give this information.

Set a target date. In the case of fresh-cut herbs, this is usually the date that you want to begin harvesting. Average length from seeding to first harvest can be found in the production time charts (see pages 160 and 178). Your target date can also be the date for transplanting to the outdoors or the date of sale for potted herbs.

In the case of outdoor planting, you need to know how old or big a plant should be before you transplant it outside. Your target date most likely depends on the last frost-free date in your area or, even better, the average nighttime low temperatures.

When your target date is set, simply count backwards, by weeks, until you arrive at your seeding date. It is better to start a little late rather than too early if you will be transplanting to outdoors. The weather may be unsettled and your ideal planting date can be set back a week or two. Plants that are too big become rootbound and often stop growing. This can really set them back when transplanted.

Starting Herbs from Seeds

There are a number of reasons that you should start plants from seed. There are many more varieties available by seed than there are by plants. With good timing and proper growing conditions, your plants will be ready when *you* need them and not on another grower's schedule.

Seeds are relatively inexpensive compared to buying plants, and you have the opportunity to try new varieties with less cash outlay. Growing plants from seed is so satisfying: It's like watching the miracle of birth each and every time a seed germinates.

Seed Physiology

Seed physiology is a complex subject. For our purposes it is important to understand that seeds perceive the conditions surrounding them and intervals of time. Most seeds germinate only when their environment is correct and after a period of ripening has occurred. Purchased herb seeds will have completed this dormancy — or after ripening — and will be ready to germinate when the proper conditions are present.

Hybrids versus Non-Hybrids

Hybrids are the results of crossbreeding two genetically different plants to produce a new variety. This is done to improve yield, quality, growth patterns, and sometimes to add disease or insect resistance. The crossbreeding may take many years to complete, as is the case with the "F_1" hybrids. F_1 means that the seeds are the first generation after the crossbreeding was completed. Hybrids have some advantages, especially for the grower of edible flowers. The yields and quality are generally better, and the flowers are more uniform.

The seeds of hybrid plants generally won't produce similar plants, which means that you must purchase the seeds when you need more plants, usually at a higher cost. Some hybrid plants can be propagated by stem cuttings. Hybrids are usually a bit more finicky to grow and require more-exact control over their culture.

There are always new varieties of herb cultivars in development. These are not usually true hybrids but are developed by varietal selection. A single plant may have a genetic variation on a single stem — a "sport," as this is called — that has a desirable characteristic. This stem will be

Seed-Starting Information

Herb	Direct Sow or Transplant	Temperature (degrees Fahrenheit)	Light or Darkness for Germination	Seed Depth	Germination Time (days)
Arugula	Direct	68	★	¼"	6 to 8
Basil	Transplant	70	Darkness	¼"	3 to 6
Chervil	Direct/transplant	60	Light	¼"	7 to 14
Chives	Direct/transplant	70	Darkness	¼"	10 to 14
Cilantro	Direct	65	Darkness	½"	7 to 10
Dill	Direct	60	★	¼"	10 to 14
Epazote	Direct	65	Light	¼"	7 to 14
Fennel	Direct/transplant	70	Darkness	¼"	7 to 14
Lemon balm	Transplant	70	Light	★★	7 to 14
Lovage	Transplant	70	Darkness	¼"	8 to 14
Marjoram, sweet	Transplant	70	Light	★★	7 to 14
Oregano, Greek	Transplant	70	Light	★★	7 to 14
Parsley	Direct/transplant	65	Darkness	¼"	7 to 21
Sage	Transplant	70	★	½"	10 to 21
Salad burnet	Direct/transplant	70	Light	¼"	5 to 10
Sorrel	Direct/transplant	70	Light	¼"	5 to 10
Savory, summer	Direct/transplant	65	Light	¼"	4 to 6
Savory, winter	Transplant	70	Light	★★	7 to 14
Thyme	Transplant	70	Light	★★	7 to 14

★ Not a significant factor

★★ Scatter seed on surface

propagated by rooting and through this process a new variety is developed. It may take years of this selective vegetative propagation for enough of the new variety to be ready for the trade.

Most herbs are open-pollinated, which means that they are naturally pollinated by wind and insects. Herbs can be cross-pollinated with other plants of the same species but of a different variety, such as marjoram and oregano. This is another way that new varieties develop. This also accounts for why seeds in the same packet will grow two or more very diverse plants.

Some herb seeds will remain viable (germinate) for up to three years. Some herb seeds will not germinate, or will germinate very poorly, after one year — you must purchase them fresh each year. See part IV: Successfully Growing More Than 20 Herbs and Flowers for which seed should be purchased fresh each year.

Seed that is 1 year old, or older, may germinate, but the seedlings will probably have less vigor and less quality. When planting older seeds, compensate for these tendencies by planting more and culling out the smaller and less vigorous seedlings.

Ordering Seeds

It is best to purchase seed from seed houses or nurseries that supply commercial growers. The quantity per packet is larger, prices are cheaper, and they usually supply more cultural information than the seed houses that cater to home gardeners. Commercial-grower catalogs are sent out much earlier than those for home gardeners because the grower usually must have plants ready for sale at the beginning of the growing season. This may mean starting some seeds in the middle of winter.

Many seed houses offer herb seeds, but not many will have all the varieties you need or the latest in the trade. Some nurseries have exclusive varieties that are available only from them. Get catalogs from as many nurseries as possible. See Resources for sources of seeds and plants.

If you grow only fresh-cut herbs, you'll need just a small assortment of seeds but in larger quantities. You'll still want the best varieties and the lowest prices. If you also grow potted herb plants for sale, you will probably want more variety of plants. This number can reach into the hundreds.

When to order. Place your seed order at least two months, or more, before you need to plant it. It is not uncommon for some seed to be back-ordered. Once I had to prepay an order of a special variety of chives a year in advance to be sure that I could get it. Try to plan ahead for your seed needs for the entire growing season. It is frustrating to run out of seed in the middle of the season and not be able to reorder that same variety.

It is also a waste to have a substantial amount of seed left, especially if it is a variety that must be purchased fresh each year. This job is getting easier now because seed companies are beginning to sell seeds by count rather than by weight, as they did in the past.

production lead times for plants grown from seed

Herb	Days from Germination to Potting Up	Days to Transplant-Ready	Days from Sowing to First Harvest	Days from Regrowth to Second Harvest
Arugula	Direct sown	Direct sown	35 to 45	4 to 10
Basil	18 to 21	21	84 to 90	7 to 14
Chervil	10 to 14	12 to 14	42 to 56	14 to 18
Chives	★	25 to 30	105 to 120	14 to 21
Cilantro	Direct sown	Direct sown	49 to 56	10 to 14
Dill	Direct sown	Direct sown	38 to 50	10 to 18
Epazote	Direct sown	Direct sown	28 to 35	14 to 21
Fennel	14 to 21	21	56 to 70	14 to 21
Lemon balm	21 to 28	15 to 30	90 to 120★★	14 to 18★★
Lovage	21	21 to 28	84 to 98★★	14 to 21★★
Marjoram, sweet	21 to 28	21 to 28	120 to 150	14 to 21
Oregano, Greek	28 to 35	21 to 28	90 to 120	14 to 21
Parsley	14 to 18	14 to 18	42 to 76	7 to 14
Sage	14 to 18	14 to 21	85 to 98★★	7 to 14★★
Salad burnet	14 to 21	28	68 to 75★★	21 to 28★★
Sorrel	14	14 to 21	60 to 70	7 to 10
Savory, summer	14 to 21	14	49 to 56	7 to 10
Thyme	21 to 28	28 to 35	84 to 98	14 to 21
Winter savory	30 to 45	30 to 60	120 to 150★★	21 to 28★★

★Usually transplant directly from flat to growing bed
★★Small yields in first year
Production times are for herbs grown under optimum conditions.

How many. Determine how many plants will fit in each growing bed and plan the number of sowings you will need to make. Some seed companies list in their catalogs the germination rate for each variety. This figure is the percentage of seeds that have germinated in their trials under optimum conditions. If they don't include this, be sure to ask for this figure because it will help you determine the volume of seeds you need. Remember, your growing conditions won't be as exact as theirs.

Planning your seed orders in not an easy task, especially if you order a wide assortment of herbs. Take the time to lay open all the catalogs and compare varieties, quantities, and prices from each company. Many seed companies offer a discount if you order more than a certain quantity, dollar amount, or pay with your order.

Many seed companies want to make substitutions if they are out of a certain variety. Some growers don't mind this, but I always preferred to get the variety I wanted, not what someone has left over. If you send your order by mail or fax, be sure to hand-write "No substitutions without customer's approval" across the order. Don't rely on the little check mark in the box used for this designation on most seed order forms; it can easily be overlooked by the order picker.

Receipt. Always make a copy of your seed orders. When your seeds arrive, check them against your copies of the orders. Make sure each order is correct and write down the lot numbers, date, and any coding on the packet. This may be helpful in getting a refund if the seed does not perform as advertised. Keep the orders for next year, so you know what you ordered from whom. Note which seeds worked well, and which you were not satisfied with.

Storing Seeds

Store the seeds in a cool, dry place. If they are not packaged in hermetically sealed foil packets, place them in moisture-proof plastic containers. Store them in a refrigerator or cool area away from direct sunlight. Allowing seed to lie on the greenhouse bench for a small amount of time, even while planting, is the surest way to a low germination rate.

Seed packets have a way of becoming disheveled in a very short time. It is really helpful to organize and store the seed of like varieties together, perhaps in plastic freezer containers. Ideally you should keep last year's seeds separate from the new shipment to prevent an accidental planting of old seed when you need a high percentage of germination.

Testing for Viability

Before sowing, you may want to test your seeds to see if they will germinate and what percentage of seed is viable. This can be especially useful if it is an important crop or a large planting. It is rare, but sometimes a batch of seeds is "dead," or will not have completed the after ripening or dormancy process. Viability

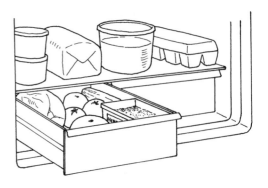

The refrigerator is an ideal place to keep seeds cool and out of direct sunlight.

testing is a simple way to avoid having to utter the phrase "crop failure."

Scatter a few seeds on a moist paper towel, cover with a moist towel, then place in a plastic bag. Keep the seeds at room temperature and inspect them daily after the third day. Germination may take a week or more, depending on the herb.

Some dead seed may be coated with mold or mildew. Viable seed, though ungerminated yet, can usually withstand this attack. The seeds should sprout within a few days of their usual germination time. This will give you an idea of the germination rate as well as the seeds' viability. You can plant these sprouted seeds if you are gentle. Just make sure to place the pointed "tail" downward in the soil.

Sowing Seed Outdoors

Sowing seed directly outdoors is practical for a number of herbs, including cilantro, dill, chervil, chives, and parsley. Direct seeding is especially useful for those herbs that do not transplant well and for successive sowings of arugula, cilantro, dill, and some basil varieties. Many growers start the first crop indoors, transplant it outdoors when the weather and soil conditions are right, then direct sow the second crop at the same time.

The initial sowing of herb seeds outdoors early in the spring can be difficult because most herbs require warm soil to germinate. Weather conditions and insect pests are other factors that make early outdoor sowing impractical for most herbs. Most growers choose to start seeds indoors, under controlled conditions, and then transplant the started plants outdoors (see part IV: Successfully Growing More Than 20 Herbs and Flowers).

Starting Seeds Indoors

Herb seeds require certain conditions before they will germinate: proper moisture levels, warm temperatures, soil or a planting medium, and, in some cases, light. There are a number of ways to provide these conditions. Experiment with different methods and you will find the system that is most convenient and successful for you.

Seeds can be planted just about anywhere where they can receive the proper conditions. This area can be in a basement, on a windowsill, in a separate building, or in a small greenhouse used for just this purpose.

Once your seeds are sown, try not to move the container until the seedlings are up. Seeds will germinate better if they are not jostled about; this can cause them to lose contact with the soil.

When germinating seeds, there are several important things to consider.

Lighting and heat. The germination area need not have natural sunlight. You can supply light with fluorescent strip fixtures, usually 4 feet long, with cool white or warm white bulbs. There should be at least one 25-watt lamp per square foot of growing area. This should supply 500 to 700 foot-candles at the plant surface. The bulbs should be suspended 4 to 6 inches above the leaf surface. If you hang them on chains, you can move them up as the plant grows. After germination, the seedlings should have 14 to 16 hours of light a day. Put the lights on a timer to make this job easier.

These lights usually produce enough heat, in a small area, to maintain proper soil temperature. Supply another source of heat when the lights are turned off. Heating cable under the containers works well.

In a small enclosed area, it may be necessary to ventilate out excessive humidity and heat. A small fan should supply the necessary movement to keep the air circulating and the temperatures even.

 creating a place for germination

Many growers set aside a certain area of the greenhouse for germination. It can be difficult to keep temperatures even in the germination area. One way around this is to build a mini-greenhouse within the greenhouse to keep temperatures and the humidity high. This can be as simple as a small enclosed area with heat and light added or as complicated as a multilevel racking system where each level has its own lights.

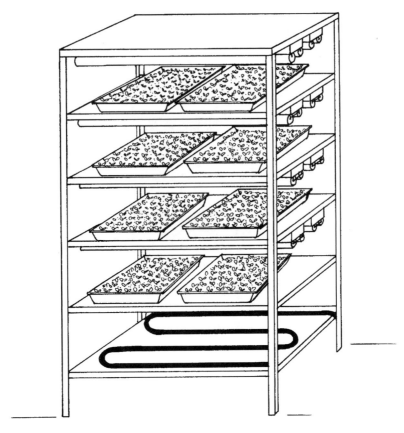

A rack germination system with grow lights and bottom heating cables ensures plenty of heat when placed in a contained area.

Providing bottom heat. The soil temperature of germination flats should be between 65° and 80°F, depending on the herb. This is most often provided by heating the bottom of the flats rather than the air temperature above them. Bottom heat speeds germination and root development.

There are several methods for providing the proper soil temperatures for germination of most herb seeds. The most common is using heating cables on benches. There are several different types of heating cables. Thermostatically controlled rubber propagation mats are the most expensive, but they also work the best. Plain heating cable, even the type meant to defrost ice on roofs (though not recommended for this purpose), also can work well if you monitor the soil temperature in the flats.

A simple and inexpensive method is to wind the cable back and forth on the bench, about 4 inches apart, making sure the cable does not overlap. Hold it in place with duct tape or tacks. Be careful to not pierce or crunch the cable if tacks or nails are used to hold it in place. Measure the flats to make sure that the cable extends all the way to the ends of the containers.

The flats should be suspended an inch or so above the cable rather than sitting directly on it. You can use pieces of wood or heavy steel mesh to support the flats. Old hog panel sections worked well for me.

Ideally, the soil temperature should be monitored. Soil that is too cool delays germination and may even cause seeds to rot. Soil that is too warm can "cook" the seeds or cause the soil to dry out quickly. An inexpensive soil thermometer will be helpful.

Containers. Herb seeds have different germination times, from three days to three weeks or longer, so it is best to use separate containers for each herb. This will eliminate the situation of part of a flat with seedlings up and the other part still needing to be covered to retain moisture.

You can use just about any type of container for starting seeds as long as it has good drainage. Shallow containers are better for starting small seeds because they allow better air circulation around the seedlings and moisture control is easier. The most common and efficient method is to sow seed in shallow flats and later transplant the seedlings into larger pots.

In days gone by, commercial growers used wooden boxes for starting seeds, and some still do. While this is an "earth-friendly" thing to do, wooden boxes are difficult to wash and sterilize to remove disease organisms. Most growers today use plastic flats, sometimes called trays, for starting seeds. The most common sizes are 10 by 20 inches and 11 by 21 inches. It may be necessary to nest two plastic flats for stability because of the added weight.

Seeding is sometimes done directly in plug trays. These are plastic inserts that fit inside the 10 by 20 flats. They are divided into separate cells in varying numbers and sizes, from 36 to 144. "Paks" are rectangular or square thin plastic containers that fit inside various-size plastic flats. These are sometimes used for seeding and are especially useful for small quantities of a variety. You can also sow small quantities of a variety directly into pots where they are to grow.

Used containers should be disinfected before you reuse them. Soak them in a 10 percent bleach solution for 20 minutes and then rinse clean. There are also special disinfectants made for greenhouse use. New flats do not need to be sterilized before use.

Growing medium. The principal roles of any growing medium are to support the seedling and to supply moisture, nutrients, and oxygen. Oxygen and drainage are very important in the germination process and health of the young seedlings. Seeds will germinate in ordinary garden soil or commercial potting soil, but because this soil compacts easily, the seedlings may be deprived of nutrients and oxygen. They may also be exposed to disease organisms and competition from weeds in unsterilized soil.

The choice of most commercial growers for seed germination is a soilless mix. These mixes are sterile and consist of peat moss, vermiculite, perlite, and sometimes other natural ingredients, and can be used as they are right out of the bag. These are available with or without a small nutrient charge. Some growers like to add a bit of sterilized soil, especially if the seedlings will be large. The soil adds more moisture-holding capacity and more stability for bigger seedlings.

When filling containers or flats with a soilless blend, do not pack down the mix. Fill the flat nearly to the top and even it off. The mix will settle as you wet it. The color of these mixes is much lighter when dry, so it is easier to see when moisture must be added. (See chapter 7, Interiors, for more information about soil mixes.)

Moisture. The most important component in the germination process is water management. Too much or too little water can interfere with the process and even cause seed or seedling death. And cold water inhibits seed germination.

It is crucial that seeds have constant, uniform moisture, but the medium you use should not be saturated or leached. Most seeds absorb all the moisture they need for germination during the first eight hours after initial contact. After that, they must have a constant supply of a small amount of moisture to prevent them from drying. If the seeds are allowed to dry out, germination will be delayed or the seed may die.

Seeds drown in a soil medium that is too wet or the tiny root may not easily penetrate the medium. Many problems can arise for the seedling whose root system is stunted from a soil mix that is too wet and compacted. Fungal diseases, such as damping-off, can easily get a foothold in overly wet soil.

After the seeds germinate and covers are removed, monitor moisture levels in the flats. Young seedlings can be seriously damaged by watering from above. Set the containers in a tub of water and allow them to absorb moisture from below. The water in the tubs should not come more than three-quarters of the way up the side of the flat or the flat might sink and the seedlings will be swamped. When the soil on top of the container is wet, remove it and allow the excess water to drain away for a few minutes.

Some supply companies sell tubs or containers to set your flats in to soak up water from below. Many growers use a little ingenuity by coming up with all sorts of containers for this purpose. Alberta Roberson, owner of the Dirty Thumb Greenhouses in Zumbro Falls, Minnesota, uses children's plastic toboggans, and they work quite well!

After the seedlings are well established, usually when they have at least 2 sets of true leaves, you can water them from above. Use a gentle spray so the "babies" are not knocked over. Be sure to water enough so that the soil is moistened thoroughly at the root level.

Sowing seeds. Sowing seed properly ensures a good stand of plants. Seedlings that are crowded may elongate or become stunted and are more susceptible to disease. It is also more difficult to transplant crowded seedlings.

Seeds that are too deep may not germinate at all. A good general rule is to cover a seed to a depth of twice its diameter. Each herb has its own requirements for planting depth, so give seeds individual attention.

Seed should be sown on top of the soil rather than pushed into it. After sowing is complete, cover the seeds with a medium or leave them without covering. They also can be scattered, or "broadcast," over the soil or you can plant them in rows, or furrows. Each grower has his or her own preference: Some say that air circulation and transplanting are better with seedlings planted in rows; others feel that planting in rows is a waste of precious space and that seeds spaced properly in the flat do better and are just as easy to transplant. Again, experiment to see what works best for you.

Seed can be sprinkled right out of the packet. Open the corner of the packet and gently tap the edge to sprinkle the seeds directly onto the soil. You can also pour the seeds into a folded piece of paper to scatter them.

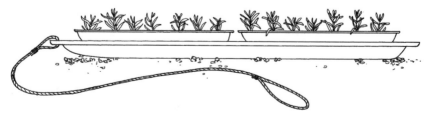

A plastic toboggan makes an inexpensive tub for soaking flats of seedlings.

Fill the flat with growing medium nearly to the top. Level off the mix but do not pack it down. At this point there are several ways to proceed; every grower has his own method and you must find out what works best for you. The one thing that you should *not* do is water from above the flats after the seeds are planted. This could cause the seeds to float, bunch up, or wash away their covering. These are your choices:

- Water the flat from above and allow it to drain well before sowing seeds.
- Place the flat in a tub and let it soak up water from below. Let it drain well before you sow seeds.
- Moisten the growing medium before filling the flat.
- Sow the seeds on the dry growing medium, cover them, and let the flat soak up water from below.

My preference is to water the flat from above (it's quick this way), let it drain well, sow the seeds, then cover them, if needed. The covering will absorb moisture from the soil and remain damp when the flat is covered.

Sowing one at a time with your fingers is an option, but is time-consuming. There are also several types of seed sowers available. Some types are battery powered and they gently vibrate, causing the seeds to drop off the end of a trowel-shaped container. Another type has a metal frame that fits over the standard 10- by 20-inch flat. This has bars, with holes that drop the seeds onto the flat at regular intervals; you simply move the bar across the flat and turn the bar as you go. It comes with several bars with different hole sizes to accommodate even very tiny seeds such as marjoram.

When using the row method, mark the rows about an inch apart or more, depending on the size the seedling will be. Make a shallow furrow with your finger or a tool and scatter the seeds into it. In a 10- by 20-inch flat you can sow anywhere from 40 to 100 seeds per row, depending, of course, on the size of the seed and seedling.

When using the broadcasting method, figure on more seeds to the flat than to the row. Starting at one end or corner of the flat, sprinkle the seeds as evenly as possible throughout the flat. Try to leave a little space

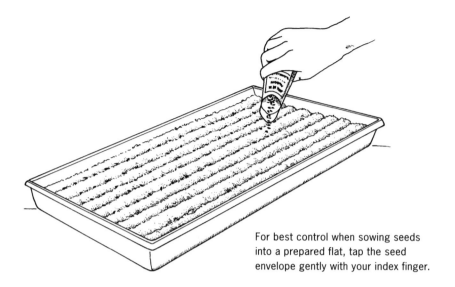

For best control when sowing seeds into a prepared flat, tap the seed envelope gently with your index finger.

between seeds. If a bunch dump out all at once, simply move them around with your finger or a tool. The exceptions to sowing thinly are chives and garlic chives — sow these quite thickly.

Covering seeds. Small seeds, such as marjoram and oregano, need not be covered at all. The seeds slip down into the tiny crevices of the soil and have enough contact with moisture there to germinate. Seeds that require or prefer light to germinate, such as chervil, dill, lemon balm, marjoram, oregano, savory, and thyme, can be covered very thinly. They will still receive enough light to germinate. As some seeds — basil, in particular — swell with water, they may poke through the covering. If this happens, add a little covering to keep them in contact with the moisture.

The growing medium, if sifted finely, can be used to cover the seeds. Finely milled sphagnum moss and vermiculite, both said to discourage damping-off, are commonly used seed coverings. I often use dry sterilized sand to cover seeds. The sand should be fine, with no large grains. If the sand is very dry, it is easy to spread this evenly over the seeds. Put the sand in a trowel or a small container and gently tap the container or trowel to spread it.

The sand should be strained to eliminate larger grains. Sterilize and dry it by cooking it in a large old pot on the stove over medium high heat for an hour or so. Stir the sand thoroughly every so often to expose all the sand to the heat. Cook it for 20 minutes longer after all the sand is dry, turning often. Cooking sand on the stove may seem unconventional, but it can lead to interesting conversation with guests!

G. Basil 4/14
Johnny's Lot #1234S

Label your seed flats with plant name, date, seed
source, and lot number.

After the flats are seeded, mark them with the variety, date, seed
source, and lot number. You can write on plant labels with a permanent
marking pen for flats that won't be covered. For flats that are to be covered, a bit of masking tape placed on the edge of the flat works well. Fold
over the edge of the tape to make removal easier.

Covering the flats. After seeding the flats, it is necessary to prevent
them from drying out. Flats that are placed in a germination chamber or other
small enclosed humid area may not need to be covered to retain moisture.

Many growers prefer to cover the flats with plastic or glass to maintain constant moisture after sowing. This way there is no doubt that the
growing medium will stay moist until you remove the covering, and it
also eliminates the need constantly to check the flats.

These containers can also be placed inside a large plastic bag, but this
method makes it a little harder to check for germination. It is also incon-
venient to remove the flat from the bag with tender seedlings in it. Plastic
wrap can be used to cover the flats, but the kind purchased for kitchen
use is not quite large enough to cover the sides of the flat. You'll need two
pieces, but even then they just don't seem to stick well. Commercial plastic wrap — the kind used in restaurants — is the best thing I have found
for covering flats. It is larger than the home type and will cover your seeds
with room to spare. It is thicker, sturdier, and adheres well to plastic flats.
Ask one of your accounts, or someone you know who works in a commercial kitchen, to purchase a package for you.

Those herbs that require darkness to germinate should also have a
dark cover placed over the plastic covering. You can use anything that will
provide total darkness as a cover. Heavy black plastic works well, and it
can be reused several times because it does not come in contact with the
seed or growing medium.

Those flats that are covered with plastic, but not a dark cover, should
not be placed in direct sun. The clear plastic covering concentrates the
sun's rays and causes overheating and the possible demise of the seeds.

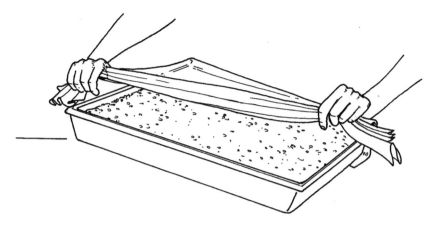

A large sheet of commercial plastic wrap works well for covering a flat of seeds.

Watch the flats closely and remove the covers at the first emergence of a seedling to allow air to circulate and to decrease chances of disease.

Light. As previously stated, the seedlings should have light as soon as they emerge. If your flats are in a lighted germination chamber or under lights already, nothing more is necessary until the plants are too large for the area or ready to harden off for transplanting outdoors. See the section above about germination areas for further information about fluorescent lighting.

Herb seedlings need as much light as possible, so when you remove the cover, place the flats where they will receive as many hours of direct sun as possible.

Fertilizing seedlings. The seedlings will grow much faster and be of better quality if they are given something to boost their growth. Herb gardeners who use a commercial growing medium that contains a "starter" nutrient charge will not have to fertilize the seedlings until they have three or four sets of true leaves.

When the growing medium does not contain a nutrient charge or much soil, the seedlings will benefit from a half-strength fertilizer given once a week when watering. Fertilizing should begin when the seedlings have their first set of true leaves.

The fertilizer can be an organic type, fish emulsion, or a water-soluble commercial recipe from 12-12-12 and up. (For more information about fertilizers, see chapter 10, Growing Herbs.) Some growers prefer to use a fertilizer high in phosphorus, such as 9-45-15, at this time because this nutrient aids the plant in developing its root system.

Transplanting seedlings. The first leaves that grow on a seedling are called cotyledons and their appearance is much different from the plant's regular leaves. These usually dry up and drop off after a few weeks.

Seedlings can be transplanted when they have their first set of true leaves, but waiting until the second set of leaves appear — if the seedlings are not too crowded — will be easier on the little plants. They are easier to handle and better able to handle the shock of transplanting at this stage. Their root systems are more developed, so they must be separated with care to avoid damaging the roots. At this young age the seedlings should be transplanted into pots to continue their growth, preferably in the greenhouse or a cold frame, where they will have some protection from the weather.

Seedlings started in flats should be trans-planted before they get their third set of leaves. Allowed to stay crowded, they will stretch out and the roots will become hopelessly entangled. The stems can mature and harden and the plants might not recover.

Prepare for transplanting by having plenty of soil mix at the ready, so you don't run out before the job is done. You can use a soil-less mix out of the bag or add some soil to increase moisture-holding capacity and provide more stability for the growing plant. Used pots should be disin-

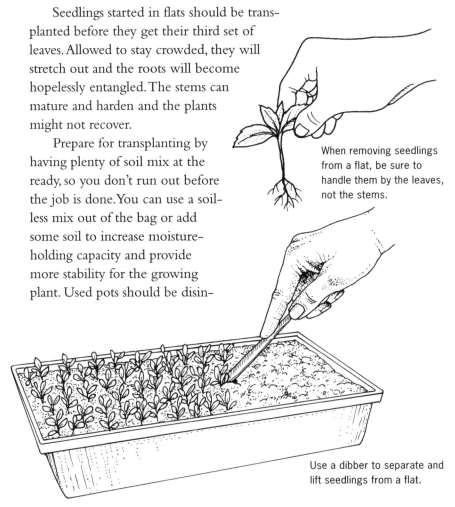

When removing seedlings from a flat, be sure to handle them by the leaves, not the stems.

Use a dibber to separate and lift seedlings from a flat.

fected, or use new pots. It is helpful to fill as many pots as will fit into a flat at one time.

Fill the pots to the brim with dry soil mix and don't pack it down. Three-inch pots are suitable for those plants that will be transplanted to the beds when they are large enough. Use your finger to make a hole in the middle of the soil mix to receive the plant.

To remove seedlings from the flat, place your fingers along the edge of the flat and push them down and under the roots along the bottom. Gently lift out a group of seedlings at a time. You can also use a dibber to lift out the plants. This is a knifelike tool used for lifting seedlings from the growing medium.

Gently separate seedlings that are bunched together. Handle seedlings by their leaves rather than by the stem — young stems bruise when you hold onto them, and this can result in stunting or death of the seedlings.

Place the root end into the hole in the pot to the same depth that it was growing in the flat. Gently press the soil around the stem. Continue until all the pots in one flat are filled. Some of the seedlings may wilt a little but don't worry; they should recover when you water them. Place the newly filled flat in a tub of water to provide moisture from below. Remove when the topsoil is moist and let it drain. You can place the flat in direct sun, but if the seedlings were very wilted, give them a few hours in the shade to help them recover.

If these plants will be for sale, clumps or groups of seedlings can be put into one pot. This will make a fuller-looking pot faster, but the plants may not grow as big in these crowded conditions. As a fresh-cut herb grower, you want each plant to grow to its fullest capacity.

Do not reuse the soil the seedlings were germinated in to start more plants. Do add it to the compost pile, though.

Vegetative Propagation

Many of the perennial herbs should be propagated vegetatively because their seeds, if they produce any, may not grow true to the type of the mother plant. Plants that are produced by vegetative means will be exact copies of the parent plant. These methods of propagation produce usable plants much faster than those started by seeds.

Although there are several ways of vegetative propagation, the most common method employed by commercial growers is stem cuttings.

The other means are certainly easy, and are useful in situations such as increasing the number of stock plants. See part IV: Successfully Growing More Than 20 Herbs and Flowers, for detailed information about how each can be propagated vegetatively.

Some perennial herbs grow roots whenever a stem comes in contact with moist soil. This is useful to the grower, but it can also be a negative situation in beds where overcrowding is a problem.

Simple Layering

There are two methods of propagating plants this way, simple layering and mound layering. Both will produce small numbers of plants without constant attention from you. Herbs that can be propagated by simple layering include marjoram, rosemary, sage, winter savory, and the thymes.

Choose young and flexible stems that will bend easily to the ground. Remove a few inches of leaves from the section of the stem that you want to have contact with the soil. If the stem has thin bark, with a sharp knife scrape off some of the bark from the section that will be in contact with the soil. This wounding helps the plant to set roots more quickly.

Simple layering can be done by anchoring a wounded stem to the soil with a rock.

Remove a small amount of soil from where the stem will be placed in the soil. Put the stem in the depression and cover it with a little soil. You may have to hold the stem in place with a U-shaped wire or a small weight such as a rock. Keep this area moist.

Check the plant in six to eight weeks. You can tell if the stem has set roots by giving it a gentle tug. If you meet with resistance, it is most likely ready to be separated from the mother plant. Make sure that roots are well developed before you cut the new plant away from its parent.

Cut the stem from the mother plant with a sharp knife or scissors. It is now ready to be transplanted to its new home.

Mound Layering

This method is good with herbs that have many branches at the soil level, such as the thymes and winter savory. This is perhaps the easiest way to propagate these herbs. When you do this outdoors in the early fall, new plants should be ready by midspring.

Mound up soil around the plant stems, encompassing the lower branches. Be sure to replace any soil that is washed away by rain or wind. If this is done during the growing season when temperatures are warm, the covered stems should grow roots in six to eight weeks.

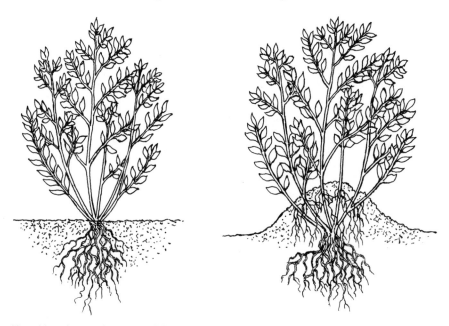

Mound layering can be accomplished easily in plants that have branches at soil level.

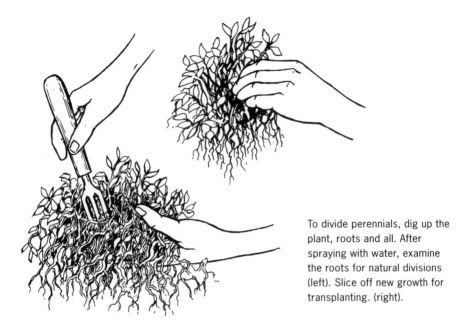

To divide perennials, dig up the plant, roots and all. After spraying with water, examine the roots for natural divisions (left). Slice off new growth for transplanting. (right).

Division

Many perennial herbs can be propagated by dividing the root system. Division is suitable for chives, lemon balm, lovage, marjoram, oregano, sorrel, tarragon, and the thymes. Division is easier on the plant if it is performed during the cool weather of spring or fall.

Lift the plant with a sharp spade, taking care to dig deep enough to get as much of the root system as possible. Wash the soil away from the roots with a spray of water. This allows you to see where the natural division should take place. Those who are in a hurry can do this division without lifting out the entire plant. With a sharp, well-placed spade, slice off the outside new growth from the mother plant. This may seem cruel, but it will yield new plants quickly. It is especially effective with old, big clumps of chives. I had some huge chive clumps that I had to divide with a saw because they were so tough! The shovel method is certainly easier.

Pull or cut apart the plant with a sharp knife or scissors, depending on the variety. In most cases, the new growth around the outside edges of the main plant is what you'll want for new plant production. Discard the old-growth mother plant, or replant it to allow it to grow more plants for you.

Trim off some of the top growth from the newly divided plants. A little trimming of the roots will encourage the roots to grow quickly.

Runners

Mint deserves a special section here because it is the only culinary herb described in this book that can be propagated by runners. Mint sends out stolons, or runners, both above and below the ground, that take root very quickly. This can be to your advantage when you want to increase your mint plants; however, it can be difficult to keep these plants where you want them.

The runners have nodes on them spaced an inch or two apart. It is below these nodes that roots will form. To start new mint plants, simply pull up the aboveground runners, cut them in sections that contain one or two nodes, and plant them — it's as simple as that! New growth will appear in a matter of a week or two.

Stem Cuttings

This method is the choice of most commercial growers for propagating a large volume of plants in the least amount of time. It does, however, require some attention to detail to be successful. All perennial herbs can be propagated this way, and even some annuals, such as basil, will set roots under the right conditions.

Stock plants, from which the cuttings will be taken, should be vigorous, healthy, and disease-free. If they are not in active growth, bring them

Mint naturally propagates by runners, with roots forming at each node.

production time for plants grown from cuttings

Herb	Days from Sticking Cuttings to Potting	Days to Transplant-Ready	Days from Sticking Roots to First Harvest	Days from Regrow to Second Harvest
Lemon balm	24	21	70 to 80	14 to 18
Marjoram, sweet	26	35	75 to 85	14 to 21
Mint	23	27	65 to 75	10 to 14
Oregano	29	35	75 to 85	14 to 21
Rosemary	26	50	120 to 140★	21 to 28
Sage	26	45	70 to 85★	7 to 14
Tarragon	30	50	108 to 115★	14 to 21
Thyme	26	50	80 to 90	14 to 21
Winter savory	28	50	98 to 112★	21 to 28

★Small yields the first year
Production times are for herb cuttings grown under ideal conditions.

out of dormancy far enough in advance of your target date to have several inches of new growth to take the cuttings from.

Tip cuttings, also called softwood cuttings, will root much faster than will old woody growth. Older woody growth — on rosemary, for instance — will not set roots at all. The first new growth of the spring has the highest carbohydrate levels. Because of the higher carbohydrate levels, cuttings taken from this new growth will set roots faster and be of higher quality than those taken later in the growing season.

You can take cuttings just about any time there is active growth of the herb. Those that are taken later in the growing season, as fall approaches, will be less apt to strike roots: The plant is preparing for dormancy, and the growth has slowed.

Tip cuttings can be taken from any branch with new growth. The cuttings should be 2 to 4 inches long. Do not take tip cuttings from plants that are flowering, as these stems won't root well. Cuttings taken from the middle or heel end of new growth stems will also root. Use a sharp scissors or a knife to cut the stems. Be sure to sterilize these tools with a bleach solution or other disinfectant to reduce the possibility of spreading disease. If lots of cuttings are being done at one time, the tools periodically should be dipped in a disinfectant. Be sure to rinse off the disinfectant with clean water.

The rooting medium should be light, to allow for fast drainage and good aeration of the cutting ends. It should also retain some moisture. Regular potting soil is too heavy; cuttings rooted in it usually won't take.

The medium should be sterile to decrease the chances of fungal diseases. Peat mixes, sand, perlite, and vermiculite are common choices. Many growers mix a combination of these ingredients and use different mixes at different times of the season. My favorite is a blend of sterile sand and peat mix. Perlite can be used alone. It is light and provides good aeration, but it is hard to keep moist. It is also difficult to stick cuttings into. The rooting medium should be moistened before sticking the cuttings. Keep it moist, but not waterlogged, during the entire rooting period.

Cuttings may be started in water, but this is not an efficient method when large amounts are needed. Tom DeBaggio, in his book *Growing Herbs from Seed, Cutting & Root,* suggests that many herbs can be started this way. The water must be changed daily, he says, and the cuttings should be potted as soon as the roots are ¼ to ½ inch long.

Containers for rooting cuttings can be anything that is strong enough to support the moist medium, has drainage holes, and is sterile. Shallow containers are best because it is easiest to control the moisture levels in them. Most growers use the standard 10- by 20-inch plastic flats. Some growers, if they have the space, "direct stick" cuttings in the pots in which they are to grow. They use a very light soil mix for the direct-stick method.

Rooting hormones are (usually) made from naturally occurring plant hormones and sometimes an antifungal agent. These greatly decrease the length of time it takes for a cutting to strike roots and increase the number of cuttings that root. Many growers who do not use these hormones report equally good results. Try it both ways to find what works best for you.

Most rooting hormones come in powder form. My favorite brand is Dip 'N Grow, which is in liquid form. A concentrate, it should be mixed with water before you use it. It can be mixed in different strengths — a lighter mix for easy-to-root herbs and a stronger mix for those that are hard to root. Because it is a liquid, a number of cuttings can be treated at one time, so it is a great time-saver.

Light is an important component of rooting cuttings successfully. They should have 12 to 14 hours of bright light daily. They should be kept out of direct sunlight, however; the sun causes them to dry and wilt quickly before they have struck roots. Shade the cuttings from direct sunlight to keep them cool and help prevent wilting. If the cuttings are in the greenhouse, build a tent over them using porous material to allow air circulation. You can buy shade cloth for this purpose or, as I did, use old washed sheer curtains as a tent.

Bottom heat will stimulate the cuttings to set roots. The temperature of the rooting medium should be between 70° and 80°F. The air temperature should be a little cooler, in the upper 60s, to help avoid wilting.

Mist the cuttings frequently to keep them from drying out. You may have to mist them every 5 to 10 minutes if the temperatures are very warm, especially after just sticking the cuttings. The number and frequency of mistings can be decreased after a few days. By observing the cuttings closely, you will be able to tell when to cut back on the misting. The cuttings will stand erect and appear adequately watered. Misting can be done with a hand-held spray bottle; with a hose end or Fogg-it nozzle; or with an elaborate, computer-controlled system.

Enclose the cuttings in a plastic dome to help them retain moisture. This cuts down on air circulation, though, so be sure to monitor the cuttings for any signs of disease.

Some growers allow the cuttings to dry and wilt slightly between mistings, believing that it aids the cuttings in striking roots faster. The danger in this method — and it can be fatal — lies in not misting the cuttings often enough. Others choose to provide misting practically as soon as the last droplets dry on the leaves. Experiment to see which method you prefer.

Fertilizing the cuttings is a relatively new concept. Cuttings seem to set roots faster and are of better quality when they are fertilized regularly during the cutting stage. One method employed by some commercial ornamental growers is to use a water-soluble fertilizer during the initial wetting of the medium before sticking the cuttings. Half-strength fertilizer is applied every time the medium needs watering until the cuttings are transplanted. Supply bottom heat when using this method. If you do not use this method, apply fertilizer as soon as the cuttings start to develop roots.

Procedures for Stem Cuttings

Before you begin to take cuttings, have all your equipment gathered and ready to use. The containers should be filled and moistened and set to receive the cuttings. This avoids delays in sticking the cuttings, so they don't have a chance to dry out or wilt. Work in an area out of direct sunlight.

Cutting. Using sharp scissors or a knife, cut 2 to 4 inches off the tips of the new growth. Try to cut at or just above a branch or leaf to encourage the mother plant to branch. Place the cuttings in a plastic bag, small cooler, or between wet paper to prevent them from drying out. If you limit the number of cuttings you take at one time to 25 or 50, there will be less chance of the cuttings drying out.

Cut off the bottom of the stem ¼ inch or so below a leaf node. (The node is the swollen area of the stem where the leaf is attached.) This cut can be straight or at a slight angle. Use a very sharp knife or scissors so the stem is not crushed. Remove at least half the leaves from the lower end of the stem. Allow at least four leaves to remain on the tip of the cutting. Less leaf surface is exposed to dry out, but enough leaf is left to feed the cutting.

Strip off the leaves by running your fingers downward on the stem. With very tender stems, such as new-growth rosemary, cut off the leaves using a small sharp scissors. Be careful when removing the leaves so that the stem is not broken. The new growth on tip cuttings can be very soft and fragile. Try not to leave any leaf fragments on the stem; these will rot when in the medium.

Prepare the medium. Using a small pointed tool or dibble, poke holes in the rooting medium at a 45-degree angle to receive the cuttings. They should be just far enough apart so that the leaves don't touch. Some growers prefer to crowd the cuttings with their leaves overlapping. This makes for poor air circulation, but usually results in more cuttings to the flat taking root, if they stay healthy.

When using perlite or vermiculite alone, this can be a tedious job because the

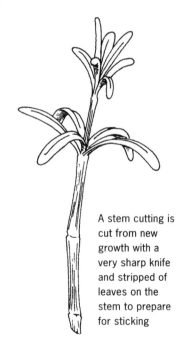

A stem cutting is cut from new growth with a very sharp knife and stripped of leaves on the stem to prepare for sticking

granules tend to fall back into
the hole you've just made. It is
easier to make sloping furrows
of the perlite, lay the cuttings
in, and then cover them with
more perlite.

Dip tips of stem cuttings in a
rooting hormone before planting,
if desired.

Dip the cuttings, if you
choose to, in rooting hormone.
When you use a powdered product,
dip them one at a time and gently
shake off the excess. When using a
liquid formula, a bunch of cuttings
may be dipped at one time, but they
must remain in the liquid for a few
seconds.

Insert the cuttings, stem ends down, in the holes that you have
made. Place them so that one half or so of the stem is in the medium. It is
not necessary to firm the medium over the cuttings unless the holes you
made were a lot bigger than the stems. It is helpful to have a small spray
bottle filled with water to mist the cuttings as you work.

More than one herb variety can be placed in a flat. Be sure to label
the container with the variety, date, and any other special information
you want to keep track of. Mist the cuttings well. Place them over the
bottom heat in a shaded area. Mist them every 10 or 15 minutes, if you
can, for the first few days. Misting is usually not required at night unless
it is very warm.

Water. Keep a close eye on the medium to make sure that it remains
moist, and water when needed. Perlite used alone may need watering
more than once a day. Water perlite very gently because the dry granules
are light and will blow away when watered with force.

Watch the cuttings closely, give them tender loving care, and
within a few weeks they should be striking roots. A gentle tug on the cutting, if met with some resistance, will tell you that roots have begun to
grow. A rooting medium that contains a lot of sand may hold more tightly
onto the cutting. You can gently remove a cutting and look for a callus
forming on the bottom or node of the stem. (A callus is a swelling from

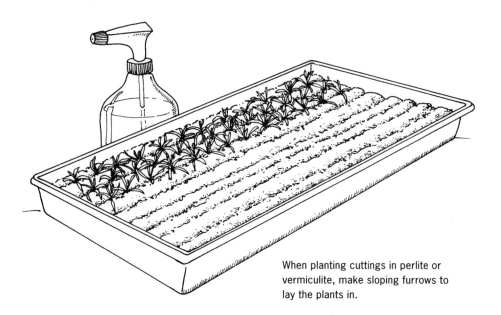

When planting cuttings in perlite or vermiculite, make sloping furrows to lay the plants in.

which the roots will grow.) Remove any cuttings that turn brown, look diseased, or are clearly dead. This will lessen the chance of the whole flat becoming infected.

When the cuttings have small roots, ¼ to ½ inch long, they usually begin to put on some top growth. This is the time to give them a dose of half-strength fertilizer and remove them from the bottom heat. Gradually expose them to more direct sunlight or place them in the greenhouse in a partially shaded area.

The cuttings can be transplanted when the roots are ½ inch to 1 inch long. They should be planted into 3- or 4-inch pots so that they can "grow on" until they are ready for sale or their final growing bed.

growing herbs

Herbs are actually many different plants from a number of families, each with its own cultural requirements and preferences. Some of these vary a great deal from each other, but if we provide good growing conditions, all should do well.

Most herbs grow best when conditions are similar to the climates of the countries where they originated. Many herbs of commercial importance are native to the Mediterranean region and other places where the climate is hot and dry. (The exceptions to the hot and dry conditions are mint, chervil, parsley, and lemon balm.) It is not always possible to duplicate these conditions, but that should be our goal as commercial growers.

Specialized information is included in part IV: Successfully Growing More Than 20 Herbs and Flowers. The charts on pages 136–137 provide quick access to this information.

Most herbs can be grown successfully in the ground.

Direct Seeding

Some herbs — mostly annuals — do not transplant well and should be sown directly where they are to grow. Arugula, chervil, cilantro, dill, and nasturtium should all be direct-seeded.

Direct Seeding Outdoors

When direct seeding outdoors, expect losses due to uneven germination, bird and animal scavengers, and weather conditions. Heavy rains will wash away seeds or float them out of the soil; muddy runoff may cover them too deeply to allow germination. Lack of rain may dry the seeds out before germination occurs. Always locate your direct-sown-seed beds close to a water supply.

Direct seeding can also result in uneven stands (the number of seeds that germinate and grow to maturity) of plants. This is a waste of valuable space and inefficient as well. Plants grown by direct seeding may need thinning. Leaf number and size are reduced when plants are overcrowded. This increases the number of leaves needed to fill bunches and packages. Smaller leaf size is less attractive to the consumer, too.

To thin, gently pull the seedlings from the soil so as not to disturb the roots of nearby plants you wish to keep. Or cut them off at soil level before they have their second set of true leaves. If the seedlings are cut at a later stage, they may regrow.

Transplanting

For the commercial grower who needs a large volume of plants, seed should be started indoors and transplanted to the outdoor beds. This allows better control over the germination rate and early growing conditions. You'll save money in seed costs because the plants are spaced exactly where they should be in the beds and no thinning is required. Transplanting ensures a 100 percent stand of plants, uniformity, and that bed space is used efficiently.

Some herb seedlings, such as marjoram and thyme, are tiny and tender and can succumb to the elements if directly sown outdoors. They require an investment of time and effort to ensure their survival, and that is not cost effective for the commercial grower.

In addition, many herbs are slow growing and need a head start of several months in order to produce enough foliage to harvest in one outdoor season. This is especially true in cooler northern climates with a short growing season. Most of the slow-growing perennial herbs are usually transplanted twice. For a large volume of plants, start seeds in flats. Transplant the seedlings into bigger pots, 3 or 4 inches in size, to grow in until they are big enough to transplant to their final beds. If only a few plants are needed, they can be seeded in the pots in which they will grow.

Choosing an Outdoor Location

The location and design of your outdoor growing fields depends on what you intend them for. Display gardens for public viewing should be located close to the main sales area. If your business will not be open to the public, the fields can be in any area that is convenient for you.

The most important consideration is that the area receive full sun. Herbs can grow with as little as 6 hours of direct sunlight a day; for maximum production, however, they should have at least 10 hours. The exceptions are those herbs that like partial shade and cooler conditions. Even mint, though, grows better in full sun if it has moist soil conditions.

The other main consideration for herb beds is air circulation. Most herbs are sensitive to fungal diseases. When air circulation is inhibited, these diseases will attack outdoor herbs. This is especially true where the climate is hot and humid. The herb beds should be away from walls, windbreaks, buildings, tall fences, and plants that would impede air flow.

Ideally the herb beds should be located close to a water supply. The fields should have easy access for tractors, tillers, and people. Consider that workers will be entering and leaving the fields carrying equipment and harvests.

Controlling Pests and Wildlife

Many of the typical wild animals such as deer, raccoons, rabbits, and groundhogs are attracted to vegetable gardens more than to herbs. If you grow a vegetable garden, plant it some distance from your herb fields because animals, too, might like to spice up their meals with a garnish of fresh herbs. These creatures can also cause harm by trampling the herbs on the way to dinner.

Rabbits find many herbs tasty and will eat them to the ground. Groundhogs, moles, and mice do their damage by burrowing in the ground around the herbs. The deer in my area are quite fond of basil.

Methods of keeping these marauding animals out of your fields abound. A 6- to 8-foot-tall fence, buried a foot in the ground, will do much to keep animals away. It'll work even better if it's electrified. However, even an electric fence won't stop a hungry bunch of raccoons. One moonlit night I watched a raccoon family cheer on a determined male as he climbed over the tall electric fence, screaming in pain all the way, to help himself to my sweet corn!

A large dog on patrol, especially at night, is the best way I have found to protect crops from animal damage. Dogs themselves can cause damage, of course, by trampling plants while on duty, but it is easy to train a dog not to enter an area of bare ground. Most dogs, in the heat of a chase, may momentarily forget this lesson but the harm done at this time will probably be considerably less than what could have been if they were not on the job.

Preparing the Soil

Despite some differences in needs, most herbs will grow fairly well in the same soil as long as it provides good drainage, proper nutrition, and appropriate pH level. Reduced yields and poor quality are the results of poor soil.

Good soil preparation is especially important for perennial herbs. These herbs will stay in their beds for a long time. Renewing the soil there can be difficult because many of these herbs have a wide and deep root system.

Determining drainage capacity. The most important element of soil for growing herbs is good drainage. Here's a test to tell whether your soil has adequate drainage: Fill a planting hole, 6 inches wide and deep, with water. If the water disappears in less than an hour, that's good drainage. The ideal soil will contain enough large particles to ensure that oxygen reaches the roots of the plants. These particles also create conditions that allow the soil to drain quickly. Sand, fine gravel, and perlite are amendments you can add to soil that lacks aggregates. Perlite has a tendency to crush when worked or tilled repeatedly, so use it sparingly.

Organic matter, which breaks down to become the beneficial decomposed matter called humus, supports the plants and provides nutrients and water-holding capacity. It also acts as a bonding agent to keep soil components together. Compost, rotted manure, peat moss, leaves, grass clippings, and cover crops all provide organic matter.

These materials should be tilled into the soil and allowed to decay before planting. It is impossible to add too much organic matter as long as it has time to decay. Fresh manure will burn plant roots and should not be used because it may contain organisms that could cause illness in people.

Compacted soil can seriously undermine plant growth because it drains poorly and does not allow enough oxygen to reach the root zone. Take care not to over-till. This has a tendency to break up the large particles in the soil, which results in less oxygen and water movement. Over-tilled soil compacts easily. Try to reduce trips over the field with machinery; even walking on it compacts the soil.

Determining pH. Every region has different soil types, which can vary from field to field as well. Deficiencies and imbalances in the soil can be determined by soil testing. This is especially helpful with new fields. You can buy test kits and do this yourself, but more accurate results will be obtained by using a professional soil test laboratory. Most state universities offer soil testing through the local Extension service. They'll supply you with soil sampling bags and instructions. Private labs offer more in-depth testing, but at a higher cost.

Some local fertilizer companies also offer limited soil testing. All services should supply you with guidelines to interpret the results and suggestions for correcting problems. Be forewarned that many labs will suggest using chemicals to rectify any problems.

The pH of soil (and water) is, simply put, a measurement of acidity and alkalinity. The scale runs from 0 to 14. The low end is defined as acid (sour) and the higher numbers, sometimes referred to as "sweet," are alkaline. Neutral on the scale is 7. Many plants prefer a slightly acid soil in the range of 6.5 to 6.9. Most herbs grow best in soils with a pH between 6.2 and 6.8. Rosemary, dill, basil, and parsley prefer the slightly higher (more alkaline) soil.

If the soil pH is not in this range, the plants can't take up the necessary nutrients. If the herbs are failing to thrive, despite proper growing conditions, an out-of-balance pH may be the problem.

The pH levels, in the field and greenhouse, should be checked every year or two. Check new fields several months before planting. The water used in the greenhouse should be tested yearly. Water with a low or high pH can gradually change the pH levels in soilless mixes.

Many types of test kits are available with different degrees of accuracy. A complete soil test from an accredited laboratory is best. The composition of the soil, organic matter, nutrients, micronutrients, secondary elements, and pH are all important. If these elements are not considered together, measures taken to correct a single pH test may not solve the problem and may even cause serious damage to plants. Having said this, the following are general recommendations to adjust problems associated with an individual soil pH test.

To raise the pH in acid soils, add some agricultural lime. There are several types and particle sizes of lime available. The finer grinds work faster but also tend to blow away in a strong wind. Because quicklime and hydrated lime are more soluble in water, they work faster. Till the lime into the soil to a depth of 4 inches. This should be done three months to one year before planting. Fall applications are preferable.

Wood ashes, marl (a mixture of chalk and clay), and ground shells will raise pH but they take longer to react with the soil. Application rates vary with each material and type of soil. These rates should be supplied by the manufacturer or supplier.

In perennial beds, scratch lime into the surface or spread it on the soil surface. It will take some time to raise pH levels with this technique. Lime moves down through the soil at the rate of 1 inch per year. Lime should not hurt plant foliage, although wet leaves may be burned if quicklime or hydrated lime is dusted onto their surfaces.

To lower soil pH, try adding pine needles, oak leaves, peat moss, or sulfur. Sulfur is the fastest to interact with the soil but it will still take months for a reduction in pH.

Sulfur for soil applications is usually supplied in granulated form. It is available also in liquid form, which is most often used to treat water. Powdered sulfur is a fungicide. It has been suggested that sulfur, when incorporated into the soil, can have a detrimental effect on soil microorganisms. Some growers believe that sulfur also injures beneficial insects. Application rates vary with the type of soil and the pH adjustment needed. Rates should be furnished by the supplier.

There are some chemical fertilizers that help raise or lower pH; others provide nourishment to plants while the pH level is being corrected by other means. However, regular application of organic matter into the soil can buffer the acidity and alkalinity without the need for additional treatments.

Field Layout

There are many ways to design the layout of growing beds for herbs. The field design is dependent on your climate, soil type, perennial or annual herbs, type of weed control, the volume of production, and personal preference. Experiment with a mixture of bed types and designs to determine what is best for your situation.

Many herbs grow well and are more attractive if planted in clusters. This type of planting is more suited for formal display gardens and where only a small amount is needed of individual herbs. For the grower needing large volumes of herbs, fields should be designed to facilitate cultivation and harvest. This can be done in a variety of ways.

Flat-Row Cropping

This is an age-old method of growing used by farmers and gardeners. It is appropriate for large-scale production because it allows the use of mechanical tilling, planting, and weed control. Plants are grown in straight rows in freshly tilled soil that is level with the surrounding area. An area around each bed or row should be left unplanted for machinery to turn around in. This should be at least 5 feet wide for rototillers, wider for tractors with cultivators.

Make the rows as straight as possible. Besides giving an attractive appearance, straight rows make cultivation with tillers or machinery easier. Mark rows with furrow markers, either homemade or purchased. The old method of stretching rope between two sticks works well, but it is time-consuming.

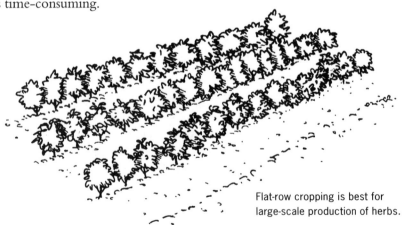

Flat-row cropping is best for large-scale production of herbs.

Paths between the rows should be spaced to accommodate people harvesting by hand. There should be enough space for a worker to squat or sit without damaging the plants in the next row. The space between the rows should be at least 8 inches wider than the tiller's tine width. Weed control in the paths can be achieved with one pass of the tiller. The weeds within the rows can be hand-pulled or hoed. The use of a small 1.5 horsepower cultivator makes this job easier.

Some growers use grass or other growing plants in the paths between the rows. This has the benefits of neat appearance and easier and cleaner walking after rains. On a large scale, though, living paths make for more work. The grass must be mowed and edged frequently. Many types of grasses spread by underground runners that will invade the crop. Living paths can also harbor insects and diseases.

The Benefits of Raised Beds

Growing in raised beds has many benefits, but may not be economical on a large scale. Although raised-bed-shaping machinery is available, it is expensive. Each grower must weigh the advantages of committing a large area to raised beds against the expense involved with purchasing the machinery or creating the beds by hand.

The labor and costs of installing raised beds are substantial. Costs can include soil amendments and border material. If a large area will be made into raised beds with the use of shaping machinery, be prepared to expend a great deal of time and energy.

Soil in raised beds, especially in those beds from the previous year, warms faster in the spring. This can be a definite advantage in cold-winter climates because you can plant crops earlier. Raised beds offer better drainage and soil aeration, which produces better-quality plants. Hand-weeding and harvesting are also easier because you don't have to bend quite so far to accomplish these tasks. Plants can be spaced closer together, helping to shade out weeds as the herbs mature.

Raised beds can be used in areas that were previously unacceptable for planting because of high erosion potential. On slopes, for example, beds should follow the contour of the land and operate like terraces.

Soil compaction is less of a problem with raised-bed culture because it discourages foot traffic. People may be tempted to step over plants growing in flat rows and cause damage. Growing in raised beds usually eliminates this problem. Soil amendments, water, and work are all directed to the growing area. Nothing is wasted on the surrounding unproductive paths. The paths between the beds are usually well traveled and lacking in nutrients, which has the effect of keeping weed problems to a minimum.

Preparing the Beds

The soil in the bed area should be double dug: First till the soil to a depth of 12 inches. Remove this soil and till the soil in the trench to another 12 inches deep, adding soil amendments as you go. Replace the top layer of soil and till in the amendments. Some experts recommend allowing the soil in the bed to rest a few days before repeating the entire process. This second tilling is not always necessary, depending on the soil in your location.

The beds should be elevated to a height of 4 to 6 inches. It is not necessary to enclose the beds with solid borders. Soil on the edge that is sloped at a 45-degree angle should hold in place. Materials for solid borders, if you choose to make your beds this way, are brick, masonry blocks, wood, or the new plastic planks made of recycled material.

Your beds can be a single row wide, as many large commercial vegetable growers have them. These beds are changed each year using bed-shaping machinery. Semipermanent beds, usually used for perennial herbs, should not be wider than 36 inches to allow easy access from each side. Leave at least 24 inches for the paths between beds.

The soil in beds that will contain perennial herbs needs special care because amendments will not be tilled in each year. The soil may become compacted and hard from the elements. Aerating and top dressing the beds each year will help to keep the plants healthy and productive.

caution on wood frames for beds

Take care in choosing wood borders. Many chemicals used as wood preservatives are toxic to plants. Do not use old railroad ties or any timber coated with creosote or penta (Pentachlorophenol). Wood coated with green cuprinol (use only the green type) or chromium, copper, and arsenic (CCA) pressure-treated wood is said to be safe for use with plants. CCA-treated wood should be washed with water six to eight times before it is considered safe to use. Use caution and space plants at least 12 inches from these wood borders.

Aerate by poking 4- to 6-inch-deep (¼ to ½-inch in diameter) holes in the soil between the plants every 4 inches. Take care not to damage plants or roots. There are tools just for this purpose or you can design your own. Top dress compost and other amendments after aeration, working it down into the holes.

Planting Seed Outdoors

Sowing seed outdoors by hand usually results in an uneven stand of plants; plants are overcrowded in some areas and sparse in others. This means wasted space or spending time thinning out crowded seedlings. If only a few plants are needed, however, sowing by hand is the easiest.

For planting a large volume of seeds, mechanical equipment will result in more even stands. There are many types of planters available, although not all are suitable for herb seeds.

The most inexpensive is the planting wheel. Place the seeds inside the wheel, set the opening to the seed size, and roll the wheel by hand down the furrow. The seeds are placed at regular intervals. This takes a little practice, but it is well worth the effort.

Many types of tractor-drawn or self-propelled seeders are available. Some will even lay plastic mulch and seed multiple rows in one pass over the field. These machines are expensive: Evaluate the expense compared with the payback time before making a purchase. Equipment manufacturers can be located by checking the buyer's guide published by most grower trade magazines or monthly editions of trade magazines such as *The American Vegetable Grower*.

Timing

Your eagerness to get an early start on the growing season must be balanced with the seeds' reluctance to germinate before conditions are right. If the soil is too cool, seeds will rot before they can sprout. Most herb seeds require warm (60° to 75°F) soil to germinate.

If early-planted seeds of heat-loving herbs do germinate before the weather is settled, the plants might fail to thrive. They are more susceptible to disease, insects, and late frosts. Direct seeding should be done after the weather is settled, or protect seeds against frost and cold winds. The chart on page 158 gives ideal air temperatures for seeding of individual herbs; see also Part IV: Successfully Growing More Than 20 Herbs and Flowers.

Plan your seeding around the last frost-free date in your area. Keep good records of planting and harvest dates for each plot. This will help you in seasons to come to determine the earliest date you can plant safely. You will learn which areas warm early and which are in frost pockets.

The early-spring sun warms the soil even when the air temperature is cool. Soil that has a loose and porous texture and soil in raised beds will warm sooner than that on level ground.

Sowing

Whether sowing seed by hand or by machine, the same principles apply. Seed depth should not be more than twice its diameter. Plant most herb seeds no more than ¼ inch deep. Cover the seeds with a fine, light medium such as sifted soil, dry sand, or medium- or coarse-size vermiculite. Small-size vermiculite should not be used because it can pack down hard over the seeds. Sand and vermiculite have the added advantage of marking the rows while allowing light to reach the seeds.

The seeds must be kept moist, which means they must have constant contact with the covering medium, even those that require light to germinate. Water or mist the seeds twice a day. In dry climates, an evening watering will help germination. In humid areas, avoid evening watering because soggy soil could cause fungal diseases or drown the seeds.

When the seeds germinate, weed meticulously. Competition from weeds will stunt or even kill the seedlings.

Fertilizer should not be needed if the seedbed has been prepared with plenty of organic matter. If the soil is marginal in this respect, a side dressing of compost or half-strength fertilizer (such as 12-12-12) will help the plants grow quickly.

Transplanting Outdoors

Transplants into the field may be potted herbs you've grown or those bought in. They can also be bare-root plants bought in or grown in another area of the field, a greenhouse, or a cold frame. The techniques for transplanting are the same for all types.

There are mechanical transplanters that will plant potted plants through plastic mulch, in bare soil, in single or in double rows. Many of these machines will water the transplants at the same time. Some can even side-dress fertilizer after planting. For those growers planting many acres, the cost savings can pay for the machine in one season.

Timing

The plants should have a developed root system and at least two sets of true leaves before you transplant them to the field. Starting seeds or cuttings yourself so the plants are ready for transplanting when the weather conditions are appropriate requires planning. Most herbs have differing growth rates and lead times from each other. Many variables, such as seed or cutting, temperature, sunlight hours, and soil fertility, all have a bearing on whether the plants will be ready on time. The lead time charts on pages 160 and 178 list guidelines as to when to start seeds or cuttings. The times are based on the average frost-free date in your area.

Keep careful records noting variety, date started, conditions, date transplanted, general weather trends for that year, and whether the plants were ready. This will be a big help in years to come.

Weather, of course, plays a big role in when to transplant. Each herb has its preference for weather conditions. Parsley and sorrel, both usually transplanted rather than direct seeded, prefer cooler weather. Basil and many other herbs must have warm day and night temperatures. Weather is unpredictable, so be prepared for unexpected frosts and ready to cover tender transplants.

Hardening Off

Greenhouse-grown plants should be hardened off before being transplanted to the field. This means they are gradually acclimated to the weather conditions where they are to grow. Without this step, plants may succumb to the elements or their growth can be stunted.

Begin this process one week before the planned transplant date. Set the plants outside in a sunny location but protected from strong winds. Bring them in at night. After a few days, the plants can be left out at night unless freezing temperatures are forecast. At this point a little wind will only serve to make the plants stronger.

Spacing

Each grower wants to get as much crop out of each section of land as possible. Often plants are spaced closer than recommended in an effort to increase yield. This can result in less foliage growth and, thus, diminished return on investment.

When herbs are overcrowded, they first reach out to open areas for expansion and become somewhat leggy. Next, the leaf size is reduced. The plants may produce more but quality suffers. Because herb sales depend on attractiveness as much as aroma and taste, it is important to allow the herbs enough room to produce large leaves.

Plants can be spaced in the odd/even pattern. This allows more light and air circulation to reach each plant. This pattern is useful in raised beds outdoors or in the greenhouse. Herbs are susceptible to fungal diseases, especially in areas that have hot, humid summers. By allowing enough room around each plant for good air circulation, you'll avoid many diseases.

Perennial plants will occupy the same bed for several seasons — and grow larger each season — so it is extremely important for them to be properly spaced. It is better to allow more space for perennials than too little. Annuals are a bit more forgiving because they can generally be pruned more radically than can perennial herbs, thus eliminating some crowding. If severe symptoms of overcrowding occur, pull some plants to make room. Correct this spacing error with succession planting or in the next season.

Most herbs used for fresh-cut sales are constantly being pruned by harvesting. By careful harvesting to eliminate leaves and branches touch-

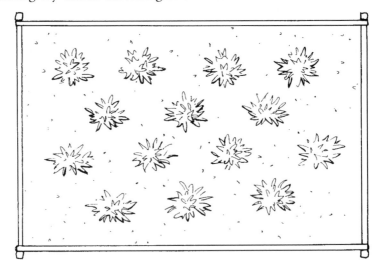

Spacing plants in an odd/even pattern in a bed maximizes the amount of light and air that reaches each plant.

ing each other, the plants can be spaced a few inches closer than recommended. This spacing should be based on your climate conditions and sales of each particular herb. See part IV: Successfully Growing More Than 20 Herbs and Flowers and the chart on page 137 for spacing recommendations.

The Transplanting Process

Transplant on a cloudy day to lessen the stress on your plants. If you must transplant on a sunny day, protect the plants from direct sunlight or plant them during the early evening. Water the herbs the day before you plan to transplant them. Plants that are overgrown or leggy should be pruned before transplanting. This is easier on your back and saves time.

Prepare the soil in the beds by tilling and adding amendments. Mark out the rows and where the plants will be placed: Make a small indentation in the soil with your hand, trowel, hoe, or any other plant-spacing device. Some growers dig the holes for the plants as they mark them; others like to make sure the marks are evenly spaced before they dig the holes. The holes should be a little larger than the rootballs of the plants.

Planting options. The plants can be "dry" transplanted or "puddled in." Dry transplanting simply means that the plants are placed in the holes and watered afterward. This requires more water and time. The foliage will likely get wet; don't use this method in the evening. If the soil is already wet from rain, this method is good. The advantage is that watering after transplanting eliminates any air pockets around the rootballs.

With the "puddling in" method, the holes are filled with water or a fertilizer mix before transplanting. The water is allowed to drain before you place the plants in the holes. Pack the soil firmly around the rootball to eliminate air pockets.

Pull dry soil from the surrounding area to cover around the plant. This method places the water where the plants need it and prevents crusting of the soil around the plants. Less water is required with this technique — especially useful when the soil is dry.

To remove a plant from its pot, place the stem between your fingers and turn the pot upside down. If the plant does not fall out, knock the bottom of the pot with your hand or a trowel. The rootball can be loosened in plastic pots by gently squeezing the pot on each side or pushing the plant out from the bottom. Don't try pulling the plant out by the stem — this may tear the plant from its roots.

Rootbound plants in clay pots may refuse to leave the pot because the roots have attached themselves to it. To remove, water the plant and the outside of the pot well. Allow the water to soak into the pot and try again. If this fails, run a thin-bladed knife around the inside of the pot. Still no success? Just break the pot.

Spread roots. Very gently, pull on the roots on the bottom of the rootball with your fingers. This helps the roots to spread out faster. If

Gently slide a plant from its pot; tugging might tear the plant from its roots.

roots are growing in a circle on the sides of the rootball, loosen these as well. Otherwise the roots may continue to grow in a circle. Many commercial growers of ornamental plants score or cut the roots ¼ inch deep on two or four sides to hasten the root spread. This is not recommended for herbs.

Place the plant in the hole just a little lower than its level in the pot. Cover the roots with soil and gently firm it around the base of the plant. By leaving a slight indentation in the soil around the plant, water will be funneled downward to the roots.

Monitor the plants closely for two weeks for wilting. It takes at least this long for the roots to spread into the adjoining soil. When watering, give the plants a boost by using half-strength fertilizer. The newly transplanted herbs will not usually produce any top growth for several weeks — the plants are hard at work developing a root system. When the roots have spread and begin to take up nutrients and moisture, new top growth will start up.

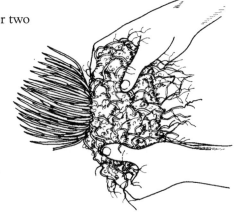

Once the plant is out of the pot, gently spread the roots on the bottom of the rootball with your fingers.

A row cover raises the soil temperature, protecting seedlings in early spring.

Extending the Season

There are several methods you can use to warm the soil early and provide protection to young seedlings during cold nights. Not all may be practical for large-scale plantings.

Row Covers

Black plastic laid over the ground will do much to raise soil temperature early. Many commercial growers use this technique. The plastic is pulled off for planting or left on during the growing season to act as mulch.

With direct seeding, the seedlings can be covered with a spun-polyester row cover. One good brand is Remay. These lightweight covers do not need to be supported above the plants. They provide at least a five-degree temperature buffer along with protection from wind and insects.

Making Your Own Tunnels and Coldframes

My favorite way to get a head start in the spring begins in the fall. The beds are emptied of the current season's crop, soil amendments are added, and then I till them. Afterward, 10-foot lengths of ¾-inch white PVC water pipe are pushed into the ground, at least 6 inches deep, every 4 feet. Push the pipe into the ground on the opposite side to form hoops over the bed. Secure the hoops lengthwise on the tops with tape, rope, or wire. The ground inside this mini cold frame is then covered with black plastic. (Do not cover the hoops.) The bed stays this way all winter.

In the very early spring, 10 to 12 weeks before the average frost-free date, heavy clear plastic is stretched over the hoops and anchored to the ground with boards. Brush any snow off this structure, as it is not strong enough to support much weight. The clear plastic concentrates the sun's rays onto the black plastic covering the ground. This melts the winter snows quickly and the black plastic warms the soil. When the soil is warm enough to plant, roll back the black plastic to expose the amount of soil needed. Planting holes can be cut in the black plastic for transplants or seeds and the plastic can be left in place.

Unless you like to crouch, the clear plastic covering should be rolled back when you plant and then be replaced. The end flaps should be opened or the sides rolled up slightly during the day to allow for ventilation. This structure can be removed when the weather has warmed or left on to provide extra warmth. This cold frame will provide extra growing time on both ends of the season.

This same concept can be used to make walk-in cold frames or tunnels. Longer PVC pipes are used and slipped over ½-inch metal pipes driven into the ground. Lightweight doors can be made for the ends. You can even make provisions for warmth by using heating cable. This type of structure is more permanent, relatively inexpensive, and can be used to grow crops well into the winter in most parts of the country.

There are many different ways that cold frames or tunnels can be constructed. Building these on a small scale does not have to be expensive. Do some research, look at the materials on hand, and design your own. The payoff will be worth the effort and money expended.

Irrigation

Most herbs grow best in dry soil conditions, but although the plants may survive in dry soil, they still need adequate moisture to produce the abundant foliage needed by the commercial grower. Moisture is the vehicle by which the plant roots take up the nutrients that produce new growth. By withholding adequate moisture, plant growth will be diminished, especially after the first cutting.

Excessive moisture can create problems as well. Consistently soggy soil causes root rot and promotes fungal diseases. Strive for frequent moisture to allow the plants to take up nutrients, then let the soil dry. By making sure the soil provides good drainage, this cycle of moist and dry is easily accomplished.

Each region of the country differs in its rainfall patterns and soil conditions. Sandy loam soils, windy zones, and raised beds dry out more quickly than do other areas. You must observe diligently the microclimates of your growing area, which can vary dramatically from one spot to another.

Whatever the rainfall pattern is in your region, water must be available when you need it. Your livelihood depends on having crops ready for market in a timely fashion. Depending on nature to supply adequate moisture is a gamble the commercial grower cannot afford. Setting up an irrigation system should be just as important to the grower as finding markets and planting seeds.

Watering Supplies

Sources of water can be anything from city water, to wells, lakes, and streams. There are many systems available to tap into the water supply and transport it to the fields. Your local Extension agent or plumbing contractor should be able to help you set up an irrigation system. Either will also be knowledgeable about any local regulations governing water use.

You'll have to decide how to supply the water to your plants. The simplest system is hand watering, with either a watering wand or a hose-end spray nozzle (see illustration on page 212). This may work for small areas close to the water supply, but it's not efficient for large fields and beds far from the water supply. Long sections of hose or pipe tend to decrease water pressure, and dragging hose around can hurt both hose and plants.

Sprinklers and overhead irrigation systems cover a lot of ground. They usually require strong water pressure. These systems are not very efficient because they wet areas not useful to the plants, and some evaporation occurs while the water is in the air. The foliage is also wet during this process and this promotes fungal diseases. Mud splashes onto the undersides of the leaves and stems, necessitating washing after harvest.

The most efficient way to supply water to the plant roots is to water at soil level. Vinyl or canvas soaker hoses are inexpensive. Many kinds of trickle or drip irrigation systems are available, through a variety of sources. These systems place the water only where it is needed and operate well with low water pressure. Injectors are available that connect at the source and mix fertilizer in measured amounts into the water flowage. A drip irrigation system is costly to purchase, but should pay for itself in one dry growing season.

If overhead watering is done, use a gentle spray so that the tender seedlings are not knocked down. At an early stage of growth they are easily captured and held down by mud after a heavy rain or watering. If so, gently lift the tops of the seedlings off the soil. This may not be possible in large fields, but don't despair! The seedlings are stronger than they appear and most will survive just fine.

When to Water

When to water can be determined only by your understanding of your plants and soil conditions. Check the moisture level a few inches below the topsoil with a finger, trowel, or moisture sensor. Provide water only when it is dry several inches below the soil level.

Water early in the day, especially if you use an overhead system, to allow the foliage to dry before evening. Wilting plants should be watered promptly. If watering of wilting plants must be done in the evening, water at soil level to keep the foliage dry.

Water moves through the soil slowly, so you want to supply water slowly over a long period of time. Short, heavy rains wet only the top few inches of soil, and this water evaporates quickly. The plants' roots stay shallow in the soil, where they will dry out even more quickly.

When irrigation is done, it should be enough to soak through the soil to a depth of 6 to 8 inches. A general rule for dry soil is to supply ½ gallon of water per square foot of soil. If applied slowly, the soil should be adequately moistened. To be sure that enough water has been supplied, wait a half hour after watering and check the soil 6 to 8 inches down. This is especially helpful with new fields. In time you will know how much water is needed, and when, to ensure vigorous growth.

Weed Control

Weeds in the field rob the crop of light, nutrients, and water. Weeds harbor harmful insects and disease. They cause the most damage during the seedlings' first six weeks of growth. Transplanted herbs usually manage to compete with weeds better than seedlings because of their larger size and more-developed root system.

Weeds, which are simply native plants that want to grow in your fields, are not all bad. They provide a safe haven for beneficial insects, break up hard, compacted soils, and serve as indicator plants (which can tell you much about your soil), among other benefits. A happy medium is to allow these native plants to grow around the perimeter of the fields, but keep them cut to less than a foot in height to allow for air circulation. This encourages beneficial insects and gives them access to their prey.

Perennial grasses that spread by underground runners are most important, and difficult, to control. Each time a runner is cut, it grows a new patch of grass — it will even take root in the compost pile. Always try to remove all the runners.

Keep weeds to a minimum so your herbs can grow to their full potential. Harvesting is much more difficult if weeds are growing among herbs that grow in clumps. These herbs — chives and cilantro, for instance — can be harvested by the "grab, cut, and rubber band" method. If weeds are in the bunch, precious and expensive time is lost in removing them.

There are many ways to control weeds. It may be necessary to use several methods in combination.

Hand Weeding

This is labor- and time-intensive but also satisfying and aromatic, especially in small areas. It is also the most costly for those who hire employees to do the job. Pulling by hand may be the only way to control weeds in seedling beds; raised beds usually must be hand-weeded. Hand weeding within the row may be necessary after cultivation with a rototiller.

If weeds are crowded close to young seedlings, take care when pulling the weeds, or the seedlings may be pulled out also. Put two fingers on top of the soil on both sides of the seedling before pulling the weed. This usually keeps the seedling where it belongs. If the seedling is pulled out, replant it right away. It will usually survive this procedure.

To protect a seedling while you weed, hold it down with two fingers while pulling up an adjacent weed with the other hand.

Pulling the weeds out with their roots intact is easier when the soil is slightly moist. If the soil is dry, the weed stem will break off at soil level and, depending on the plant, may even regrow. If dry soil weeding is necessary, loosen the soil with a trowel or hand fork. Don't attempt to pull weeds when the ground is very wet because too much soil will remain attached to the roots.

Cultivation

Cultivation kills weeds by cutting the roots, uprooting, or burial. There are several ways to use cultivation to control weeds. Most methods use a rototiller or tractor-drawn cultivator. Human-powered cultivators or hoeing will produce the same results, just with more work.

The rows must be straight and spaced properly for the equipment to be used efficiently. The soil should be dry. Delay irrigation for a few days after cultivation to prevent weeds from rerooting.

Preplant cultivation involves irrigating the already prepared seedbed before planting in it to encourage weed seeds to germinate. After the weeds come up, cultivate the area to a shallow depth. This can be done a second time before planting. The second tilling is very effective at killing weeds.

Use care if cultivating when the herb seedlings are still small; they can easily be buried by the side soil worked up by the tiller. When the herbs are established, they are better able to withstand competition from weeds. Weeds should still be controlled to prevent them from producing seeds, however. The next time to cultivate depends on how tall the weeds are. Tall weeds often become entangled around the tines of the tiller; these can be a headache to remove. Knock them down when they are small — then only shallow cultivation is required.

Repeated tilling or cultivating is not good because it breaks up aggregates in the soil, enabling it to pack down easier. Alternative weed-management methods should be used whenever possible.

Water Management

Weeds, like other plants, need water to germinate. By withholding water, most annual weeds will be practically eliminated. Deep-rooted perennial weeds are not affected as much by this technique, and this method works best in areas with a dry growing season.

The easiest way to use water to suppress weeds is to use subsurface drip irrigation. Drip lines or tapes are buried in the ground and plants grow over the lines. The lines should be deep enough that water does not reach the soil surface, which would allow weed seeds to germinate. Buried drip irrigation systems usually run about $1,000 per acre, including installation. Most of these lines or tapes should last five years or more.

Soil Solarization

This technique entails stretching clear plastic, with UV stabilizers added, over the soil surface. Heat from the sun intensifies and is trapped below the plastic. This has the effect of cooking and killing weed seeds, seedlings, and some fungi and nematodes. The root-knot nematode is not seriously affected by this method of soil sterilization, however.

Soil solarization is most effective during the period of maximum solar radiation, usually May through August. Till the soil to seedbed texture, level it as much as possible, then irrigate. Stretch the plastic over the bed and secure the edges with soil. The plastic should be left on a minimum of four weeks — six to eight weeks is better. Repair any tears in the plastic immediately to keep the heat enclosed.

Ideally the surface soil temperature should get up to 134°F and 100° at a depth of 10 inches. After the solarization period is over, remove the plastic and immediately plant the crop. The clear plastic can also be painted white or dark and used as a mulch by planting through holes cut into it. Do not till the soil, because this brings untreated weed seeds to the surface, where they then germinate.

Some organic growers in California use a very effective "biofumigant" technique that includes solarization. In the spring, they plant a crucifer crop, which includes plants in the broccoli and cabbage family. They then till it into the soil and cover the rows or fields with plastic to let them "bake." As the crucifer crop breaks down, it releases isothiocyanates, which are similar to the toxic compounds found in chemical soil fumigants.

Mulching

Mulches work by blocking the light. This inhibits weed germination and growth. However, mulches also provide a perfect hiding place for insects, slugs, voles, and a variety of other destructive critters and disease organisms. They also hold moisture in the soil, which is not conducive to the health of many herbs. For these reasons I do not recommend using most types of mulches for herbs unless your climate is very dry. The exceptions to this are landscape fabric, coarse sand, and light-colored gravel or rocks. These are useful in beds that contain perennial herbs. Because these materials are porous, soil moisture can evaporate and weed growth will still be suppressed.

In hot, dry climates, mulch can be beneficial. There are many materials you may use for mulch. New products are being introduced regularly to meet the needs of commercial growers, such as colored plastics

Bark mulch should only be used for winter protection.

that allow only certain wavelengths of light through to the soil. Pressed peat and compressed paper mulches are available in differing-size rolls that degrade by the end of the growing season.

Organic mulches. Many organic materials can be used to mulch plants: straw, hay, bark, wood chips, grass clippings, and compost, to name just a few. These mulches degrade in time and supply nutrients to the soil. Materials such as pine needles, bark, and grass clippings can alter the pH of the soil and deplete nitrogen over time. All mulches should be free of seeds, which can germinate and defeat the purpose of using mulch. Make sure the mulch material has not been treated with herbicides or pesticides.

Organic mulches should be 4 to 6 inches deep in order to block all sunlight from the soil. As the mulch degrades, new material must be added to maintain the necessary thickness. Till these mulches into the soil at the end of the growing season to add organic matter.

Plastic mulches. Plastic mulch is available in a variety of widths and thicknesses. It is usually sold in rolls that can be used with machinery or placed by hand. Black is the most common color because it completely blocks light and warms the soil. A new clear plastic treated with a wavelength inhibiter blocks wave-

Gravel mulch provides good drainage, but also should only be used in the winter.

lengths needed for seed germination but allows most light to pass through. The colored plastic mulches work in much the same way.

Spread out the plastic mulch, hold down the edges with soil or another heavy material, and plant through holes cut in the plastic. Direct seeding can be done either before the plastic is applied or after. Use drip irrigation under plastic

Straw is commonly used as winter mulch.

mulch to provide the plants with moisture. Machinery is available that will lay the irrigation lines and plastic simultaneously.

In some areas, clean plastic mulches, along with greenhouse poly glazing, can be recycled. In other places, though, plastic mulches can present disposal problems.

Use only plastic manufactured for mulch. Some plastic will degrade with exposure to sunlight, causing it to tear or break into small pieces.

Flaming

Burning has been used since antiquity to remove crop residue and is still practiced today. Burning effectively kills most weed seeds on, or just below, the soil surface. It is also destructive to soil organisms. Flame weeding uses fire to kill living weeds by rupturing the cell walls of the plant. It can be done with anything from a hand-held propane torch to large, tractor-drawn propane equipment.

A technique used by large vegetable growers is to prepare the seedbed for planting and then irrigate to encourage weed seeds to germinate. The field is then flamed and the crop is planted without tilling. With crops that are slow to germinate, the crop rows are again flamed just before emergence. While this is an effective way of controlling weeds, it is expensive in labor, time, and equipment.

An ordinary hand-held propane torch can be useful to kill tough perennial weeds. It is a safe way to selectively control weeds in established beds. Because the flame from these torches is small, you can direct it to the base of the weed without damaging the crop.

Heat is also used to kill weeds in other ways. A new system uses hot water that is injected into the soil a few inches below the surface. It is

quite effective at killing even
tough grasses. This new machine,
though big and probably expensive, demonstrates that technologies are being developed that can
provide good weed control without using dangerous chemicals.

Herbicides

This term is generally used to
describe inorganic chemicals that
are used to kill weeds. It is important to note that many are not registered (legal) for use with herbs.
Some are currently being with-

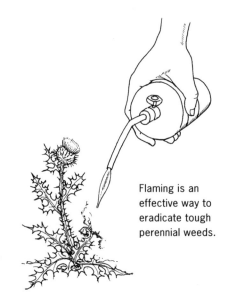

Flaming is an
effective way to
eradicate tough
perennial weeds.

drawn from the market as the Environmental Protection Agency requires
reregistration of these chemicals.

It is costly for chemical companies to certify an herbicide for use with
any crop; herbs, classified as a minor crop, are treated on an individual
basis. Because this area is rapidly changing, check with the supplier or
your Extension agent to determine whether an herbicide can be used
with a particular herb.

Most growers choose not to use these chemicals even if they are legal
to use with a particular herb: They add still more chemicals into our
already polluted world, and most herbicides must be applied by a person
trained and licensed to do so.

Herbicides come in three basic types: preplant, applied prior to the
planting of the crop; preemergent, applied prior to the emergence of the

 watch for roadside spraying

In rural areas, many counties apply herbicides to the roadsides to
control weeds and overhanging trees. The drift from these chemicals
can cause serious crop damage in nearby fields. Check with the
highway department in your county to find out how you can prevent spraying from happening close to your property lines.

weeds; and postemergent, applied after the emergence of the weeds. The safest way to control weeds with herbicides is preplant. Depending on the herbicide, there is a waiting period of a week to four months before planting seeds or transplanting can begin.

Fertilizers

Herbs, like vegetables, must be well nourished in order to achieve optimum growth. The soil in your fields is subject to intensive production and repeated cropping, and it is necessary to replace the nutrients regularly. The organic matter that is tilled into the soil each year may not be enough to supply the needed nutrients all through the growing season.

While it is true that the flavor of some herbs is slightly diminished by too much fertilizer — especially an overabundance of nitrogen — it is nearly impossible for this situation to occur in field-grown herbs. Some herbs absolutely require additional fertilizer after cutting in order to regrow with vigor. Among these are chives and parsley. Annual herbs, such as basil, dill, fennel, and cilantro, profit from additional feeding. Many of the perennial herbs are not heavy feeders but will benefit from periodic fertilization, especially in their second and subsequent years.

The science of fertilization is complex and beyond the scope of this book. Many materials and books are available that will educate you about the intricacies of this subject. Each grower must learn the symptoms of nutritional deficiencies and excesses. The chart on page 211 lists the symptoms of nutritional problems.

Synthetic fertilizers and many commercially prepared organic fertilizers express the percentages of their three main components with numbers relating to nitrogen (N), phosphorus (P), and potassium (K). These are the main nutrients for plant growth. Trace elements, also called macro- and micronutrients, are required for good plant growth. Many fertilizers include these in their formulas in specific amounts to correct soil deficiencies or for specific plants. These fertilizers are easy to use and are faster acting than are organic compounds. For large-scale production, their ease of use has made them the first choice for many growers.

What fertilizer should be used for herbs? That depends on the growth stage of the plants, fertility of the soil, and personal preference. Many growers use fertilizers with equal percentages of the three main nutrients such as 12-12-12, or higher, through the entire life of the plant. These can be diluted to half strength for seedlings.

As the herbs approach midlife, I prefer to use a formula lower in phosphorus (21-5-20), because this nutrient can hasten maturity, causing flowering and seed development. This is something the fresh-cut-herb grower wishes to delay. Phosphates are important for good root growth. A fertilizer high in phosphates (9-45-15) is excellent for young seedlings and transplants because this nutrient is not available to plants when the soil temperature is below 60°F.

When to fertilize depends on the herb and soil conditions. Some herbs, such as chives and parsley, should be fertilized after each cutting. Most other herbs do well with a preplant feeding for transplants or a half-strength feeding during the seedling stage and a full-strength feeding in midseason. Thin, sandy soils may need fertilizing more often. The ideal situation, of course, is to have soil that is very fertile with lots of organic matter so that fertilizing is not necessary.

Soil testing and/or leaf culture can tell you exactly what the soil and plant fertility is. Soil testing should be done yearly, or even more often, especially if the soil is not amended regularly with organic matter. New fields should always be tested before planting. Leaf culture may be necessary if severe problems exist.

There are many fertilizer options; you might have to use more than one to supplement your soil.

symptoms of nutritional deficiency

Nutrient	Deficiency
Major Nutrients	
Nitrogen	Yellowing or light coloring, starting with lower leaves. Drying and shedding of older leaves. Growth is stunted, stems are spindly, yields are severely depressed.
Phosphorus	Stunted growth. Leaves dark green developing toward red or purple, beginning at leaf margins or tips. Leaves may have necrotic spotting, and older leaves may turn brown.
Potassium	Stunted growth, tendency to wilt, weak stems. Older leaves may have bleached spots; margins may appear scorched. Leaves may curl.
Minor Nutrients	
Calcium	Soft growth, deformed young leaves. Decreased growth of leaf and root tips.
Magnesium	Bleaching between the veins of older leaves; may be pale, mottled, or tinted purple at the margins. Leaves may be stiff and brittle, or may appear withered. Usually worse in soil with low pH.
Sulfur	Stems are thin and brittle. Young leaves are pale or yellow. Slow, limited growth.
Micronutrients	
Boron	Young leaves deformed, wrinkled, or thickened. May have blue-green color, brittle leaves and stems. Decreased root growth. Terminal buds may be brittle or die.
Copper	Leaf tips bleached, twisted, or curled. Leaves may turn dark blue-green and wilt and drop. Plant appears withered and seed stalk may wilt.
Iron	Bleached or yellow new growth. Leaf tips and margins appear dry. Usually seen in soils with high pH.
Manganese	Bleaching between the veins, dead spots at the base of young leaves. Young leaves may drop. Usually more severe in soils with high pH. Symptoms vary by species.
Molybdenum	Stunted growth. Leaves are pale or yellow, withered, may have dead spots or rolled margins. Begins with middle leaves, then moves to older leaves.
Zinc	Bleaching between the veins along the midrib and leaf margin beginning with older leaves. Leaves may be thickened, and premature leaf death can occur. Sometimes short internodes are present, and decreased bud formation. This is the most common mineral deficiency.

Many growers choose not to have a soil test each year. They simply go with the odds and feed the plants a balanced fertilizer. This approach can work just fine — or it could result in serious plant damage.

Soil testing is very important if synthetic fertilizers are to be used. Excessive synthetic fertilizer in the soil can burn young seedlings and injure beneficial soil microbes over a short period. It also can lead to salt buildup and reduce the plants' ability to take up water, both of which will suppress plant growth. Soil that is rich in organic matter has the ability to buffer many excesses and will rarely have deficiencies.

Fertilizer Types

There are three types of fertilizers on the market today, varying in form and the way they are administered.

Dry powdered or granular fertilizers. These were the first types developed and are still in use. They are broadcast onto the soil surface through the use of spreaders, incorporated into the soil, or can be used for sidedressing. They are sometimes mixed with water but require frequent agitation to stay evenly dispersed through the water.

Water-soluble fertilizers. These newer fertilizers are easy to mix and don't need constant agitation to stay blended with water. They are easy to apply with sprayers, machinery made for this purpose, or can be used with a hose-end-type sprayer. Water-soluble fertilizers can also be used for foliar feeding. They are available in many strengths and a wide variety of formulas for different kinds of plants.

Water-soluble fertilizers are usually in powder form and require exact measurements for mixing, usually by weight or expressed as parts per million (ppm). If mixing instructions are difficult to decipher, call the manufacturer for directions. Most companies have horticulturists or technicians who can supply you with mixing instructions in plain language. Take care to mix these exactly per instructions. It is simple to dilute these to half strength but just as easy to mix them too strong, which can burn plants.

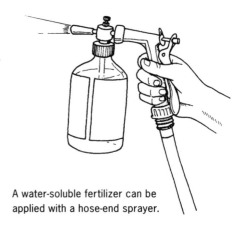

A water-soluble fertilizer can be applied with a hose-end sprayer.

These fertilizers are most often available, and are more cost efficient, in 50-pound bags. The bags readily absorb water, so you must protect them from moisture. Anyone who has tried to break apart a rock-hard bag of this product will tell you to store it sealed tightly in plastic.

Resin-coated fertilizers. Beads of fertilizer are coated with resin or some other material that slowly breaks down to release nutrients continuously over a period of time. They are the easiest to use because they must be applied only once or twice during the growing season. This type is most often used in potted plant production but some growers do use it in, or on, the soil in fields or greenhouse beds. It is usually used in conjunction with a liquid feed program.

There are several brands available; Osmocote is the most well known. As with other fertilizers, these are available in many formulas and release rates. Some may last as long as nine months under optimum conditions.

The release of the fertilizer is generally controlled by temperature. This can cause a problem when the air, water, or soil temperature is too cool to allow the fertilizer to release. If the temperature is too warm (as in early-spring greenhouses), the fertilizer may release too quickly. Close observation, as with other fertilizers, is the key to success.

Resin-coated fertilizers can be incorporated into the soil or top-dressed. The application rates must be exact. Too much can injure or even kill plants with an overabundance of soluble salts. Pots or raised beds (with gravel underneath for drainage) can be leached with water to remedy this. (Leaching is nearly impossible in fields, however.) Too little fertilizer will require other fertilizer applications.

 ## synthetic vs. organic fertilizers

The debate over synthetic versus organic fertilizers rages on. Each grower must make her own decision about which direction to take. Synthetic fertilizers are not in the same dangerous category as chemical pesticides and herbicides. The problem with relying solely on synthetics is that they are often used in place of organic matter. Fertilizers feed the plant and do nothing to enrich the soil. If you opt for synthetics, be sure to add plenty of organic matter to the soil.

To calculate the amount to incorporate, measure the total number of cubic yards of soil. This is easier to do with potted plants and with completely enclosed raised beds, of course.

Top dressing is somewhat simpler. The beads are scattered evenly around the plants on top of the soil. The application rates are based on the size of the pot the plant is growing in. Rates for both types of applications will be supplied by the manufacturer.

Organic Fertilizers

The basic philosophy of organic growing techniques is to refrain from using any chemicals, including pesticides, herbicides, and fertilizers. Instead, the grower uses natural materials to maintain soil fertility. This approach focuses on feeding the soil and not the plant. Fertile, well-balanced soil produces healthy and abundant plant growth.

Building soil fertility to the point where it will provide nutrients for intensive cropping can take years, depending on the soil structure. The materials incorporated into the soil to build fertility, usually organic matter and rock powders, are slow to interact with the soil. Deficiencies in pH or nutrients may take a year or more to be corrected by using natural materials. Because of this, soil testing is even more important for the organic grower.

Fortunately, there are many types of organic fertilizers now available. They are made from a variety of natural ingredients, from seaweed to

 ## success with synthetic fertilizers

If you plan to use synthetic fertilizers, these basic guidelines will help ensure success.

1. Get a soil test each year in the spring or fall and follow the recommendations supplied with the results.

2. Add plenty of organic matter and soil amendments annually.

3. Use fertilizers in moderation. Resist the temptation to overfeed.

4. Fertilize before rain or irrigation to work the fertilizer into the soil.

5. Follow the manufacturer's directions exactly when mixing.

6. Closely observe the plants for signs of deficiencies or salt buildup.

composted manures. These provide nutrients in smaller percentages and are usually slower-acting than synthetic fertilizers. They are available in powder, liquid, and granular form and are administered just as you would synthetic fertilizers without fear of burning roots (except for raw manure) or foliage.

The market for certified organic produce, including herbs, is growing steadily. In the past, a grower could become certified as organic by complying with strict regulations governed by state or independent certification organizations. As I write, new national standards are being proposed that will be monitored by the federal government. The organic grower must work continuously to build soil fertility. Supplementing with organic fertilizers not only feeds the plant, ensuring an abundant crop, but feeds the soil as well. Check with your local Extension service for the most up-to-date regulations.

Preparing Field-Grown Herbs for Overwintering

In all but the southernmost areas of the country, field-grown herbs will at some time experience freezing temperatures, which will cause them to go into dormancy. This is a time of rest and allows the plant to recoup its energies. Many of the perennial herbs will continue to grow, though sometimes poorly, during the reduced-daylight winter season in southern climates or in the greenhouse. Tarragon and some types of spearmints require dormancy in order to flourish each growing season.

Some herbs are tender perennials and will not survive frigid temperatures. Most of these herbs do not require a period of dormancy, but will benefit from it nonetheless. Marjoram, rosemary, bay, and lemongrass will perish if exposed to frigid temperatures for an extended period. They must be overwintered in a warm atmosphere or treated as annuals. See part IV: Successfully Growing More Than 20 Herbs and Flowers for information on overwintering individual herbs.

We can do everything in our power to see that our herbs are protected properly for winter. However, there are some plants that just will not survive, for one reason or another, especially where winters are severe.

Plants are conditioned to harden off for winter by the normal progression of lower temperatures and shortened day length. But as we all know, Mother Nature does not always behave "normally" and can provide conditions that are less than perfect for plants approaching dormancy. We can help the herbs prepare for winter by employing good cultural practices at this time.

Watering

During dormancy, herbs must have a normal amount of sap to survive. You can ensure this by providing the plants with ample moisture before the onset of winter. If the autumn has been very dry, regular watering should continue until the first killing frost.

A wet growing season and an autumn with saturated soil can have adverse effects on herbs approaching dormancy. If the soil does not provide good drainage and becomes waterlogged, the plants may be weakened and not survive a severe winter. Tarragon is especially vulnerable to wet soil over winter. Newly planted herbs may not have deeply set roots during a wet growing season, and they are especially susceptible to heaving out of the soil during alternating freezing and thawing. A thick layer of mulch will help to prevent heave.

This is the time to make sure that weeds are not growing within the crown of the herbs. Weeds trap moisture and make the plant more vulnerable to freezing temperatures.

Pruning

Most perennial herbs should be cut back after the first killing frost or when the foliage begins to die back; this helps to prevent mice, insects, and disease organisms from taking refuge in the dead foliage. Cutting them back too soon, before they begin the natural dieback, encourages new tip growth, which you don't want. These new growth tips are especially vulnerable to winterkill.

Herbs with hollow stems should not be cut before winter. Water can seep into the stem and freeze all the way to the crown or roots. Florence fennel is one herb that fits this category. It is a tender perennial, however, and thus will not survive frigid temperatures. In climates with a longer growing season, it will usually have reached maturity and been harvested before frost occurs.

Do not cut lovage before winter. Allow it to die back naturally and prune it after new growth begins in the spring. Chives have hollow stems, but the plants are very hardy and will survive most winters whether or not you cut them back.

Some growers recommend not pruning off the dead top growth after it dies back and allowing it to remain on the plant during winter. They believe the dead foliage protects the plant from severe cold, so they prefer to cut it back in the early spring. Try both ways and see which works better in your climate.

Pruning amounts will vary with the individual herbs. Some plants should be cut back to the ground, while some need only partial pruning. (See part IV: Successfully Growing More Than 20 Herbs and Flowers for guidelines on how much to prune from each plant.)

Growers with large stands of perennials may find it time-consuming and tedious to cut each herb by hand. Pruning by hand must be done on some herbs with irregular growth patterns, such as the sages and oregano. Some growers use a string trimmer, sometimes called a weed wacker, to cut large stands of chives, mints, and sorrel because they can be cut back at ground level. Although this is certainly quicker and easier than pruning by hand, it may be painful to see your beloved herbs treated this way!

Mulching

Snow is nature's best winter protection for perennial herbs, but in most parts of the country it cannot be counted on to be there when you want it. Mulching with organic materials is the next best winter protection. The purpose of mulch is to prevent alternate freezing and thawing of the soil, not to exclude the cold. If plants are not protected from this cycle, they might heave out of the ground. First-year perennials are especially prone to this because their root systems may not be deeply set.

A thick layer of mulch helps to prevent the herbs from breaking dormancy prematurely during a midwinter thaw. If the plants begin to grow during these thaws, and the temperature plunges again, many of them will be lost.

Do not apply mulch until the ground has frozen slightly. By waiting until after three or four hard frosts, you can be assured that the plants are completely dormant. Mulch can be applied over snow if necessary. A 6- to 8-inch layer of mulch material should be sufficient in most areas. Use something that won't pack down much under the weight of heavy snow. Oat straw, hay, oak leaves, chopped corn cobs, and shredded cornstalks or corn leaves/husks are all good mulching material.

Gradually remove the mulch as the weather warms in early spring. If the spring is very wet, remove the mulch from around the base of the plants rather than risk giving a toehold to mold or fungi. Keep the mulch close by the plants in case the temperatures dive again.

Overwintering Container-Grown Herbs

Stock plants grown in pots are more susceptible to winterkill because their roots are more exposed to the cold. They need special treatment in areas where winter temperatures plunge.

The ideal temperature for ensuring total dormancy of container-grown plants is 32° to 36°F. This range helps to inhibit the growth of fungi. Keeping the plants in total darkness also helps maintain dormancy. While this temperature range is ideal, most often your area is far colder. Most plants will survive, though, given proper care.

The same procedures apply to preparing container-grown and field-grown plants for dormancy. The plants should be outdoors so the natural elements of colder temperatures and shorter days will harden them off. Make sure that the soil is moist but not wet. Allow the plants to die back naturally before you cut back the foliage.

Having had very little experience with overwintering dormant container-grown herbs (I grew year-round in the greenhouse), I enlisted the aid of a horticulturist friend, Beth Tidwell. Beth owns and operates Perennial Design Flower Nursery in Wabasha, Minnesota, and, as the name implies, grows perennial flowers and herbs. She overwinters her perennial stock plants in the following manner.

Her container-grown plants spend the spring, summer, and fall in the outdoors retail sales area. As the plants die back, she cuts them back and allows the soil in the pots to freeze. This is usually completed by Thanksgiving in southern Minnesota. Beth waits to move the plants into the greenhouse until the weather is continuously cold in hopes that the mice have found other places to live besides her greenhouses.

Transferring the Plants Indoors

Beth then prepares the plants for overwintering by watering them until just slightly moist a day or two before moving them to the unheated greenhouse. The floor of the greenhouse is covered with black landscape fabric. The plants are placed on the floor in large blocks with the pot sides touching. She keeps the pots a foot away from the edge of the glazing, where the temperature fluctuates more. Mouse bait is placed around and in between the plants because mice can do a great deal of damage. The plants are then covered with white, ¼-inch-thick poly foam made specifically for this purpose. (Some growers use poly-bonded blankets or similar materials.) This protects the plants from harsh cold and provides some degree of darkness. The plants spend the winter in this environment.

Because dormancy must be broken in late winter so she can take cuttings early for plants to be ready for sale in the spring, Beth overwinters her stock plants in the greenhouse. Her greenhouses have automatic heating and venting systems. When Beth wants to break dormancy, she sets the thermostats to begin heating when the temperature reaches 30°F and the automatic venting at 45°. This allows the plants to break dormancy slowly. She waits to uncover the plants until she expects a two-day cloudy spell to prevent them from being sunburned. She does not water the plants until the soil in the pots has thawed completely.

Beth's approach to providing dormancy to her container-grown perennials is basically the same as that of other growers around the country. Some report having problems with fungal growth on the plants (and the covering), so they use a fungicidal drench before establishing dormancy. Beth does not have a fungal problem now — her biggest problem comes from mice. Mice and fungal growth were big problems when she previously used straw to mulch the pots. She also warns that plants in small pots (2 to 3 inches) do not usually survive dormancy because the small size does not hold enough moisture, so they dry out quickly and die.

Guidelines for Dormancy of Container-Grown Herbs

Your situation may differ from Beth's. You may not have a greenhouse or a hoop house. You may not need to force plants out of dormancy early for propagation purposes. But whatever your situation, the principles are the same for providing dormancy for container-grown plants:

- Allow the plants to harden off by keeping them outdoors
- Cut back the foliage after it dies back naturally
- Keep the soil moist but not saturated
- Allow the soil in the pots to freeze
- Group the pots together and place rodent bait or traps around them
- Cover the pots to protect them from extreme cold and to provide darkness (lightweight foams are best)
- Try to keep the temperature below 36°F
- Break dormancy *slowly*, either naturally or controlled
- Remove the covering during cloudy days to prevent sunburn
- Water only after the soil has thawed

controlling

insect pests

Insect pests are a fact of life for all gardeners. But for the commercial grower of fresh-cut herbs, they present major problems because they can cause cosmetic damage to the foliage, thus rendering it unsalable. Of course, fresh-cut herbs cannot be sold with insects on the foliage. Some herbs will withstand rinsing forceful enough to wash off the bugs; others will be seriously damaged by this process.

Insects can spread disease, stunt the growth of herbs, and mar the foliage with blemishes and "honeydew." (Honeydew is the sticky sweet excretion given off by many insects.)

In the field, large stands of a single herb are an open invitation to an insect invasion. Small plantings or mixed plantings are not as attractive to insects, and damage is usually kept to a minimum by nature's own checks and balances. It is, however, important to remember that most insects are attracted to more than one of the culinary herbs. A wide variety of insects, some neither pests nor beneficials, may take up residence in the herbs outdoors and have to be removed.

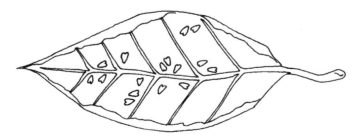

Whiteflies, one of the most common greenhouse pests, feed on the underside of leaves.

Insect pests are more of a problem in the greenhouse because it is an enclosed environment. But this closed environment also makes it easier to deal with an insect problem. Pests are a big problem in the greenhouse that is operated year-round. If the greenhouse is allowed to freeze in winter or become highly overheated in summer, many insect problems can be avoided.

More than likely, pests will be a constant nuisance for the year-round greenhouse grower. In southern Minnesota, insect infestations are much worse during the fall, winter, and spring. The timing of insect "explosions" may be different in other climates.

How do we deal with insect problems? For many years the "cure" was routine spraying with chemical pesticides. Now, of course, we realize that these chemicals have been poisoning both us and the world around us. Insects build up a resistance to chemicals very quickly, and the pesticides soon become ineffective. Most synthetic pesticides are not legal to use on herbs anyway, so we must find other, more environmentally friendly ways to deal with insects.

Types of Insect Pests

The first step in controlling pests is to recognize them. Purchase a reference book that describes in detail all pests, their life cycles, and the available control options. Make sure that the book shows detailed photographs of all the life stages of pest insects. This will be of great aid in their identification. Many different insects can attack your herbs. We will focus on the pests that most often cause problems, especially in the greenhouse.

Aphids

Many types of aphids (sometimes called plant lice) are attracted to herbs. The most common are the green peach aphid (green in color) and the melon aphid (these are black). They are small, soft-bodied insects shaped like a pear. They have long antennae and two small "tubes" that protrude from the back end.

Most aphids are wingless. It is only when they become overcrowded that some will develop wings and fly off in search of more food. Once winged aphids develop, they cause serious problems because they spread rapidly to other plants and start more colonies. Winged aphids, which are frequently darker in color and hold their wings vertically when at rest, are a sign that a serious aphid infestation exists already. They can often be seen in large numbers resting on the greenhouse glazing that faces south.

Wingless aphid

Ants are attracted to aphids and "tend" them as we do domestic animals. Ants feed on the honeydew excreted by the aphids and "milk" them by stroking their undersides. They some-times carry aphids to other plants in order to start new colonies. If you have a large population of ants and aphids, controlling ants will make it easier to control the aphids.

Winged aphid

Preferred environment. Aphids can be a problem at any time of the year in the greenhouse or in climates that have no frosts. Their populations increase dramatically as the temperature rises. In late winter or early spring, as temperatures climb in the greenhouse, be on the lookout for a rapid increase in the number of aphids. Their num-bers tend to decrease at temperatures above 80°F, but they seem to thrive in all humidity levels.

Because aphids like to feed on the tender new growth of plants, they can be found on the undersides or backsides of leaves and buried deep in the crevices where new branches sprout from the main stems. You'll often find them in clusters.

Feeding habits. Aphids feed on almost all of the culinary herbs, even those said to repel them. Their favored herbs are arugula, basil, dill, mint, parsley, rosemary, sorrel, tarragon, chervil, and salad burnet. They also love nasturtiums and pansies and can often be found hiding deep within the flowers.

Aphids insert their mouthparts into the plant and suck the juices. Their feeding can cause deformities of the new growth, dieback of the stems, and general stunting. They excrete honeydew, which makes shiny, sticky spots on the leaves. This is hard to wash off and attracts the forma-tion of black sooty mold, which is impossible to remove. Huge clusters of aphids feeding on a plant also can cause the leaves to lose color. This cer-tainly makes the foliage less marketable. Aphids are known carriers of sev-eral plant diseases as well.

Life cycle. Aphids have a very high rate of reproduction and can give birth to three to six young every day. These insects have the ability to produce at least two generations of live young without fertilization (mating). In the fall aphids mate to produce eggs that will overwinter and hatch in the spring. Aphids in the greenhouse are almost always female and give birth to live, fully formed young. They give birth as soon as they mature, which is usually a week after they are born. The newborn nymphs start feeding immediately. Outdoors more males are born, probably because conditions are less favorable due to the abundance of natural enemies.

Monitoring. Sometimes you'll see yellow sticky cards, which are used to trap insects, in a greenhouse: These are not useful for monitoring aphid populations because feeding aphids have no reason to leave the food site. Winged aphids do go to the yellow cards but they are not the ones that cause problems. It will, however, be helpful to know when there is a big enough population for winged aphids to form. Look for aphids on the growing tips of plants and the undersides of leaves.

Controls. The time to begin aphid control is when you find the very first aphid because of their rapid rate of reproduction. You can simply crushed them or brush them off plants. A hard stream of water will wash them away. Be sure to crush any that are brushed off onto the floor because they can and do walk to other feeding sites.

Since they're soft bodied, aphids are susceptible to a variety of botanical sprays. These include insecticidal soaps, pyrethrum, rotenone, neem, and horticultural oils. Spraying should be done at least three times, five to seven days apart.

There are numerous beneficial insects that will eat aphids. These include ladybugs, lacewings, aphid parasites, and the aphid midge. There are also some specialized predators that will attack only certain types of aphids. It is important to note that if winged aphids are present in the greenhouse, it is too late to use beneficial insects for their control.

Naturally occurring disease and fungal pathogens are available that help to keep the aphid population in check without affecting the plants. There are more biological controls being developed on a regular basis. Insect-growth regulators can regulate the growth stages of aphids. Check with your greenhouse supplies company to see if there are any that are registered for use on herbs in the greenhouse.

Fungus Gnats and Shore Flies

Although these small black flies do not have the same bad reputation that other pests do, they can cause considerable damage to plants, since they are said to carry and transmit root-rot disease organisms.

Fungus gnats are very common in the greenhouse. They often fly around plants and soil mix in great clouds. Insects of any kind are undesirable to the general public and to other growers. For those growers who sell potted herbs or are open to the public, fungus gnats must be controlled.

Adult fungus gnat; top view

Shore flies are often mistaken for fungus gnats but they are actually very different. It is important to distinguish between the two for purposes of control. Both insects are black, about ⅛ inch long, and fly around the soil and plants. The larvae of both can be found in the top inch of the soil mix. Fungus gnat adults are more delicate, though. They have long antennae, long legs, and a slender body. Their wings are clear and have a distinctive Y-shaped vein. Their larvae are slender, off-white in color, and have a black head capsule.

Adult fungus gnat; side view

Shore fly; side view

Shore flies are slightly larger and have a heavier body. Their antennae are very short and they have red eyes. The wings are darker in color with white spots on them. Their larvae are darker colored, maggot-shaped, and do not have a distinct head capsule.

Preferred environment. Both insects thrive in high moisture and high humidity. Puddles of water standing on the greenhouse floor or after a rain outdoors are a perfect breeding ground for both insects. There is evidence that fungus gnats, in some stages of their life cycle, can live for extended periods in dry conditions. They begin development again when moisture reappears. This may account for the many reports of "spontaneous generation" of clouds of fungus gnats where before there were none.

Feeding habits. Fungus gnat larvae live in the top inch of soil and feed on fungi and decaying organic matter. In the absence of organic matter, as in many potting mixes, they feed on plant roots and the fine root hairs. Large numbers of feeding larvae can cause serious damage, especially to rooted cuttings and young plants. Damaged plants will wilt and show leaf distortions.

Shore fly larvae feed on algae, not on plants or plant roots. The only damage caused by shore flies is that the adults rest on plants and leave dark excrement spots on the foliage. A large population of shore flies is an indication of a recent algae bloom somewhere in the greenhouse or field.

Life cycle. Fungus gnat females lay their eggs in clusters on top of the soil or in crevices. Each female can lay up to 300 eggs at a time. The eggs begin hatching in 10 to 14 days and it takes 4 to 7 days to complete the hatch. The larvae feed in the soil for several weeks before emerging as adults to start the cycle again.

Monitoring. The gnats can often be seen running across the top of the soil. They will sometimes rest on the lower stems of plants. They fly quickly around the plants and land back on the soil.

Fungus gnats and shore flies are attracted to yellow. The use of yellow sticky cards placed face up on top of the soil or hanging vertically with the bottom resting on the soil will quickly attract the adult flies. You can monitor larvae by placing raw potato slices on top of the soil. After several hours, you'll see the larvae under the slices.

Controls. Prevention begins with careful watering to avoid waterlogged soils and puddles on the greenhouse floor. Store your unused growing medium in a dry area and keep it covered. Keep the greenhouse weed-free. Decomposing organic matter, such as compost piles and refuse from weeding, attracts egg-laying females.

Botanical sprays will kill adults but won't do anything against the eggs or larvae in the soil. Large numbers of yellow sticky cards will help to trap many adult fungus gnats and shore flies before they lay eggs.

There are a several biological controls available commercially that are effective at controlling the gnats. Beneficial nematodes (microscopic soil-dwelling roundworms), mixed with water and applied as a soil drench, prey on fungus gnat larvae in the soil. Use this treatment too on gravel floors in greenhouses and to treat pots and unused growing medium. There are several strains of beneficial nematodes, marketed commercially under different names.

Gnatrol is a strain of *Bacillus thuringiensis* (Bt) that is effective against some species of flies, including fungus gnats. This is also applied as a soil drench, and it kills only the feeding larvae. Gnatrol remains effective for only 48 hours, so you must reapply it weekly to kill the next stage of larvae.

Geolaelaps is a predatory mite that is being used successfully to keep fungus gnat populations low. It offers some control of thrips because they also pupate in the soil. The very best control for fungus gnats is most likely a combination of all of the cultural and biological methods.

Spider Mites

There are more than 150 species of spider mites. The most common, especially in greenhouses, are the two-spotted and the red spider mites. They are not insects but rather members of the family that includes spiders, ticks, and scorpions. Mites are serious pests in the greenhouse because they are very tiny and tricky to control.

Spider mite

Spider mites are so small that many will fit on the head of a pin. They have four sets of legs. The two-spotted variety is tan to greenish yellow and the biggest ones have two dark spots on their backs. It is necessary to use a 10X to 20X magnifying hand lens to correctly identify this pest.

The first sign of a spider mite infestation is stippling (very small yellow spots) on the leaf surface. As the stippling increases, the leaves develop a bronze appearance. This is caused by the mites feeding on the underside of the leaves, which leads to chlorophyll loss in the leaves. As the population increases, very small silken webs show up around the leaves and the new growth tips. These webs are used for their dispersal to other areas as well as protection for the eggs and immature mites. Webbed areas on plants often turn brown and die. Mites can also spread by dispersal on air currents. And they very easily hitch a ride on clothing.

Preferred environment. Spider mites prefer hot, dry conditions. Populations can explode at temperatures above 85°F and humidity levels below 60 percent. These conditions are often found in greenhouses in late spring and summer.

Feeding habits. Mites feed on plant cells on the underside of leaves. When there's a heavy population, they will feed also on the upper surfaces of leaves. They usually are on the tops of plants where it is warm and dry. You may notice the first stippling as a roughly circular pattern toward the center of the leaves or along the main veins.

Mites have a wide host range and can be found on all of the culinary herbs. Rosemary, mint, sage, and basil are among their favorites.

Life cycle. The life cycle of spider mites can be as short as eight days. The typical life span of a female is only three to four weeks. During this time she'll lay as many as 200 eggs. The number of days for an egg to develop into an adult is greatly dependent on the temperature. It can take 30 days for this cycle at a 60° average. If the average temperature is 90°F, this cycle can be completed in four days! It is easy to see how the spider mite population explodes under optimum conditions!

Spider mites are also sensitive to day length and they are capable of hibernating as the days become shorter. They find cracks and crevices in the greenhouse for overwintering and reemerge when conditions improve. In moderate climates, or in greenhouses operated year-round, they may continue feeding all winter.

Monitoring. Scouting must be done by looking carefully at the plants, because mites are not attracted to yellow sticky cards. A simple way to detect mites is to hold a sheet of white paper under the leaf and tap the plant. Mites will fall off the leaf and can be seen on the white paper.

A better way to spot mites is to use a 10X to 20X hand lens. Turn over the suspect leaf and look for mites crawling under the stippled area on the leaf surface. Be sure also to look for unhatched eggs, which are clear and round. Empty egg cases and shed skins may also be present. This will give you an idea of the extent of the infestation. Additionally, look for webbing on the young growing tips of plants.

Controls. Sanitation is the first control for spider mites. Keep the greenhouse, surrounding area, and the field weed-free. Weeds are a haven for mites because of their wide host range. Remove all old plant debris. Make sure workers are aware that mites latch onto clothing and are thus transferred to other host plants. As soon as you spot a mite infestation, limit access to the area. Cut or move infested plants last.

Spider mites are difficult to control by spraying because they can hide along the main leaf vein. The spray must contact the mite to kill it. These pests are notorious for their ability very quickly to develop resistance to most pesticides. The use of pyrethrum seems actually to increase their numbers, perhaps because this substance kills off any beneficial insects that may be present.

Mites have shown little ability to build up resistance to insecticidal soap and horticultural oils, though. For the grower of fresh-cut herbs, insecticidal soap sprays are the best choice because oils leave a coating on leaf surfaces. If you spray, thoroughly coat the undersides of leaves. Spraying should be done twice a week if live eggs or webbing are present.

There are many different predatory mites available from commercial insectaries for control of spider mites. Many of them are targeted at specific environments and/or prey. Most of the predatory mites are more suited to the greenhouse environment because they require specific temperature and humidity levels to be effective. Some of these mites are also well adapted to outdoor conditions or are starvation-resistant, tolerant to specific pesticides, or host specific. In my experience, predatory mites have been very effective in the greenhouse.

Thrips

There are many kinds of thrips, but the two that cause the most damage to plants are the western flower thrips and the onion (sometimes called tobacco) thrips. A few thrips species are beneficial and eat other thrips, mites, and other small insects. These beneficial thrips usually have markings on their wings that distinguish them from the damaging types.

Nonflying thrips

Thrips are minuscule (⅟₅₀ of an inch), slender insects. The western flower thrips are black or brown and the onion thrips are lighter tan to yellow. They both have two sets of feathery wings. Although the adults are capable of flying, they are not particularly efficient at it. They get up into the air usually just enough to be carried along by wind currents.

Flying thrips

Severe cosmetic damage to foliage and flowers is caused by thrips. A most serious threat from these insects is their ability to carry and transmit viral diseases. They are known vectors of tomato spotted wilt virus and necrotic spot virus. Because of their life cycles, feeding habits, and small size, these insects are difficult to control in the greenhouse.

Preferred environment. Thrips prefer hot, dry conditions. When they're happy, their numbers can increase quickly. However, they can live, feed, and complete their life cycles at almost all temperatures and humidity levels. This makes them a serious threat in greenhouses operated year-round or in areas that do not have subfreezing temperatures. Adult thrips will usually not fly when temperatures fall below 60°F.

Feeding habits. Because they have a very wide host range, thrips will eat almost all plants, including weeds. All of the culinary herbs are susceptible, and thrips favor the new growth tips. The onion thrips is attracted to chives. It particularly likes our edible flowers, and mars the blossoms by burrowing deep inside the flower bud. When the flower opens, it is scarred with white marks and the petals may be distorted.

Thrips feed by scraping the leaves or flower buds with their raspy mouthparts to induce the flow of plant juices. They then pierce the leaf and suck out the juice. Their feeding results in white trails and mottling of the leaves, along with black fecal specks. Some damaged areas may actually enlarge and open to become a "window" on, or through, the leaf. New growth tips may become distorted, turn brown, and die back. The damage caused by these insects almost always leaves the foliage unsalable.

Life cycle. In ideal conditions — 80° to 85°F — the life cycle from egg to adult can be as short as 10 to 13 days. The female thrips cuts a slit in the plant and lays her eggs inside. She can lay between 150 and 300 eggs during her lifetime.

The nymphs hatch in two to four days, and begin immediately to feed on the plant. Thrips have two nymphal stages that last between three and six days combined. The larvae then drop to the soil, where they burrow in to pupate. They will sometimes pupate in leaf litter or in tight flower buds. When this stage is completed, usually in about six days, they emerge as adults to feed, mate, and repeat their life cycle.

Monitoring. You can detect the tiny yellow thrips larvae by holding a white sheet of paper under the plant and gently tapping the leaves; this should shake loose some of the young.

Adult thrips are attracted to yellow and blue. Special sticky card traps are available with the exact shade of blue that thrips love. These are especially helpful if you suspect lots of thrips. Yellow sticky cards are attractive to a number of pests, so you can use them while scouting for other insects as well.

Controls. Physical exclusion with screening is the first line of defense against thrips in the greenhouse. Cultural controls include keeping the area weed-free, disposing of plant refuse away from the greenhouse, isolating new plants, and keeping propagation benches separate from the main growing area.

Though not good fliers, thrips are mobile and will easily travel from one area to another. Greenhouses with wide open spans can be separated with curtains (plastic or screen) to prevent the migration of thrips.

Control of thrips with sprays has always been difficult because in almost all of their life cycle they are protected from exposure to sprays. The eggs are inside a leaf or stem, the adults and larvae feed inside flower buds or new growth tips, and in the pupal stage they are in the growing medium. You can try a spray, but droplets should be small, 100 microns in size or fewer. Small droplets perhaps can penetrate into the holes made by the thrips if coverage is thorough.

Thrips are vulnerable to insecticidal soap, horticultural oils, pyrethrum, and rotenone. Sprays should be used sparingly and classes should be rotated because thrips rapidly develop resistance to pesticides. Spray every three to five days for two to three weeks with the same insecticide for control of one generation of thrips before switching to another class of sprays. Diatomaceous earth spread on the soil surface is effective for those thrips entering or emerging from the growing medium.

Predatory mites offer only limited control of thrips because they do not feed inside flower buds, where the thrips hide out. There are two types (currently) of aboveground predaceous mites that attack immature thrips, *Amblyseius cucumeris* and *A. degenerans. Geolaelaps,* or *Hypoaspis* mites, are soil-dwelling mites that will eat thrips pupae in the soil as well as fungus gnat larvae. Minute pirate bugs, *Orius* spp., are useful for control of adult and immature thrips because they will feed within the flower and on thrips in all active life stages.

Whitefly

Whiteflies are one of the most common and serious pests for the commercial grower. They can infest plants in large numbers and fly about in white fluttering clouds when plants are disturbed. They are not only annoying but also destructive. Hordes of feeding whiteflies cause yellowing of the leaves and loss of plant vigor. The sticky honeydew they excrete makes a mess of the foliage and attracts black sooty mold.

Whitefly

There are many species of whitefly and it seems that more are being identified regularly, such as the silverleaf and banded-wing whiteflies. Some species have been known to attack and nearly destroy huge fields of a single crop, especially in warm climates.

The two most often found in greenhouses are the sweet potato white-fly and the greenhouse whitefly. In some parts of the country the sweet potato whitefly has taken over and the greenhouse whitefly is no longer found in abundance. However, both species may be present at the same time.

It is important to identify the species for purposes of control. The differences between the sweet potato whitefly (SPWF) and greenhouse whitefly (GHWF) are important because the SPWF has a greater egg-laying capacity and is more difficult to control; total egg-laying capacity of the SPWF, up to 65 per female, is nearly seven times that of the GHWF. During optimum conditions the SPWF can lay twice as many eggs per day as the GHWF, and has a longer egg-laying period.

In general, whitefly adults are white and $\frac{1}{16}$ long with white powdery wings. The GHWF adult is slightly larger and holds its wings flatter over its body than does the SPWF. The SPWF holds its wings in a peak over its body and is more of a yellow. It is easier to note the differences between the two species when the insects are in their pupal stage.

Preferred environment. Whiteflies prefer warm temperatures, between 65° and 90°F. Their egg-laying capacity increases and development time decreases at the higher end of the temperature range. They cannot survive freezing, so in cool climates the flies will enter greenhouses or hitch a ride indoors on plants to overwinter.

In my experience in the greenhouse, humidity has no effect on their life cycle or feeding habits.

Feeding habits. Whiteflies at all stages of the life cycle feed on the underside of leaves. Adults will sometimes feed on upper leaf surfaces. The adults and newly laid eggs are most often found on the new growth on the upper part of the plants. Older nymphal stages can be found feeding on the lower and older leaves. In the second through the fourth nymphal stages, they attach their mouthparts onto a leaf and feed in the same place without moving around.

Whiteflies are attracted to the strong-flavored herbs such as basil, sage, rosemary, marjoram, and oregano. They are especially fond of mints and will lay many more eggs on these than on other herbs.

Life cycle. The entire life cycle of whiteflies depends on temperature. At temperatures in the 50°F range it may take four months for a life cycle to be completed. When temperatures are in the 80s it can be completed in as few as 18 days. They go through several nymphal stages, or instars, before becoming adults.

Whitefly eggs are very small and spindle-shaped, and are on the end of a short "stalk" on the undersides of leaves. They may be in a crescent pattern or placed singly. When they are first laid they are pearly white to pale yellow. As an egg ages, it turns darker. GHWF eggs change to a tannish brown and those of SPWF become a dark gray.

Greenhouse whitefly pupa

Sweet potato whitefly pupa

The eggs hatch into the first nymphal stage and are then known as crawlers. These white crawlers move around for a short time before attaching themselves to a leaf to feed for the next three instars.

These nymphs are flattened, oval, and have a white, green, or yellow color that is almost translucent. They look the same through the next life stages except that they grow bigger with each molt. During these stages, the nymphs are difficult to kill with insecticides because they are covered with a protective shell. Both species remain in each of these nymphal stages for two to seven days, depending on temperature.

At the end of the fourth instar stage, the nymph changes to a pupa. This stage usually lasts five or so days with each species. During this period it is easier to see the differences between the GHWF and the SPWF. The pupa of the GHWF is oval-shaped with straight, flat sides and a level top. It has a fringe of short hairs around the edge of the top. The SPWF pupa is oval but without straight sides. When viewed from the side, it is more dome-shaped. It does not have the fringe of hairs but it does have several longer hairs rising up.

At the end of the pupal stage, the adult whitefly emerges and begins feeding and mating almost immediately. Adult females lay 4 to 10 eggs daily, usually within two days of emergence.

Monitoring. Adult whiteflies of all species are attracted to yellow. Place yellow sticky cards just above the canopy of their favorite herbs. Adults can be trapped this way because they fly up to the cards when the plants are disturbed.

Also check for immature stages of whitefly on the lower leaves. Use a 10X to 20X hand lens to look for eggs, all the nymphal stages, and to determine the species of whitefly present.

Controls. The best control is exclusion through screening. Avoid wearing yellow clothing because whiteflies can hitch a ride from one greenhouse to another or into the greenhouse from outdoors. Screen all incoming plants. If you notice any stage of whitefly, spot-treat these plants with sprays before bringing them into the greenhouse.

If possible, allow the greenhouse to freeze during the winter. An alternative is to close up the greenhouse tightly during the summer and allow it to become overheated for at least a week. Try to remove all plants during this time. Adult whiteflies will die after a week without food.

Whiteflies are susceptible to some botanical sprays, including pyrethrum, rotenone, ryania, tobacco, neem, insecticidal soap, and oil sprays. They will quickly build up a resistance to these agents, however, except for the horticultural oil spray.

Biological control using the tiny beneficial wasp *Encarsia formosa* is well researched. These wasps are especially effective in the greenhouse. *Encarsia* are usually sold as parasiticized whitefly pupa stuck on small cards. Use these beneficials at the first sign of a whitefly infestation. If no whiteflies are present, *Encarsia* will die of starvation.

Encarsia lay their eggs in whitefly immatures, which turn black or brown as the young *Encarsia* develop. Different species of parasiticized whitefly nymphs will show slight variations in color. *Encarsia* also kill nymphal stages of whiteflies by poking holes in them and then feeding on the juices.

The insectary from which you purchase *Encarsia* will provide you with the proper amounts of predators to release for your situation. The usual rate is one wasp for every four plants per week for four to six weeks. Sweet potato whitefly is more difficult to eradicate with *Encarsia* and normally requires many more wasps to achieve control.

Monitoring for whitefly adults and parasiticized nymphs should continue after the *Encarsia* are released. This will help you determine when the releases should stop. Limited spot spraying with insecticidal soap can be done because *Encarsia* are somewhat tolerant of it.

Recent research is bringing about other whitefly predators. Some are host-specific for a certain species of whitefly. *Delphastus pusillus* is a very small black lady beetle that has a ravenous appetite for whitefly in all life stages. There are other species of *Encarsia* available as well as whitefly fungal diseases.

Other Pests to Know

There are other insect pests that can be problems for your herbs. They may not be found in as great numbers as the ones previously described, but they can still damage your field or greenhouse herbs, and should be controlled.

Earwigs

These creatures are ¾ to 1 inch long and black or brown in color. They have a pair of pincers protruding from the back end and, yes, they do sometimes bite people when provoked. Some adult earwigs can fly. They eat decaying plant material and other insects (including beneficial insects), which sometimes classifies them as beneficial. They do eat living plant material, including your herbs and flowers, and can make a mess of the foliage.

Earwig

Earwigs are communal and can be found together in dark hiding places. They will often enter greenhouses when the weather cools by coming in on plants (usually in the soil) and supplies. They lay eggs in the soil and often feed there.

Controls. Control earwigs by attracting them to crumpled-up newspaper or rags, or sections of bamboo or hose placed on the soil surface and work areas. Collect these materials and destroy the earwigs by crushing or drowning them. One species of tachinid fly lays its eggs in earwig eggs and eventually decreases their numbers.

Flea Beetles

These little (⅟₁₆- to ¼-inch), hard-shelled beetles are more of a problem outdoors than in the greenhouse. They are usually dark and shiny but the pattern and color vary by the species. They are quick moving, often jumping away when the plants are disturbed. The adults do the most damage early in the growing season.

Flea beetle

They eat "shotgun"-type holes in the leaves, which certainly makes the plants and leaves useless for sale. They are especially fond of arugula. Flea beetle larvae eat plant roots before they emerge from the soil. Serious damage to young seedlings, even death, can result from great numbers of feeding adults.

Controls. Flea beetles lay their eggs in the soil, so shallow cultivation will expose and kill the eggs before they hatch. A thin layer of wood ashes or diatomaceous earth spread on the soil also helps to deter these pests. Garlic sprays are said to repel flea beetles, but I have tried both homemade and commercial types and found them to be utterly ineffective.

Flea beetle adults are attracted to white sticky traps. Place them just above the foliage and secure them from moving around in the wind. Rain causes flea beetles to stop feeding and hide. Check a heavy infestation by frequent misting or overhead sprinkling.

The best way to prevent damage from flea beetles is to cover the plants with lightweight row covers immediately upon emergence. Cultivate shallowly around the row to expose any eggs that are close to the seeds. Carefully secure the edges of the row covers with soil or boards to prevent beetles from crawling underneath to reach your plants.

Grasshoppers

These insects need no description, and everyone is familiar with the damage they can do. They are included here because of the serious destruction they cause when they are allowed to become established in the greenhouse. No plants,

Grasshopper

except those in high hanging baskets or those covered with row covers, are safe from these pests. Big populations can devour your entire crop in a very short time.

Grasshoppers enter the greenhouse attached to the clothing of workers. They lay their eggs in the soil and may even be present when a new greenhouse is erected. The eggs sometimes arrive with unsterilized topsoil or soil mixes.

I can say from experience that, once established in a greenhouse, grasshoppers are very hard to get rid of. They are quick to move when approached and nothing short of strong chemical sprays will kill them, especially the larger adults. I have tried, with no success, many kinds of traps and controls.

Controls. One technique that does control them is to walk through the greenhouse with a long, sharp pair of scissors. Each time you see a grasshopper, carefully and slowly approach from behind and cut it in half. Although I love all living things, this somewhat time-consuming method gave me a curious feeling of "sweet revenge" with each one killed!

Another way to control grasshoppers is Nolo Bait. This microscopic spore kills only grasshoppers. After it is spread around, often mixed with wheat bran, the "hoppers" ingest it. It produces a disease in the insects that is passed on when they eat each other and also in their eggs. This treatment is effective, but it does need some time to work.

Leafhoppers

There are many species of leafhoppers. They are ⅛ to ¼ inch long and narrow in a wedge shape. Their color changes with the species but they are usually light colored, and some have spots. They are notorious for spreading viral diseases. Their feeding causes stippling of the leaves and stunting of plant growth. They like the strong-flavored herbs: sage, mint, oregano, and especially rosemary.

Leafhopper

They seem to be more common in greenhouses now, and when they become established they are extremely difficult to eradicate. Most species of leafhoppers live part of their life cycle in the soil, inside chewed holes in the stems, and on the leaves. This, and the fact that they are good, fast fliers, makes them hard to control with sprays. Insecticidal soap has no effect on them at all — they just fly away, laughing.

Controls. Control leafhoppers before they become established, especially in the greenhouse. Pyrethrum, rotenone, ryania, neem, and horticultural oil sprays are some of the botanical sprays you can use. I once paid more than $400 for several different beneficial insects that were supposed to control leafhoppers, both adults and the immature in the soil. Nothing worked, perhaps because the leafhoppers were too numerous. The result was a very thorough spraying over the entire greenhouse with pyrethrum, spaced several days apart. That eliminated most of the problem.

Scout carefully after the initial spraying to catch any that eluded you and to watch for the development of immatures that were missed because they were in the soil or within the plant. Diatomaceous earth spread on the soil surrounding plants can help control the immatures as they enter the soil.

Leaf Miners

Leaf miners rarely cause extensive stunting or loss of vigor. They are usually not found in great numbers, unlike the case with many other pest insects. They do cause serious cosmetic damage to plant foliage, however, and that makes for lost profits for the grower of fresh-cut herbs.

Adult leaf miners are tiny, ⅟₃₂- to ⅟₁₆-inch flies. They are black and yellow, with yellow heads and brown eyes. The eggs are white and are found in clusters on the undersides of leaves. The adults rest on the foliage and fly about when the plants are disturbed.

Leaf miner

It is the leaf miner larvae that cause the problem. As the eggs hatch, the larvae burrow into the leaf to feed, tunneling through the leaf between the upper and lower surfaces. This causes white to light green streaks on the leaf surface. There can be many larvae feeding in one leaf at a time.

Leaf miner damage

Controls. The adults are vulnerable to most botanical sprays but usually only spot spraying is necessary. Horticultural oils are effective against leaf miner larvae. The best control is to pick off leaves that have leaf miners at work. Place the leaves in a plastic zipper-lock bag and keep them sealed tightly as you work through the infested area. Then burn the bag.

Mealybugs

In some parts of the country, mealybugs are serious pests in the greenhouse; elsewhere, they are almost nonexistent. These slow-moving insects are white to tan in color, soft-bodied, and oval in shape. They have a cottony substance on their bodies and are covered with many leglike hairs. The adults are ⅕ to ⅓ inch long.

Mealybug

One species, the long-tailed mealybug, has a long tail protruding from its back end. This species gives birth to live young, whereas the other species lay eggs in a cottony sac on the undersides of leaves. Another species of mealybug lives in the soil and feeds on plant roots.

Mealybugs suck plant juices and cause the same type of damage as aphids do. They produce copious amounts of honeydew. You'll find them on the undersides of leaves, on tender stems, and on new growth.

Controls. Mealybugs are vulnerable to most botanical sprays, insecticidal soap, and horticultural oil sprays. Small numbers can be dabbed with a cotton swab dipped in rubbing alcohol. Beneficial insects such as ladybugs and lacewings will eat mealybugs. Some insectaries offer *Cryptolaemus* (a predator) and *Leptomastix* (a parasite) for control of mealybugs.

Scale

There are many different species of scale insects. Some are host-specific and only attack trees or citrus, for example. The most common in greenhouses are the soft brown scale and the armored scale. They have not been reported to be serious pests of herbs.

Scale look like small (mostly legless) bumps on a leaf or stem. They are black, brown, or gray and ⅛ to ¼ inch long. Some have ridges or a spot on the back. They may be hard- or soft-shelled. They are immobile during most life stages. Scales suck plant

Scale

juices, produce honeydew, and some inject toxic substances into the leaf.

Controls. Small numbers of scale should be hand-picked and crushed. Control larger infestations with horticultural oils. There are two biocontrols available: *Aphytis melinus* and *Metaphycus helvolus*.

Slugs and Snails

These two slimy creatures are not insects but rather are members of the mollusk family. They do their damage to crops by eating holes in the foliage. They can destroy a whole batch of seedlings overnight — they are nocturnal feeders.

Slug

Slugs and snails are dark gray to black with soft, wet-looking bodies. Snails have a hard shell covering that may be multicolored; they retract into the shell for protection. The average size of the adults of the most common species is about 1

Snail

inch long. Both slugs and snails leave a silvery trail as they travel.

Slugs and snails prefer warm, dark, moist conditions. They hide under debris, boards, anything that would give them protection during the day. The bottom drainage holes in pots and the undersides of flats are perfect hiding places.

Controls. There are many home remedies and commercially available traps to control these pests. The saucer of beer, set out where you see a trail, is a common trap. I have found this, and most of the other traps, to be ineffective, but they may work for you.

Hand-picking works. Place boards, potato slices, or other heavy dark objects on the soil where slugs have been. In the morning, lift these items and hand-pick the critters. Drop them into a jar with rubbing alcohol in it, cut them in half with a scissors, or sprinkle them with salt. The salt kills them quickly but large amounts can harm foliage, and salt is detrimental to the soil.

Slugs and snails die if they eat tobacco. Some growers control slugs and snails by scattering tobacco around their crop plants. Take care, though, because tobacco, if infected, spreads the tobacco mosaic virus to your plants. It's best to grow your own tobacco for this purpose and carefully monitor it for symptoms of this disease — yellow mottling of the leaves — before using it.

Slugs and snails are reluctant to cross rough surfaces. Create barriers around your plants of diatomaceous earth, sharp or silica sand, wood ashes, or ground lime. Use these last two in limited amounts because they can change the pH of your soil.

Tarnished Plant Bugs

These bugs are members of the *Hemiptera* family, which includes chinch bugs, harlequin bugs, and squash bugs. The tarnished plant bug is a serious pest of basil, especially in the field.

Tarnished plant bug adults are ¼ inch long, oval, and greenish to brown with a brassy hue. They have mottled darker markings on their backs. Their two sets of wings cross over the bodies to form an "X" on the back. The immatures resemble their parents but are smaller.

Tarnished plant bug

These bugs feed on new growth. As they feed, they inject a toxic substance into the plant that causes the surrounding area on the leaf to turn black. The new growth tips thus will be deformed.

Tarnished plant bugs first appear in early spring and their numbers gradually increase until their peak in late summer. They are very active and will fly away or hide when disturbed. This, coupled with their hard-shelled bodies, makes them difficult to control with sprays. They are easier to target with sprays in the early morning, when they are cold and move a bit more slowly.

Controls. Sabadilla dust, made from the seed of a lily family plant, is said to be effective in controlling them. A very tiny wasp, *Anaphes iole,* offers gradual control of the tarnished plant bug and is available from some insectaries.

Integrated Pest Management

The term "integrated pest management" (IPM) may be relatively new but the concept is an old one. It is being embraced by growers all over the world, especially by ornamental and nursery growers who in the past routinely used synthetic pesticides exclusively for insect control.

IPM is defined as a control strategy to reduce insect populations to an acceptable level that is friendly to the environment and cost-effective. It is a commonsense approach to promoting plant health by evaluating and using *all* available techniques and combinations of methods. It could be said that IPM is a "holistic" approach to growing quality crops. These methods may include the use of some chemical, biological, or botanical pesticides. Beneficial insects also play an important role in harmful-insect control. The best control is prevention, but this rarely works completely.

IPM takes skill, planning, and perseverance. It involves setting a threshold for damaging insect populations rather than striving for total elimination. Most of all, IPM involves educating yourself about your crops and the insects that want to eat them.

It is our job as growers to know the cultural requirement for each of the herbs that we grow and to strive to create these ideal conditions. We must also learn what pests are attracted to the individual herbs, what kind of damage they do, and what their ideal environmental conditions are. Armed with this information, we can then try to create an environment that is favorable to the herbs and much less so for the insects. Most often we must settle for a balance between the two. Obviously, this is much easier in the greenhouse. Outdoors, we are subject to the whims of Mother Nature.

The strategies of integrated pest management involve making use of several methods of pest control. These multiple-control tactics can be cultural, physical, environmental, biological, and chemical.

There are some things to remember as you begin to use these various techniques. There is no magic cure for damaging insect populations. It takes work and perseverance to control them. Certain tactics may or may not be less costly than the traditional chemical sprays.

Remember, you don't need to eliminate all pests, only to manage them. Take swift action when pests appear but don't overreact and try to wipe them out at once. And don't expect these techniques to work perfectly all the time. If your first attempt at control doesn't work, don't let that sour you on the technique. Try again and be more vigilant.

Educate yourself. Keep abreast of this ever-changing field of pest management. New biorational controls are being researched and developed at a rapid pace. Subscribe to grower trade magazines and ask for help and information from companies that supply pest-control supplies and beneficial insects. A list of insectaries and related companies can be found in the Resources section of this book.

Implementing Prevention Techniques

Preventing pests from becoming established should be your first line of defense. Excluding insect pests outdoors can be achieved only by using row covers. You can also use row covers in the greenhouse. The following techniques are especially important inside.

1. Practice good sanitation. Be vigilant about weeding. Clean plant debris from the top of the soil. Don't allow weeds to grow around the perimeter of the greenhouse, on the greenhouse floor, or under benches. Insects hide in these materials, just waiting for the chance to set up a feeding and breeding station in your crops.

I once found an immense colony of whiteflies on a fugitive mint plant that was living in the small space between the growing bed and the greenhouse glazing. The escapee had tunneled through the boards of the bed and taken root where it was sure it would not be cut and consumed!

2. Quarantine new plants. Hold plants that are purchased in a separate area for a week or two before placing them in the greenhouse. Inspect them carefully for insects and disease. Spray them, if necessary, before introducing them into your greenhouse.

3. Dispose of infested plants. This can be a very difficult thing to do, especially with an important stock plant or if you are short of a certain herb. Culling these plants will help a great deal with controlling a pest invasion. It is cheaper to replace a few plants than to lose a lot due to insect damage.

Avoid dispersing the pests on these plants by carefully covering them with a plastic or large paper bag before pulling them from the soil. Burning is the best method of disposal; this prevents insects from migrating to other areas outdoors.

4. Install screens. All openings into the greenhouse should have screens. Ordinary house-type screening will keep out many of the larger insects such as houseflies, as well as bees and birds. The mesh size is too large to exclude whiteflies, thrips, and flying aphids.

Many greenhouse-supply companies now offer insect barrier screening with a mesh size small enough to block smaller insects. This screening is very effective at excluding pest insects, but it also reduces airflow. This makes it more difficult to ventilate the greenhouse.

The use of these screens may mean that you have to increase the size of air inlets and/or fan capacity. The supplier should be able to help you with a formula to determine the screening size and cooling requirements for your greenhouse.

5. Confine plants with barriers. Another method of pest exclusion is to divide the greenhouse into separate areas. Insect barrier screens or poly curtains can partition the room. It is especially useful for multiple cropping situations such as the growing of different herb species. Pest control is easier in a smaller area as opposed to a large open space. This technique allows you to try different methods of pest control. For example, beneficial insects in one area would not be endangered by spraying in another compartment.

6. Avoid wearing yellow and blue clothing. This has been said before but bears repeating. These colors attract insects, which can then easily hitch a ride on clothing from one crop area to another or into the greenhouse.

Plan Your Strategy

Educate yourself about the pests and controls before you are confronted with a pest problem. Planning in advance of a problem enables you quickly to implement controls as soon as a pest appears. With swift action you'll be able to manage the pests easily and avoid an out-of-control situation that necessitates stronger tactics, such as spraying the entire greenhouse.

Determine Pest Thresholds

Once a pest appears, especially in the greenhouse, it will more than likely always be present. If you plan on using beneficial insects for pest control, a limited amount of the pest must be present as a food source or as a method for reproduction in order for the beneficials to survive.

Decide what your allowable limits of pest damage will be before you begin a control tactic. If you can tolerate only 1 percent of your crop to be damaged, there is not much leeway for control. If you will allow 5 percent to 10 percent of damage, that gives a little more time to begin using controls. For example, if small numbers of spider mites are present only in the lower leaves of some plants and you are harvesting the new growth, more tolerance could be in order.

These numbers have much to do with the volume of an herb you have and what your market is. If you are selling to restaurants that will chop the herbs for flavoring food, a little more damage is acceptable. Explain to your accounts and chefs that you allow a little damage because of the natural controls you use and that the damage will not affect the flavor.

Garnish, however, must be absolutely perfect with no pest damage. Supermarket packs must also be blemish-free. In these instances, your threshold for damage is lower. If the market is flooded with a particular herb and prices are low, your threshold for damage should be lower. In this instance, strive for a superior product.

Scouting and Monitoring

In order for an integrated pest management program to work, you must be able to monitor pest levels. This simply means scouting for pests and recording the results. Be consistent in performing this chore in order to control the pests before an infestation gets out of hand.

In large operations, one or more employees are trained to do this job and they devote a certain amount of time each week to it. In small operations, you may train an employee or do the job yourself. If you have only one or two people who help you pick herbs part time, teach them what to look for so they can report any pests to you. All people who work for you should be able to identify pests and know where on the plant to look for them.

To scout most efficiently, carry the right equipment: a hand lens, a small notebook and pen to jot down types and numbers of pests or

Use a hand lens to scout for pests on your plants.

a plan of the greenhouse or crop area, and markers to place where infestations appear. In large operations, the scout will also carry a small counter to record the number of insects trapped on yellow sticky cards.

Some operations even have special insect-trap reports printed up on which the scout records the number and type of pests. These list the location, cultivar, number of plants checked, number of plants infested, time and date, number of sticky cards checked, pest type, and its life stage. Markers are a good idea, especially for the busy small operation, because areas of infestation can be forgotten as the busy person rushes off to do another task.

Orange flagging tape is good for marking areas where pests are seen. The small red flags on the end of wire stakes, sometimes used by utility companies, are quick and easy to use as markers. Disinfect the wire stakes before reusing them to prevent the spread of disease.

Scouting should be done on warm, sunny days because this is when pests are usually more active. Start with herbs that are known to attract certain pests and in the greenhouse near doorways and vents. Work in a zigzag pattern through the entire crop area to get a representative sample of each crop. Examine each plant carefully from bottom to top, paying special attention to the underside of leaves.

Yellow sticky cards are especially useful for detecting fungus gnats, adult leaf miners, whiteflies, winged aphids, and thrips (thrips are also attracted to blue). These are coated on both sides with a sticky substance that traps insects. Yellow flypaper will also trap some insects as well as houseflies. These work well in the greenhouse hanging up high and out of the way.

Recent research recommends placing the sticky cards vertically just below the top of the foliage. Placing the cards just above the foliage is nearly as effective, and it will be easier for you to avoid brushing the foliage (and clothing) against the cards. The sticky goo, especially that on homemade cards, can make a mess of the foliage, causing it to be unsalable.

For monitoring fungus gnats and shore flies, place the cards horizontally or on the soil surface. Sweet potato whiteflies are trapped when the sticky cards are placed at a 45-degree angle just below the crop canopy.

Space the cards around the crop area, with extras near susceptible plants, doors, windows, and vents. A general rule is to use one yellow card for every 1,000 square feet of greenhouse space.

These yellow sticky cards are available from many grower-supply companies. They are simple to use, and most have a protective covering that you just peel off. They are expensive, though; you can save some money by making your own sticky traps. Look for plastic yellow plates, 6 to 12 inches, in many paper-supply stores. Coat them with a sticky substance available from grower-supply companies. There are a number of kinds to choose from. The one with the brush attached inside the lid is easiest to use. Aerosol sprays are simple to operate, but they are expensive.

Either way, it's a messy job to coat these plates. Leave a small area on the edge uncoated for easier handling. If you plan on hanging your new sticky plates, punch a hole near the edge of the plate before coating it with sticky stuff! Attach the plates to sturdy wire or bamboo stakes with a clothespin. Use long stakes so they can be raised as the plants grow taller.

Indicator plants are especially attractive to pests. Perhaps you have noticed that a certain variety of an herb seems to attract more pests faster than others of the same species. Rather than throwing these plants out, pot them up and space them around the greenhouse or crop area. They can be used to monitor pest activity. When you see pests on these plants, it's time to begin using control methods before they infest your main crop.

These same plants can have other beneficial uses as trap crops or banker plants. These uses are explained later in this chapter.

Pest Controls

Many types of pest controls are available to the grower. Most of these can be purchased from your greenhouse supplies company or companies that specialize in natural and biological insect controls. (See Resources.)

Traps

There are traps for many garden pests, from cabbage loopers to slugs. Unfortunately, there are not many available yet for the most common pests for the commercial greenhouse grower. It would be wonderful to have pheromone-scented traps that would lure whiteflies, aphids, leafhoppers, spider mites, and fungus gnats. Someday this may be the most common method for controlling these pests, and research has already been done on a pheromone-dispersal lure that causes aphids to drop from their host plants.

At the present time, though, sticky traps are all that we've got against these insect pests. To use them as traps, place as many as you can around the infested plants. Gently shake the plants — the insects will fly off and toward the attractive sticky card, where they will be trapped. This method can catch many adults, which may prevent them from reproducing, but does nothing to catch the immatures that don't fly.

Trap crops are plants that are more attractive to insect pests than your main crop. These are usually ornamental plants that are heavily treated with systemic or residual chemical pesticides — which remain not only on the exterior of the plant, but also are absorbed by the leaves and roots — that cannot be used on food crops. The pests are lured to the trap plants and then perish after feeding on it.

Pot trap plants so they can be treated outside the greenhouse, then place them near heavily infested plants. They can also be used with newly purchased plants in isolation to detect pests. Re-treat these plants often with the residual type insecticide.

Examples of plants that can be used to trap whiteflies are poinsettia and hanging baskets of fuchsia. Rue (*Ruta graveolens*) is very attractive to whiteflies, but handle it carefully because some people have an allergic skin reaction to this herb. If edible flowers are used, take care to mark them so that blossoms are not accidentally cut and sold for consumption.

Petunias attract western flower thrips. Their feeding marks are easily seen on the leaves, especially if the flowers are removed. Fava beans attract thrips and are good also for monitoring viral infections transmitted by these pests.

Spider mites love beans. Potted dwarf beans are excellent indicator plants and trap crops for spider mites. Cereal grains are alluring to some species of aphids.

It is extremely important to monitor trap, indicator, and banker plants frequently. Employing these techniques means that you are cultivating a population of pest insects. There is a danger that the pests could escape and infest your main crop. Do not use this technique if you are releasing predatory insects.

Homemade Pesticides

Recipes for homemade insecticides abound. They range from garlic, dish soap, peppers, and nicotine all the way to tea tree oil. I have tried several homemade insecticides, but none worked well. As a very busy commercial grower, I could not afford to waste time and risk serious plant damage trying an unproven method. If you have the time to experiment, by all means do try some of the recipes. You'll find them in gardening books and magazines. Test a recipe on a plant or two before spraying the entire crop. Some growers have had good luck with these.

Commercial Pesticides

The label on the pesticide container will list on what plants and where (in the greenhouse or field) it is legal to use this product. If we are legally able to use a particular pesticide, the label should list individual herbs or the word "herbs." There are very few pesticides whose labels list herbs.

It is extremely costly for pesticide manufacturers to provide the supportive data necessary to register their products (both botanical and chemical) with the EPA. Each crop that a pesticide is used with — tomatoes, corn, beans, or grains, for example — must have data to support its safe use on that crop. Unfortunately, many "minor crops," such as fresh-cut herbs, have not and probably will not be tested because of the expense involved.

According to several EPA officials that I spoke to, there is another problem. Both the EPA and pesticide manufacturers are not sure if herbs should be classified as vegetables, or if herbs should be tested individually or as a group. The officials say that, until this issue is clarified, you may use pesticides that are labelled for use on vegetables (particularly leafy green vegetables or lettuce).

In 1996 the Food Quality Protection Act was signed into law. This law makes broad changes in the way that the Environmental Protection Agency (EPA) regulates the use of pesticides in the United States. New safety requirements force the EPA to consider several health issues when registering or reregistering a pesticide. The law also mandates that all existing pesticides be tested and reregistered within the next 10 years. This includes the pesticides listed below.

The list of what we now consider "safe" botanical pesticides may not be legally available to fresh-cut-herb growers in the future (see pages 250–252). As these insecticides are reregistered, those listed may or may not be used in the greenhouse or on herbs. There may be new greenhouse reentry times or they may be dropped from the market. Please check with the supplier regarding the status of a pesticide before you purchase it.

Current laws. Various state agencies and the EPA are enforcing laws governing the sale and use of pesticides. Vendors and users of many pesticides must obtain a permit before using or selling them. Many of these pesticides can be used only by a certified private applicator licensed by the state.

Comprehensive records of pesticide use must be kept and signs be posted where they are used. Some pesticides (including those listed here) are considered safe, and no permits for their use or sale are required.

In 1992 the EPA revised its Worker Protection Standards. The new provisions became effective on January 1, 1995. These regulations require you to take steps to reduce the risk of illness or injury from pesticide exposure to your workers and yourself. You are considered a Worker Protection Standards employer even if you are the sole employee of your business or if work is done only by your family members.

The EPA and the Worker Protection Standards consider botanical pesticides (see pages 250–252) basically the same as chemical pesticides. Therefore, if you use these pesticides, you must comply with many of the same protection standards as those using chemical pesticides.

 summary of pesticide use requirements

You must follow all the requirements on the pesticide label, which is a legal document, as well as other regulations. In general, the provisions call for:

1. A central area where information about pesticides is kept. This includes a safety poster, nearest emergency medical facility, and pesticide application records with restricted-entry information.

2. A decontamination site with soap, water, emergency clothing changes, single-use towels, and eyewash materials.

3. Personal protection equipment (see page 258).

4. Posting of treated areas. Large signs must be posted at the entry points of outdoor fields or into the greenhouse. Individually treated areas can use smaller signs.

5. A training program for employees or you. Records must be kept of employee training sessions.

Greenhouses are considered enclosed pesticide application sites, and there are special rules governing pesticide use in them. These rules involve reentry times and ventilation restrictions.

The Worker Protection Standards provisions are fairly complex, and compliance with them is mandatory. You can get complete information about this in the EPA's manual *The Worker Protection Standard for Agricultural Pesticides — How to Comply: What Employers Need to Know,* available from the regional EPA office or the EPA headquarters, whose address I list in Resources.

It is important for you to know and understand these rules and how they apply to your operation and the pesticides you use. The EPA's new priority is inspections of grower operations and enforcement of these rules. You never know when your operation will be on an inspector's list.

Botanical Pesticides

Botanical pesticides are derived from plants. Most of the pesticides described here are relatively nontoxic but they can be dangerous if misused. Even these "safe" pesticides should be used only as a last resort for pest control. Any foliage that has been sprayed should be washed before selling or using; even "natural" pesticides leave a residue on the leaves that should not be consumed.

While these are safe to use around warm-blooded beings, most are toxic to fish, birds, and beneficial insects. Use them with extreme care and avoid any runoff that could contaminate waterways.

Many of the commercially available formulations contain piperonyl butoxide (PBO). This is a synthetic substance added to enhance the toxic action of the pesticide. There are some companies that manufacture botanical pesticides without using PBO should you wish to avoid this chemical.

Most of these pesticides, purchased in large sizes from distributors, will have "reentry times" listed on the label. This is the length of time, after spraying a greenhouse, before a person can safely return to the area. According to the Worker Protection Standards laws, an owner may reenter sooner than an employee. The reentry times for the following pesticides vary from 4 to 12 hours. As these products are reregistered, these reentry times may change.

Always read and follow label directions. You are obligated, by law, to read the label and follow all the directions stated there.

Repellents. Hot pepper wax, a recent addition to the market, is a commercial version of homemade pepper sprays. It is a repellent rather than an insecticide, similar to a garlic barrier spray. A combination of paraffin wax, cayenne pepper, mustard oils, and, in some brands, concentrated plant nutrients, it binds to the leaves and is transparent when dry. It sticks to the leaves up to 30 days, even through rains. It does wash off with warm water, which may be detrimental to fresh-cut herbs.

Although they are labeled for use in greenhouses, these repellents are not usually as effective in a closed environment because the pests have nowhere else to go to feed. Outdoors, the pests may be repelled enough to seek other food.

Rick Hannigan, of J.R. Johnson Supply, Inc., a greenhouse-supply distributor based in St. Paul, Minnesota, says that several large growers indicate that hot pepper wax is effective at repelling rabbits as well as many insects! I found all commercial garlic repellent sprays to be utterly ineffective, even outdoors. Perhaps my insect pests are garlic lovers too!

Insecticidal soap. These soaps are made from certain fatty acids found in plants and animals. They are usually sold as a liquid concentrate that is diluted with water before being applied to the plants. The soap breaks down quickly and has no residual effect. It is safe to use around people, animals, and birds but is moderately toxic to beneficial insects. These soaps must contact the body of the pest insect in order to kill it.

Because it has no residual effect, you must reapply it every three to five days to control a serious infestation. It is effective in killing soft-bodied insects such as aphids, mealybugs, whiteflies, and especially spider mites. Insects do not readily build a resistance to insecticidal soap spray. It can be used up until harvest. These soaps are sometimes sold mixed with pyrethrum. Do not use in strong sunlight, in temperatures over 75°F, or on cuttings or new transplants.

Neem. This substance is extracted from the seeds of the neem tree, which is native to India, where it has been used for centuries to repel and kill insects. Some brands are now labeled for use on food crops. Neem has been commercially available for several years, but is registered only for ornamental crops. It is sold as a concentrated liquid that is diluted with water before use.

It kills, repels, and stops the feeding of most herb pests. It also disrupts their hormonal balance, causing the insects to die before they reach the next life stage. It can be used two or three times, 7 to 10 days apart. Do not use on stressed plants, cuttings, or new transplants or in temperatures above 90°F. Excessive application can cause phytotoxicity or even death to the plants. Neem is very toxic to fish.

Nicotine. Nicotine is a highly toxic insecticide sometimes mentioned as useful for growers. It is as dangerous to humans as it is to insects, and it is not recommended. Homemade versions of this toxin are even more dangerous because the concentration can vary so much. Nicotine sprays and fumigators that are marketed for use on ornamentals are federally restricted pesticides and are also restricted-use chemicals in many states.

Pyrethrum. This insecticide is derived from a type of chrysanthemum. It is highly effective at killing many insects, such as aphids, leafhoppers, leaf miners, and whiteflies. It does not seem to work as well with spider mites, and its use actually seems to increase their numbers a short time after spraying, perhaps because it kills beneficial insects.

Pyrethrum breaks down quickly in the environment and is relatively nontoxic to bees and ladybug larvae, but almost all other insects — including beneficials — are killed by it. The spray is a contact insecticide and must touch the body of the insect to kill it. It is important to mix pyrethrum correctly to its lethal dose. The insects will not die if the concentration is too weak, and the survivors quickly build a resistance to pyrethrum. Pyrethrum may be used up to five days before harvest.

Pyrethrum is sometimes sold mixed with other insecticides such as rotenone and ryania. A self-releasing aerosol spray of pyrethrum, PT 1100, manufactured by Whitmire, is labeled for use in greenhouses and on leafy food crops. This nonorganic product fogs the entire greenhouse. It has a 12-hour reentry time. It was effective the first time I used it but much less so with later applications, perhaps because of pesticide resistance.

Rotenone. Rotenone is an extract from the roots of the derris and cube plants, both tropicals. It kills many types of insects including various worms, moths, and beetles. It is effective against the typical greenhouse pests such as aphids, mites, whiteflies, and thrips. It also works as a repellent and has a short residual effect for three to five days.

Rotenone is available in concentrations of 1% or 5%, comes as a liquid or wettable powder, and is mixed with pyrethrum and sometimes copper. It is a contact insecticide but will also kill or at least sicken the pests if they eat it. Rotenone is highly toxic to fish and birds.

Ryania. Another botanical pesticide extracted from a tropical plant, ryania is a contact insecticide. Rather than killing the bugs, however, it makes them sick. They stop feeding and eventually die. Its availability in the United States is limited. It can sometimes be found in small packages, as a dust or liquid, in garden-supply stores or by mail order.

Pesticide Management

There are many factors to consider when contemplating pesticides, so that we may use them as efficiently and safely as possible. Poorly managed use of pesticides can make a product ineffective. Apply insecticides only when the pest population warrants it. Spot treatment of an infested area will save product, discourage pesticide resistance, and save the lives of many beneficial insects that may be present.

Handling and Application

Botanical pesticides will contaminate the environment and groundwater if not handled properly. It is the responsibility of each person who uses these products to learn how to protect our water sources and environment. Know the properties of your soil and how deep the groundwater is. Soil texture determines how easily a product can leach through the soil layers to reach groundwater. Sandy soil, for instance, will allow a pesticide to seep into the groundwater faster than will loamy soils.

Learn the properties of the products you will use. For example: How soluble is it in water? How easily does it adhere to the leaves? How long does it remain active? How easily is it lost into the air when sprayed? Each time a pesticide is used, read the label. Don't rely on your memory to tell you how to mix it, what it can be used on, application rates, and so on. Again, Worker Protection Standards require a grower to comply with the provisions on the pesticide label.

Some growing operations may want to use injectors to apply pesticides, fertilizers, or fungicides via an irrigation system. If you are interested in this type of system, be aware that many states require the installation of anti-backflow devices. These prevent accidental introduction of chemicals backflowing into the water source. Check with your local Extension agent regarding these regulations.

Most pesticides are applied with a backpack or hand sprayer. These must be calibrated properly to prevent over or under application of pesticides. Calibration rates and procedures are supplied by the manufacturer in the instruction booklet included with the sprayer when it was purchased.

Pesticide Resistance

Some pests develop the ability to survive pesticide exposure. When a pesticide is applied, some insects will survive because they are hardy and naturally resistant or they simply did not receive a lethal dose. They then develop a resistance to that class of material. The susceptible insects without natural resistance die.

Those resistant individuals breed and pass on these genetic traits to their offspring. Most insect pests have a short generation time and a high reproductive rate. Each time a pesticide is applied, more and more resistant insects survive to pass on these traits. Eventually, almost all of the pest population will be resistant.

The most common way to retard the development of resistance is to rotate insecticides. The pesticides should have a different mode of killing the pests. You can use the same pesticide through two generations of pests before switching to another type. The third generation will likely consist of mostly resistant insects.

Of the pesticides that we can use for herbs, insects are most likely to develop resistance to pyrethrum, rotenone, and neem. Resistance to insecticidal soap and horticultural oil is almost nonexistent because of their modes of action, or the ways in which they kill the insect.

Phytotoxicity

"Phytotoxicity" refers to the foliar or flower damage, or "burning," caused by some pesticides. Some herbs are highly susceptible to pesticide damage, and all herbs are susceptible to damage from sprays that are used improperly. Of the "natural" pesticides, insecticidal soaps and super-light horticultural oil sprays have been shown to cause the most phytotoxic damage.

Pesticides can cause phytotoxicity, or a burning effect on leaves or flowers.

The leaves and flowers may bleach white, have opaque or brown spots, or simply curl up and die. Flowers are more easily harmed than foliage, and some varieties of an herb species may be more vulnerable than others. Some will be damaged only if sprayed in strong sunlight. Damage could be worse than the blemishes caused by insects — and any injury renders the foliage unsalable.

The manufacturer may include a DO NOT USE ON warning on the label, but rarely will this specify herbs. It is up to you to test each pesticide on each herb cultivar before you use it on a whole crop. Test a crop

1. Keep good records. Make a list of herbs and the pesticides that have been shown to injure them. Make this list available to any workers who may be applying insecticides.

2. Follow label rules. Never mix a pesticide to a greater strength and do not spray more frequently than the label suggests. Damage can occur when the residue on your plants combines with new applications.

3. Don't mix pesticides. Some products become more toxic when combined and can cause twice as much damage as when used alone. Some combination products are now available commercially. These are fine to use because they have been mixed to the proper concentrations.

4. Agitate the sprayers often. If the pesticide's solution is not kept continuously mixed, the pesticide may settle on the bottom of the sprayer, where this concentrated solution, with most sprayers, is taken from. If this occurs, add more water to return the mix to the proper concentration.

5. Use separate sprayers for each product. Even a small amount of residue left in a sprayer may be enough to injure foliage, especially with herbicides. Label each sprayer and use it only for that product.

6. Avoid spraying in strong sunlight. Strong sunlight intensifies the phytotoxic effects of some pesticides, especially in the green-house. Spray on cloudy days, in the morning, to allow the foliage to dry before nightfall.

by spraying one plant from top to bottom with the full-strength mixture. Any damage should show up within nine days.

Spreader-Stickers

These are products that are added to pesticides to help them cover the leaves uniformly. They help the material to adhere to the leaves and not wash off with rainfall, irrigation, or dew. This improves the efficiency of a pesticide, so you will have to use less — thereby saving both money and time.

Most pesticides tend to run down the leaf, leaving a large area uncovered. The solution pools on the lower edge of the leaf, which could lead to phytotoxicity. It also can drip off the leaf and into the soil, where you don't want it to go. Some pesticides will bead up on the leaf and not cover the leaf uniformly. The exception is insecticidal soap, which covers and sticks better than most insecticides.

Spreader-stickers come in various formulas. Some are considered organic and safe on food crops; others are not. Be sure to check the label before purchasing. Spreader-stickers are available from distributors, garden centers, and mail-order catalogs.

Pesticide Storage

Store pesticides in their original containers in an area separate from growing, packaging, or customer activities. Do not store them in the greenhouse; temperature extremes will decrease a pesticide's potency. Keep the storage area dry and well ventilated. The storage area should be at least 150 feet from any water source. The floor should be watertight, with no drains. Inspect the containers regularly for any signs of leakage.

Any pesticide spills, even small ones, should be cleaned up immediately. Have handy the following materials close to the storage area for cleanup: protective wear, including gloves, respirator mask, and clothing; several rolls of paper towels; large plastic bags; and a bag of pet litter.

In case of a spill, stop the source immediately and contain it. Divert it away from any water source by containing it with paper towels or sweeping with a broom. Soak up the spill with paper towels and pet litter, place the materials in plastic bags, and dispose of the bag. The pesticide label, the supplier, or the manufacturer will have information regarding the proper disposal. Report any large spills or those that enter a waterway. Your local Extension agent or county health department should know whom to report the spill to. The Materials Safety Data Sheet supplied with the pesticide also will have information regarding spills.

Even our "safe" pesticides are considered hazardous waste, and you could be held liable for improper disposal. You cannot simply bury or burn empty containers or unused pesticide. The label on the container will give you directions on how to dispose of any unused portions. In general, the containers should be triple rinsed, then punctured or crushed so they cannot be reused.

Many areas of the country now have hazardous waste disposal sites, where you can drop off unused pesticides or empty containers. Some areas also have a container-recycling program. Contact the local EPA office or your Extension agent for information about these programs.

Sprayers

Today the grower has many types of sprayers to choose from. There is everything from a hydraulic, automatic, high-power fogger to a gasoline- or battery-powered backpack sprayer, from trombone-type slide sprayers to electric and hand compression sprayers. The small hand-held sprayers, quart size or less, are fine for spot spraying, but most growers need something bigger.

The most popular of these for the small- to medium-size grower is the 1- or 1.5-gallon hand-pump sprayer. These sprayers are versatile, portable, and easily operated. They are relatively inexpensive: They run anywhere from $10 at the hardware store to $50 for the superior-quality sprayers sold by grower-supply companies.

Buy the best quality that you can afford. The best ones are made with finer materials, especially the seals, and will last much longer than the less expensive models made for the home gardener. The ideal is to have separate sprayers for each pesticide you use. It eliminates any inadvertent mixing of sprays and the sprayers will not have to be cleaned so thoroughly after each use.

The sprayer you choose should have these features: the ability to pump 60 or 70 pounds per square inch (p.s.i.) of pressure; a carrying handle (usually this is also the pump handle); a long wand with an angled tip to reach the undersides of the leaves; and an adjustable nozzle. Make sure that the accessories (such as a hose extension kit) and replacement parts (gaskets and seals) will be available. This is usually the case if you buy from a distributor rather than the corner hardware store.

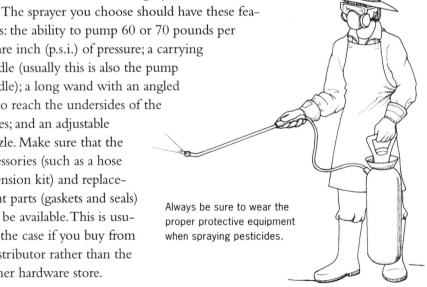

Always be sure to wear the proper protective equipment when spraying pesticides.

Personal Protective Equipment

You will find the term "personal protective equipment" listed on most pesticide labels and material safety sheets. It describes what gear must be worn while handling that particular product. These are equipment requirements according to Worker Protection Standards.

Botanical pesticides are certainly less harmful than chemical sprays, but they are still dangerous. Breathing the vapors or absorption through the skin and eyes can have damaging effects. Always protect yourself when spraying these products, even insecticidal soap.

Worker Protection Standards require that your workers wear certain protective gear if they do any spraying. You should too. Always wear long-sleeved shirts and long pants when spraying. Cover your feet completely. Lightweight rubber slip-on boots will protect your shoes from accidental spills. Plastic, rubber, or vinyl gloves, with cotton liners for comfort on long spraying jobs, will protect your hands from exposure.

Safety goggles are a must to protect your eyes from splashes. Don't rely on sunglasses, which don't protect your eyes from the side. Most goggles are reasonably comfortable, can be worn over glasses, and are ventilated. They cost less than $10 and can be purchased from most grower-supply companies.

A respirator mask is mandatory! A doctor-type dust mask is simply not good enough to protect your precious lungs. The kind you want covers your nose and mouth and has two filter cartridges. These cartridges are easy to replace. The mask must fit snugly to work properly. It should be so tight, in fact, that it can numb parts of your face on long spraying jobs. (I once told my dentist that he could replace Novocaine with a respirator mask for his patients.) Instructions for proper fitting will be included with the mask. They sell from $30 to $40 and are available from grower-supply companies.

A chemical-resistant apron is another protective item that you may find useful. These protect your clothing from splashes and from absorbing odors. Growers who have reason to buy chemical pesticides should know that all sorts of protective gear is available from suppliers, including disposable clothing, protective hoods, and full face masks.

- Before spraying, make sure that the plants to be sprayed are adequately watered and show no signs of moisture stress. If you are spraying in a greenhouse, shut off all ventilation fans to avoid "drift" of the spray to areas where it is not wanted.
- Test the sprayer with plain water before completing the mix to make sure it is working properly. Try to estimate how much spray you will use at this time and mix only enough that you won't have any left over.
- Fill the spray container halfway with water before adding the pesticide concentrate. After adding the remaining water, close the sprayer tightly, then agitate it well before pumping up the pressure. Continue to agitate the sprayer during spraying. Adjust the spray nozzle to produce the finest mist possible. Small droplets stick to the leaf better, especially the underside, and cover the area better. This saves money, time, and kills more insects.
- If you are spraying outdoors in any wind at all, larger droplets will result in less drift. This may be necessary if you are spraying near an area where the pesticide is not wanted. The larger droplet size has its drawbacks, however. The drops splash onto the leaf surface and often will run, collect on the tips, and drop off into the soil. Larger droplets also do not stick well to the undersides of the leaves, where they are usually needed the most.
- As you spray, try to cover the undersides of the leaves completely. This is more easily accomplished if you turn the spray wand over so the tip angles up. Cover each leaf completely but stop just before the spray begins to drip. Finish the area you are spraying before moving to the next. Stop periodically to pump up the sprayer, if you have that type, to maintain a constant high pressure.
- Check the sprayers often for leaks and clogging. Nozzle clogging can be a real problem with insecticidal soaps, especially if you have hard water. Be sure that the nozzles are dispersing properly. Try a toothbrush or other soft brush to clean the nozzle. Flush the sprayers with water when you are finished. The rinsate — diluted contents of the sprayer — should be sprayed over a large area of the crop rather than pouring it on the ground or down the drain.

Other Pest-Control Remedies

The following are more pest control techniques. The use of beneficial insects is most often thought of when Integrated Pest Management is mentioned. There are other methods, as well, that can be useful as you strive to reduce your pest insect problems.

Horticultural Oils

Petroleum-based horticultural oils have been used for many years as dormant oils; because they are very harmful to foliage, these oils are used when the trees are dormant. They are sprayed, usually on fruit trees, to reduce insect damage in the next growing season.

The new breed of highly refined oil is known as ultra-light, super-light, or superior oil. These oils are much less damaging to foliage and do a good job of controlling pests and diseases. See chapter 12, Controlling Plant Diseases, for more about oils.

These oils work by suffocating or smothering the insects and their eggs. The oils must contact the insect to work. They are quite effective on soft-bodied insects, which includes most herb pests. Handle oil sprays as you would any other pesticide. Read and heed the labels.

Beneficial Insects

The use of beneficial insects to control pests mimics nature's own checks and balances. In the well-integrated outdoor garden, predatory insects occur naturally or are drawn to an area because of the presence of prey. When we practice monoculture, greenhouse growing, and/or a weed-free environment, we eliminate the natural habitat for predatory insects. Luckily, we now can purchase these predatory insects and put them to work for us.

The ladybug is one of the most familiar beneficial insects.

The advantages of using beneficial insects will be seen in your crop — lush growth and the absence of damaged foliage from sprays. Insects, even beneficial ones, are not welcomed by the end user of fresh-cut herbs, so removing these insects is a must before selling. This constant harvesting and washing off of insects, both good and bad, is another reason why it may be necessary to replenish the supply of beneficials frequently.

Part IV: Successfully Growing More Than 20 Herbs and Flowers lists some beneficial insects that are specific to the pests that favor those herbs. An ever-growing field, new beneficial insects are being discovered and developed at a rapid pace. Keep informed by reading trade magazines. IPM Laboratories publishes a quarterly newsletter that is very informative about recent advances in pest control and beneficial insects. For more information, see Resources.

The first time I used predatory insects (ladybugs) to control an aphid population in my greenhouse, I was elated! They did their job so well that I proclaimed this technique the "lazy person's pest-control method!" I found out later, on my second attempt, that I was just lucky the first time because I really knew nothing about this rather complicated process.

Preconditions. The use of beneficial insects requires dedication and good management to be effective, especially in the greenhouse. It is not a one-time application, as spraying can be. This method requires understanding that it is not a 100 percent cure for pest insects and that there will always be a pest population in your crop.

Low levels of pests must be present in order for the predators to feed and reproduce. Most predators, in some life stages, will eat alternative food such as nectar, pollen, or artificial food supplied by you. Beneficial insects can be used outdoors, but much larger amounts are required because they often simply fly away. They are more effective in the captive environment of the greenhouse.

Ordering insects. If you expect to use beneficial insects to combat a pest problem, make contact with the suppliers before you need to order from them. Find out lead time in ordering and the best methods of transporting the insects to you. Suppliers can be a big help in all aspects of biological pest control. It is best to order from a biological pest-control company rather than from gardening catalogs. Insectaries have a larger assortment of beneficials, are more knowledgeable, and can give you more help and advice. (See Resources.)

Effectiveness. Many beneficial insects have specific temperature and humidity ranges at which they are most effective. Some insects must have particular conditions present before they will feed or breed. If these conditions are not met, failure is the result. The insects may leave the area or starve to death. For instance, ladybugs and their larvae are voracious aphid eaters, but if they are released in the greenhouse during winter, they immediately go dormant. They need the long sunny days of the growing season to be active.

Release beneficials at the first sign of an insect pest. If the pest population is too great, the beneficials will probably be overwhelmed, and good control cannot be achieved. If the infestation is heavy, knock down the pest population with a spray. This should be done at least one week — two or three weeks is better — before the release of beneficials. You don't want your beneficials to die from sprays or vapors; because they have not been exposed to the pesticides, they'll have no resistance. Spraying should *not* be done after the release of beneficials. Careful spot spraying of a heavily infested plant or two may be necessary.

Sticky cards are not recommended after the release of beneficial insects. Many of the beneficials that fly during part of their life cycle are attracted to yellow. Sticky cards are helpful for monitoring the balance between the pests and beneficials, but use them for a 24-hour period only, to avoid trapping too many predators.

Receipt and storage of insects. Some beneficials are supplied as live insects in vials with a carrier such as rice hulls or wheat bran. Many, predatory mites, for example, will have a small amount of the host mites included in the vial as food. These are simply sprinkled, at certain rates, on the affected plants. Others, such as lacewings, are shipped as eggs in a vial with a small amount of food.

Beneficial insects are often shipped in vials with rice hulls or wheat bran.

These are typically stored at a certain temperature until a few larvae can be seen crawling around inside the container. You then sprinkle them wherever they are needed. Still other predators are supplied as eggs attached to a sticky card. You place the cards near the prey before the eggs hatch.

Some beneficial nematodes arrive in sponges or in dry powder that is mixed in water and sprayed or drenched on the soil. Ladybugs, on the other hand, are often shipped live in small muslin bags. Mist the area where they are to be released so the ladybugs can have a drink of water when they are set free. Usually after this, rather than feeding, they will fly off to mate. While this is no doubt enjoyable for them, it does nothing to solve your immediate problem of pests that need to be eaten.

The ladybug larvae are more efficient than their parents at eating pests, so if you can afford to wait until they hatch, you eventually will see some reduction in pest population. During this waiting period, you must

not spray or destroy any eggs. The eggs are oval, usually yellow, and laid in clusters.

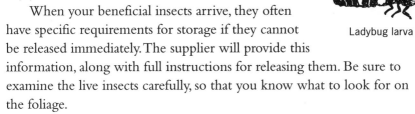

Ladybug larva

When your beneficial insects arrive, they often
have specific requirements for storage if they cannot
be released immediately. The supplier will provide this
information, along with full instructions for releasing them. Be sure to
examine the live insects carefully, so that you know what to look for on
the foliage.

Follow-up. Monitor the beneficial-to-pest ratio weekly to ensure
that the beneficials are keeping the pests under control. It is usually not
enough to make one release. In fact, many large commercial growers
release predators weekly. Establishing a good, active predator population
may take months.

During high pest insect activity time, subsequent releases of predators
may be necessary. For instance, whitefly activity increases when the temperature goes up to 80°F, while *Encarsia* slows at this temperature.

As you can see, using beneficial insects is a complicated venture. There
are many causes for failure and just as many criteria for success. When this
method works, and Mother Nature's balance is achieved in an artificial
environment, it is indeed a wonderful thing.

Banker Plants

These are the same plants that are used as indicator and trap plants. In
a sense, banker plants act as a "nursery" for predatory insects. Do *not* use
this technique if plants have been treated with pesticides.

The ideal time to use this "banking" technique is when the indicator
plant is infested with pests and before they move on to the main crop.
Concentrate releases of beneficial insects on these plants. You want the
predators to get a good start and reproduce on these plants. This technique is useful if you are not able to release the beneficial insects into the
main crop immediately due to a recent spraying or time constraints. These
plants, and their beneficial passengers, can be moved out of the greenhouse if it becomes necessary to spray insecticides.

Insect Growth Regulators (IGR)

At this time, there are no insect growth regulators legal to use on
food crops in the greenhouse, other than neem products, which have
some similar properties. As reregistration of pesticides progresses, perhaps

we will be able to use some of these environmentally friendly products on herbs in the future.

IGRs work by interfering with an immature insect's development so that it does not grow to adulthood and reproduce. Adults are not affected, so they continue to feed and reproduce. Control may not be achieved for two to four weeks with most insects.

There are several types of IGRs and each is specific to certain insects. This makes most of them safe to use when beneficial insects are present. They are unstable in direct sunlight and have little residual action. These properties make the existing IGRs useful only in greenhouses, where some of the damaging UV rays are filtered.

Keep an eye toward the future for IGRs that can be used on food crops and outdoors. Research is continuing on these new types of pesticides.

Bacterial, Fungal, and Viral Insecticides

This is a relatively new area of pest control generally referred to as biocontrols. There is much research being conducted to develop bio-pesticides using bacterial, viral, and fungal organisms. I am confident that eventually we will be able to manage insect pests using these biocontrols and nothing more.

One of the most commonly known bacterial biocontrols is *Bacillus thuringiensis,* commonly known as Bt. It has been used by home gardeners and commercial growers for years. There are several different strains of Bt sold under a variety of trade names. Each is specific to certain insect classes. *B. thuringiensis san diego* for Colorado potato beetles, *B. thuringiensis kurstaki* for many different caterpillars, and *B. thuringiensis israelensis* for mosquito and blackfly larvae are the most common.

There are many other subspecies of Bt, but, as yet, none is available commercially to control the most serious herb pests — whiteflies, aphids, spider mites, and thrips.

Beauferia bassiana is a naturally occurring fungus that is effective against whiteflies, aphids, thrips, spider mites, and mealybugs. Mycotech Corp. of Butte, Montana, markets this fungus under the trade name Mycotrol-WP. The company hopes soon to add lygus bugs, tarnished plant bugs, and several others to its label.

This fungus, and others being developed, works by means of spores that attach themselves to an insect's body. They burrow in and grow there, eventually killing the insect. Some insects turn colors as they become infected. The fungus must contact the insect's body to be effective. It usually takes two to four weeks for death to occur.

There are other biopesticides currently undergoing research and some not yet labeled for food crops that should and may be in the future. A natural viral insecticide, with a very short life, has shown promise against certain insects, and research continues on this. *Streptomyces avermitilis,* under the trade name Avermectin B or Avid, is a bacterial insecticide currently labeled for ornamental crops only.

Other Pest-Control Options

Vacuuming sucks away a large amount of insects. This can be a time-consuming, but satisfying, job. Any type of vacuum will work as long as it has a hose attachment, powerful suction, and is easy to move around. Shop-type vacuums have these requirements but often do not have bags, which will make it more difficult to dispose of the pests. It works best on those pests that don't have their mouthparts stuck securely into the plant, such as aphids. Whiteflies and leafhoppers are easily pulled out of the air with a powerful suction using a wide-nozzle end. Just disturb the plants and vacuum away!

It is a little more difficult to vacuum the pests off leaves, as this may harm the foliage. Dispose of the vacuum cleaner bag by wrapping it securely in a plastic bag and throwing it in the garbage or into a roaring fire.

Diatomaceous earth is a fine silica powder made from the fossils of prehistoric algae. It works by absorbing an insect's protective coating and then puncturing its body with its tiny sharp particles. To use it, scatter it on the soil surface. It works against slugs, snails, and any insects that crawl on, in, or out of the soil.

Use only the type intended for the garden, and not the diatomaceous earth that is meant for swimming pool filters. Wear a dust mask when spreading diatomaceous earth; the dust is very fine and it can be harmful if inhaled.

Lizards and spiders in the greenhouse will help control pest insects. Small chameleons are entertaining as well as effective insect eaters. They may not be suitable in greenhouses operated year-round in cold climates, however, or in large commercial operations.

Spiders are naturally present in most greenhouses and are an effective control. Some people do not deal well with spiders of any kind in their midst, however. I was such a person, but as time went on I took great joy in watching my many resident wolf spiders eat any bug that came within their reach.

controlling plant diseases

Diseases may infect just a few plants, stunting their growth; they can damage foliage, thus making it unsalable; or they can wipe out an entire crop. Large stands of a single species are more susceptible to plant diseases.

Because most herbs are prone to one or more diseases, it is important to have a basic understanding of the causes and measures that can be taken to control them. The available information about the many plant diseases is complex. Keep handy one of the many excellent books on the subject to consult as a reference.

Mold, mildew, blight, wilt, rot, rust, leaf spots, and scab are common names for what may infect your plants. These diseases are caused by bacteria, viruses, or fungi. The afflictions can be soil borne, airborne (or both), or be transmitted by insects. Aphids, leafhoppers, and thrips are known virus carriers. Nematodes — tiny parasitic worms that live in the soil — attack plant roots.

Keep in mind that environmental problems can cause damage to plants that is similar to many diseases. Poor cultural practices, such as moisture stress or poor air circulation, can cause stress-induced disease or weaken plants, making them more vulnerable to infection from pathogens — living organisms that cause disease in plants. They can be bacteria, fungi, viruses, or nematodes. Over or under watering, excessive pesticides, nutritional stress, excessive soluble salts, improper pH levels, and the wrong planting depth can all cause stress-induced diseases. Usually the whole bed is afflicted. Pathogen-induced diseases usually begin in small pockets. What we see on a sick plant are the results of one or more of these factors. The pathogen responsible for the damage is almost always invisible to the naked eye.

Disease pathogens are present in the soil and in the air. In order for them to infect plants, certain conditions must be present. There must be enough susceptible host plants, enough pathogens, and a favorable environment.

Scouting is the first step in disease control. Take the time each week to examine several plants in each bed. Early detection of a disease, before it infects an entire bed, can make the difference between profit and loss.

The next step in disease control is diagnosis. This might be difficult because plants can be infected with more than one disease, many diseases have similar symptoms, and insects or environmental problems can cause symptoms that mimic diseases. If you have trouble determining what's wrong, it may be necessary to have a tissue culture done of diseased plants to identify the pathogen responsible. Most universities offer tissue culture, along with suggestions for proper control measures. Your local Extension agent can give you guidance on how to send a plant for diagnosis.

If a fungus is the problem, control is difficult because few fungicides are legal to use with herbs. Most fungicides that are acceptable are for prevention rather than cure. At this time there are no commercial sprays or remedies for viral or bacterial diseases; the best method for control of diseases is prevention through the use of good cultural practices.

One means of control is to pull up and destroy severely infected plants. Care must be taken when doing so to avoid dispersing the spores to other plants. Place a large plastic bag over the plant and close it tightly around the stem at soil level. After the plant is pulled from the soil, turn it upside down and tuck the roots inside the bag. Seal the bag tightly. The bag can then be burned or placed in the garbage.

Foliar Fungal Diseases

Airborne fungi are the most common pathogens to attack herbs and the easiest to identify. The various fungi that cause disease generally need high humidity or the presence of water to grow; this can make them more of a problem in the greenhouse than in the field. There are many types of fungi that cause foliar disease. Here are just a few of the most common ones that infect herbs.

Gray mold

Caused by the pathogen *Botrytis cinerea,* gray mold is one of the most devastating diseases, especially for basil. It thrives in cool to moderate temperatures, with poor air circulation and high humidity.

The brown to gray mass of fungal growth appears on dead plant debris and over the point of invasion, such as a stem end cut during harvest. Stems turn brown, wither, and die. The fungus may progress down the stem and kill the entire plant.

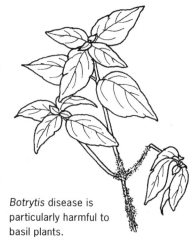

Botrytis disease is particularly harmful to basil plants.

The fungus spreads when the mass of spores is disturbed, causing clouds of spores to disperse. Watering, wind, harvesting, and moving of infected dead plant material all can cause spores to be scattered. These spores then settle on healthy plants. If there is sufficient moisture on a wound from harvest for the spores to germinate, infection will set in.

Currently there are no fungicides that are legal to use on herbs for controlling or preventing gray mold. Cultural control in a large outdoor field is difficult, because the weather does what it will.

 preventing **Botrytis** *infection*

To minimize the potential for *Botrytis* infection in the field and greenhouse:

1. **Do** plant disease-free transplants.
2. **Do** keep the foliage dry. Water only at soil level or overhead early in the day to allow the foliage to dry before nightfall.
3. **Do** keep fields weed-free.
4. **Don't** overcrowd plants; allow for good air circulation.
5. **Don't** disturb the plants immediately before and for 24 hours after harvesting to reduce spore showers that could land on wounded stems, thus causing infection.

Powdery Mildew

This airborne disease is carried by
Podosphaera, Sphaerotheca, Erysiphe spp., and, to a
lesser degree, *Oidium* sp. It is characterized by
white powdery growths that live on the stems or
upper leaf surfaces, although *Oidium* may grow
on the undersides of leaves. Powdery mildew is
most often found on the tender new growth of
rosemary and sage, and sometimes on mint, tarragon, and tansy. A severe infestation stunts plant
growth, causes yellowing of leaves, and certainly
makes the foliage unsalable.

Unlike most mildews and molds, these fungi
prefer warmer temperatures (68° to 80°F) and dry
leaves to germinate. This makes powdery mildew
more of a problem in areas with a dry growing
season. It will germinate and grow, however, in a
wide range of relative humidity.

Powdery mildew spreads rapidly when left
unchecked. It spreads easily when plants are
disturbed by wind or any air movement over
the plants. Because of this, the first place to
check for powdery mildew is near doors, vents,
and windows.

Rosemary is particu-
larly vulnerable to
powdery mildew, as
is sage.

There are no fungicides labeled for use on herbs that will control or
prevent powdery mildew, but there are measures you can take to curb it.
Removing the infected stems or leaves helps stop the spread. Carefully
cut them from the plant, place them in a plastic bag, and seal it tight
before you leave the area.

Powdery mildew spores, unlike other fungi, will drown in water.
Thus, a spray of water that washes young colonies off the leaves should
slow the spread of spores. This treatment has limited effect on heavily
infected plants. Try this water treatment only in the morning, so the
foliage can dry before nightfall. Otherwise, you may encourage yet
another fungal disease to infect the already weakened plants. I have found
that spraying the infected parts of plants with insecticidal soap suppresses
this disease. The mildew returns, however, within two or three weeks, and
then you must spray again.

Downy Mildew

Downy mildew is caused by *Plasmopara* spp., *Peronospora,* and several other pathogens. It is an airborne disease, but this fungus lives within the plant and sends out branches that create growths on the underside of leaves. These are usually red or purple blotches that, when left to grow, produce a gray, brownish, or purple downy growth. Severe infestations result in leaf death, defoliation, and the eventual demise of the entire plant. Each pathogen that causes downy mildew is host specific, so it will attack only closely related species.

This fungus has been reported on chives, dill, mint, parsley, sage, and thyme. It favors warm temperatures (75° to 85°F), high humidity, crowded plant conditions, and poor air circulation. It requires periods of leaf wetness to germinate and is more of a problem in greenhouses and humid areas of the country.

To control downy mildew, remove diseased plants. Take care not to allow the spores to become dispersed when disposing of plants.

Rusts

Rusts, usually caused by *Puccinia* spp., particularly attack mint. However, they also can be found on chives, cilantro, sage, savory, tarragon, and thyme. Rusts are so named because of the rusty-looking spots produced by the fungus. The spots are really tan to purple, and are usually first found on the underside of leaves.

Rusts favor warm temperatures and crowded plant conditions. They require water to germinate and can do so with only short periods of leaf wetness. Controls include destroying infected plants by burning; some herb growers have burned entire fields of standing mint in an effort to control rust. Other growers have used sulfur, dry or as a spray, with some success.

Leaf-Spotting Fungus

This is a generalized term that describes the effects of a great many pathogens, too numerous to list here. Most herbs are susceptible to one or more leaf-spot pathogens. The majority of these fungi produce spots on both sides of the leaves. The spots usually have a "bull's-eye" look, with an outside ring of a different color, and will cross over leaf veins. As the disease progresses, the spots join together. At this point, the disease is called a blight.

Fungi that produce leaf-spotting diseases need water to germinate, but their temperature requirements differ. Cultural controls are the same as for other types of fungal diseases.

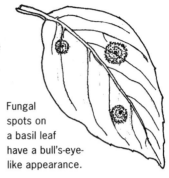

Fungal spots on a basil leaf have a bull's-eye-like appearance.

Controls for Airborne Fungal Diseases

At the present time there are no fungicides registered for use with herbs. Prevention, through the use of good cultural practices, is your best control. Keep foliage as dry as possible, remove infected plants, and provide good air circulation. Keep the fields and greenhouse free of weeds that may harbor disease. Keep insects under control. Try not to disturb infected plants 24 hours before and after harvesting to avoid spore showers. Disinfect scissors, hands, and harvesting containers regularly.

Some growers use one of several organic compounds to control fungal diseases on herbs. It should be noted that these are not labeled for use with herbs, although their labels state that they can be used with leafy vegetables such as lettuce and spinach. Copper, sulfur, and Bordeaux work by coating the leaves to prevent spores from becoming established. They are usually available as a dust, which does not mix well with water. They must be reapplied after rainfall or overhead watering.

Growers and researchers have worked for years to find natural and safe remedies to control fungal infections. Always try any new remedy first on just a few plants, covering both old and new leaves. Wait at least 24 hours and inspect the plants for damage before spraying the rest of your crop. Protect yourself with long pants and sleeves, a mask, and rubber gloves when spraying these treatments; they can be absorbed through the skin as well as the lungs.

Don't spray during cloudy, humid weather; the leaves take longer to dry and damage to the foliage could result. Shake the sprays often to keep the ingredients evenly mixed. Be sure to spray both sides of the leaves. Of course, be certain to wash thoroughly any treated herbs before selling them.

What follows are some of the treatments that have shown some success. These remedies, including the horticultural oils, are not currently registered for use as fungicides on herbs.

Ultra-fine horticultural spray oils are effective in preventing and controlling powdery mildew, as well as some insects. These are mostly petroleum based but at least one soybean oil is available commercially. Buy a type that is mixed to a 3% concentration.

Oils work by coating the pores of the leaves (and insects) to prevent fungi from infecting the plant. This can suffocate unhealthy plants, so use these only on plants that are in good condition.

Baking soda helps control foliage and soil diseases. There are several formulas that you can try. Mix 1 teaspoon of baking soda to 1 quart of water and spray on the foliage. Some growers use 1 ounce of baking soda to 1 gallon of water.

New research has found that mixing baking soda, horticultural oils, and water is even more effective at controlling airborne fungal diseases. A recipe developed by Cornell University is 1 tablespoon each of baking soda and horticultural oil to 1 gallon of water. I have also seen a formula expressed as 4 teaspoons each to 1 gallon of water. In addition, this mixture has also been shown to reduce the number of spider mites on plants.

Horsetail tea has long been used to provide disease protection both in the soil, especially for damping-off, and for airborne fungal diseases. One recipe calls for boiling ¼ cup dried horsetails in 1 gallon of water for 20 minutes. Allow this to stand overnight and then strain to remove the large pieces. Spray on foliage or soil.

Insecticidal soap is effective in controlling powdery mildew. An active infection must be sprayed every week or so. These soaps coat the leaves, thus providing some protection from other airborne fungal diseases.

Neem tree oil controls some airborne fungal diseases as well as certain insects. Currently, neem tree oil–based products are not labeled for use on herbs. Let's hope this will change in the future.

Soil-Borne Fungal Diseases

These organisms live in the soil and can affect seeds, seedlings, and mature plants. They cause root rots, aboveground damage, or both. Stunting of growth, wilting, defoliation, even death may result. Unfortunately, once you notice the damage, the disease has already become established.

Some pathogens that cause soil-borne diseases are always in the soil and attack only plants that are susceptible or weakened due to bad environmental conditions. Soil-borne fungal disease organisms can be spread by insects that live part of their life cycle in the soil, such as fungus gnats and shore flies. Ponds or surface water may harbor these pathogens. Seed can also be infected. Even commercial peat moss mixes sometimes contain disease-causing organisms. Purchased plants may appear healthy but can harbor diseases and infect your soil.

There are a number of pathogens that cause soil-borne diseases. Some will attack plants at all stages of growth. Some, such as *Pythium,* may not exhibit foliar symptoms but cause stunted growth. As you can see, it is almost impossible to exclude these organisms from your growing beds outdoors or in the greenhouse. There are, however, a number of ways to prevent infection.

There are numerous pathogens that cause root rots. This list is by no means complete but does contain some of the major fungi that cause root rots in herbs.

Damping-off

This disease is caused mainly by *Pythium* but several other pathogens also cause it. Fungal diseases are the most common cause of poor seedling stands. They can attack the seeds when they are planted, causing failure to germinate. In this case, the grower will often put the blame on poor seed.

Damping-off can cause new seedlings to topple over at the soil line.

Post-emergence damping-off causes the seedlings to topple over at the soil line, often in a roughly circular pattern. You'll see a dark brown area on the stems just at or below the soil line. The fallen seedlings quickly wither and die. This is a most distressing situation for growers and one of the reasons for my advice always to start more seeds than you think you will need. This is especially true for those herbs that need a long time to germinate.

These diseases favor cool, wet, heavy soil with excessive nitrogen. Damping-off diseases can be a serious problem for growers who direct-seed outdoors in early spring, especially with parsley, because of its long germination time.

Certain cultural practices help avoid damping-off. When planting seeds, be careful to place them at the proper depth and spacing. Seeds that are crowded or too deep weaken and become more susceptible to disease. Do not fertilize the soil until the seedlings have at least two sets of true leaves. Remove any covers over the flats or soil at the first sign of emergence and provide good air circulation. Overwatering seedbeds promotes seed decay or damping-off after germination, so keep the seedbed just moist enough to promote germination. Put finely milled sphagnum moss over seeds at planting, as it is thought to inhibit damping-off diseases.

Root Rots

The pathogens that cause root rots are numerous, but the most common are *Pythium, Rhizoctonia solani, Fusarium solani,* and *Phytophthora.* These same fungi also attack cuttings when plants are propagated vegetatively. Most herbs are susceptible to at least one of these pathogens. Many fungi are always present in the soil or on dead plant debris, and will attack plants only when the conditions are right.

Plants infected with root rot show a variety of symptoms, which may come on quite suddenly. Plants may be stunted, yellowing, dropping leaves, wilted, or dying. When plants are dug up, the roots will not have white feeder roots (the tiny roots growing off the larger main roots), or the outer covering on mature roots will slide off. The rotting areas are wet and soft.

Phythium, often called a water mold, usually strikes after a recent overwatering. If the roots were weakened by periods of severe dryness, they are even more susceptible. Excessive fertilizer, high salt levels, and high pH levels in the soil can stress plants, making them more vulnerable to root rot.

Rhizoctonia solani infects plants of all ages. It has a very wide host range and has been reported on almost all of the culinary herbs. The symptoms vary considerably on older plants. The stems will have lesions, or the entire stem will be decayed with a dry, rather than wet, rot. The leaves dry and drop off, beginning with the lower leaves. The outer covering of the roots sloughs off, as it does with other fungal infections. Eventually the entire plant dies.

Rhizoctonia favors warm soil (70° to 75°F) and does not need an overabundance of moisture to germinate. Because of this, it is often seen in the field or the greenhouse later in the growing season.

Fusarium solani, which causes fusarium root and crown rot, is sometimes seen in plants that are also infected with *Phythium* or *Rhizoctonia.*

More than one pathogen infecting plants is known as a disease complex. *Fusarium* can also cause other diseases such as wilts and "yellows."

Plants infected with this pathogen show stunted growth, appear unthrifty, and may wilt on sunny days. The foliage may yellow. The roots and lower stems usually have a reddish to dark brown discoloration on the interior which can be seen when they are cut open. This is another fungus that prefers warm soil.

There are several types of root rot that prefer high temperatures and are common particularly in the South. *Sclerotium rolfsill* causes southern root rot or southern blight. It has been reported on dill, parsley, rosemary, sage, and thyme. Symptoms include stunted growth, yellowing, and possibly wilting. White areas in the soil surrounding the infected parts contain small tan or brown spots. Texas root rot, caused by *Phymatotrichum* spp., is a disease with similar symptoms and temperature requirements.

Vascular Wilts

Wilts are caused by several different organisms but *Verticillium* and *Fusarium* are the most common in herbs. Rosemary, thyme, and especially mint are highly susceptible to wilts. These fungi invade the plant through the roots and establish themselves in a plant's vascular system. Eventually this system becomes plugged, and water cannot move from the roots to the leaves. This causes the leaves to turn yellow, or have yellow blotches, starting at the bottom. Soon the stems wither. The plant wilts, especially in strong sunlight, and eventually dies. Dark areas are seen in the stems when they are cut crosswise. These fungi have very small resting spores that can live in the ground several years without a host plant.

Verticillium infects a wide range of plants but *Fusarium* is host-specific. *Fusarium oxysporum* f. sp. *basilicum* was identified in this country in 1991. It had previously been found in Russia, France, and Italy. This is a serious threat to the basil crop. The first symptom is wilting of the leaves, although large-leaf basils may drop leaves without wilting first. The leaves usually do not yellow. You may see brown streaks running down the stems before the plant finally succumbs.

Controls for Soil-Borne Fungal Diseases

As with other fungal diseases, there are presently no fungicides labeled for use with herbs. The only controls are prevention through good cultural practices.

Make sure soil in the field is well drained. This helps to prevent the heavy wet soils that most fungal pathogens need to germinate. Avoid excessive nitrogen, especially from chemical fertilizers. Check the soil pH yearly. If the pH is above 7.0, apply soil amendments to lower it to 6.0. Fungal and bacterial disease pathogens flourish in soil with high pH levels.

Plant disease-free transplants. Seeds treated with fungicides, though not favored by many growers, can help prevent diseases when planting early in cold soil. Space transplants far enough apart to allow good air circulation when plants are mature. Water at soil level or over-head only in the morning so foliage can dry before nightfall. Avoid overwatering.

Remove diseased plants and dispose of them by burning, if possible. Situate piles of diseased plants far from the fields or greenhouses until you can dispose of them. Wind, rain, and dust can contain contaminated soil particles, which may be transported back to the field or greenhouse. Control fungus gnats, shore flies, and other insects that spend part of their life cycle in the soil. These insects are known to spread soil-borne diseases.

Alternate annual crops every two years with unrelated families of herbs. Plow deeply to remove plant debris. In fields infected with soil-borne diseases, flaming, or burning, to destroy the spores and plant debris before plowing can provide some control (see page 207).

Other techniques. Some growers routinely drench the soil in their beds with horsetail tea to prevent these diseases. This works especially well to prevent damping-off in seeds or seedlings. Soil solarization destroys some fungal pathogens.

There are several biofungicides now on the market. Mycostop is a popular one. It has been used in Europe for years and now has been approved for use in the United States. Use it on vegetable and ornamental crops in the greenhouse or in fields. Approved for use by organic growers by private and state organic certification groups, it is effective in prevention of soil-borne and seed-borne fungal diseases such as *Pythium, Phytophthora,* and *Fusarium* and has shown some suppression of *Botrytis.* Greenhouse or agricultural supply companies sell Mycostop.

Mycostop is a wettable powder that contains spores and mycelia of a strain of the naturally occurring *Streptomyces* bacterium. It works by colonizing plant roots before fungi pathogens can develop there, thus depriving them of living space. It also secretes enzymes that inhibit pathogen growth. The effect is basically preventive, so apply it immediately before or after planting.

Much research is being conducted to develop biocontrols for diseases and pests. A suppressive growing medium, now available commercially, contains carefully composted organic ingredients that help to inhibit the growth of soil-borne pathogens. Keep informed of new products and use them when appropriate.

Viral Diseases and Mycoplasmas

These diseases have been reported on some herbs, but they are not as common as those caused by fungal pathogens. There are many types of viruses and they all require sophisticated equipment to be seen. Plant viruses cause considerable loss of crops and are dreaded by growers because there are no cures. Mycoplasmas are large viruslike organisms that cause diseases with effects similar to those of viruses.

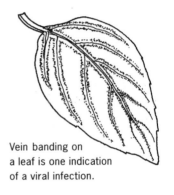

Vein banding on a leaf is one indication of a viral infection.

Most plant viruses have a wide host range and can infect herbs in unrelated families. Once a plant becomes infected, it will remain infected for its lifetime. Any cuttings taken from an infected plant will also carry the virus.

Plant viruses attack only a plant that has been wounded through cutting or by rubbing the leaves, both common activities for the fresh-cut-herb grower. Viruses spread easily when sap remains on scissors and the worker goes from plant to plant while harvesting. Sucking insects such as aphids, thrips, leafhoppers, whiteflies, and parasitic nematodes carry the virus to other plants as they feed.

The single most common symptom of virus infection is stunted growth. You may not notice this if there are no healthy plants for comparison. These diseases also cause distorted or malformed leaves and yellowing of leaves in various patterns, usually angular. These are best seen when a leaf is held up to the light. Both of these symptoms render the foliage unsalable.

The symptoms of viral disease are easy to recognize but it is important to note that similar effects can be caused by spray or cold damage. The symptoms of each disease can vary from herb to herb. Before destroying plants, make sure that it is a virus that has caused these symptoms and not a recent spell of cold temperatures or spray damage.

The following are some of the most common viruses that have been reported to infect herbs. If you grow plants other than herbs, the viruses that are common with those plants also infect some herbs.

Cucumber Mosaic Virus

Cucumber mosaic virus has been reported on basil, dill, chervil, fennel, and parsley. It causes stunting, green and yellow mottling of the leaves, and a downward curl of leaf edges.

Western Celery Mosaic Virus

This organism causes stunting, green and yellow mottling, crinkled leaves, and loss of color in the leaf veins. It attacks anise, dill, chervil, cilantro, and parsley.

Mottled leaves with curled edges may be suffering from a mosaic virus.

Curly-Top Virus

Spread by leafhoppers, curly-top virus has been reported on dill, cilantro, borage, fennel, chervil, and parsley. The symptoms include stunted growth, leaf crinkling and a downward curling, enlarged leaf veins (especially the lower leaves), and leaf discoloration.

Tobacco Mosaic Virus

Tobacco mosaic virus infects many plants, including tomatoes and peppers, and may also spread to herbs. This virus is much tougher than other types because it is easily spread by workers. Contaminated plants, infected pipe or chewing tobacco, and other contaminated materials leave a tenacious residue on clothes, workers' hands, tools, and even doorknobs. A horticultural disinfectant must be used to kill this residue because soap and water won't work. The symptoms vary with plant species but all infected plants show light and dark areas on leaves. Luckily, most plant viruses are not as resilient as this one.

Aster Yellows

Caused by a mycoplasma-like organism, aster yellows affects anise, caraway, dill, fennel, parsley, and sage. It is spread by leafhoppers. The symptoms include stunted growth, yellowing of inner leaves, downward leaf curl, red or purple coloring of older leaves, bud proliferation, and some flowers may revert to green, leaf-type structures. Aster yellows may also produce a bad taste in herbs.

Controls for Viral Diseases

There are no cures for these diseases. Plants that are infected must be destroyed. Pull the plants carefully and burn them. If infected plants are simply placed in a pile, they can still transmit diseases through insects. Because most viral diseases are spread by insects, controlling these insects will control the disease.

Seed is usually virus-free but plant material may not be. Try to buy from nurseries that certify their plants or cuttings to be disease-free. If you do buy in plants or cuttings, hold them in an area separate from your fields and greenhouse. Watch the plants carefully for several weeks for signs of infection or insects. If you suspect a problem, testing can confirm it.

Some universities and commercial labs offer testing for viruses. An on-site testing kit called ELISA — enzyme-linked immunosorbent assay — is available from several commercial testing laboratories. To do a test yourself, transfer sap from a suspected diseased plant to a wound on a healthy plant. Keep the indicator plant far away from fields and greenhouses. Watch carefully for a week or two for symptoms to show up on the indicator plant.

Bacterial Diseases

These are caused by one-celled organisms that require several time-consuming lab tests to identify. Fortunately, these diseases are not as common as fungal diseases. In the North, bacterial diseases are even less common because they cannot thrive in cold, dry winter weather.

Bacterial diseases can infect herb plants, cuttings, and scented geraniums. Bacteria multiply by cell division, and they do so quite quickly. Symptoms caused by different pathogens are similar. The disease usually strikes suddenly, beginning with wilting of the plant or cutting, often without any discoloration. After a day or two, the tissue becomes soft, rotten, and smelly.

Bacteria pathogens are spread by insects, contaminated tools, containers, hands, and splashing water. They favor wet conditions, high humidity, and warmth. High temperatures increase the development of disease symptoms. There are several bacterial diseases that can infect herbs.

Bacterial Soft-Rot

Most often caused by *Erwinia carotovora,* bacterial soft-rot affects fennel and parsley. This pathogen prefers temperatures in the high 70s to low 80s. It invades the plant through tiny wounds and then produces a cell-wall-dissolving enzyme. Invasion and growth proceed rapidly and can transform a cutting or plant into a rotten mass in one day. Suspect this disease if the lower leaves and stems of your plant are turning brown.

Bacterial Leaf-Spot

These diseases are caused by several different *Pseudomonas* pathogens. They infect the mints and catnip. Scented geraniums are also susceptible to bacterial leaf-spot and wilt caused by *Xanthomonas pelargonii.* These pathogens prefer warm temperatures and soil with a high pH. Symptoms include wet-looking spots on the leaves that usually do not cross the veins. As the spots dry, they often appear translucent. The disease progresses quickly and the plant collapses, rots, and dies.

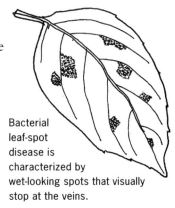

Bacterial leaf-spot disease is characterized by wet-looking spots that visually stop at the veins.

Bacterial Wilts

These diseases are also caused by a strain of *Pseudomonas.* Sage, the mints, cilantro, nasturtiums, and scented geraniums are susceptible. The plants wilt, vascular systems turn brown, the sap becomes stringy, and the plants die.

Controls for Bacterial Diseases

There are no cures for these diseases. Infected plants, and those surrounding them, must be destroyed. Bacteria can live and multiply in the pulled plants, plant debris, and the soil, so put the affected plants in an area far from fields and greenhouses. Burn them if possible.

Bacteria prefer soft tissue that is high in nitrogen. Do not *over*fertilize stock plants used for cuttings or other herbs during rapid growth periods. Soil pH should be kept below 6.5, as most bacteria favor a high pH. Provide good drainage and air circulation. Sanitation is of utmost importance in preventing the spread of these diseases. All tools and containers that come in contact with infected plants should be disinfected or destroyed. Wash hands thoroughly.

Recent research shows that treating vegetable seeds with hot water can kill some bacterial diseases that are within seeds. The temperature of the water (usually between 122° and 125°F), soaking time (20 to 25 minutes), and air drying temperature (70° to 75°F) must be exact to prevent damage or death to the seed. The temperature and soaking time vary by the vegetable. This particular research was done only on vegetables, but perhaps someday this treatment will be tried with herb seeds.

Nematode Infestations

These are microscopic roundworms that parasitize plants or insects. Some species of nematodes are beneficial and are available commercially for control of insect pests that live part or all of their life cycle in the soil. Several species of nematodes feed on stems and leaves aboveground, but these rarely cause problems in herbs.

Root-Knot Nematode

This is the most common nematode found in herbs. It lives in the soil and invades the roots of plants. The most common is *Meloidogyne* spp., which can be seen with the naked eye. These nematodes have a wide host range and have been reported on almost all of the culinary herbs.

Infection by nematodes may not kill a plant but it will limp along, barely able to supply a single cutting. Nematode infestation often mimics plant diseases. Taking even one cutting may stress the plant enough that it may never recover. The harm caused to the roots may also allow other diseases to become established within the plant.

If the soil is warm and your plants appear stunted, wilted, show yellow foliage, or are malformed, check for nematodes. These creatures enter the roots, where the females begin to feed. Their feeding causes oblong or round galls on the roots, which block the flow of nutrients through the plant. Cutting open the galls reveals very small, pearly, pear-shaped nematodes or egg masses.

Controls for Root-Knot Nematodes

Control of this pest is difficult because there are no nematicides labeled for use with herbs. Crop rotation may work, but because most herbs are susceptible, that would seem to be of little value. Soil solarization before planting may kill most of the nematodes.

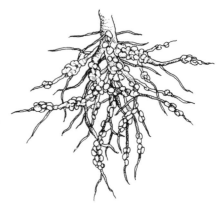

Tests show that marigolds produce in their roots a chemical that kills nematodes, both in the soil and in roots. This chemical is released slowly into the soil, so grow marigolds all season long if you have a nematode problem. The chemical can remain in the soil up to three years.

Root-knot nematode infestations can spread very quickly.

A bad infestation of this type of nematode in a field may call for planting the entire area in marigolds for one growing season. White and black mustard are also said to have a toxic effect on root-knot nematodes.

Soil that is rich in organic matter normally contains a variety of fungi that attack nematodes in differing ways. Much research is being done to develop a biological control with these beneficial fungi for commercial use. Keep informed of new remedies and products; perhaps there will be a biocontrol for this pest, as well as for many other plant diseases, eventually.

harvesting, handling & packaging

Bringing a quality fresh-cut herb to the marketplace has as much to do with growing a healthy plant as it does with the care given to it during and after harvest. A healthy, disease- and insect-free plant will remain fresher and more appealing after harvest longer than one that is moisture- and nutrient-stressed.

Fresh-cut herbs are perishable after harvest. The cut stem continues its respiration, but it has no way to renew its supply of nutrients and water. Thus, it will eventually deteriorate. We can help delay this process by harvesting only the best and strongest herbs. Proper handling and storage conditions during and after harvest will also help postpone decay. Take considerable care during harvest to prevent damage to the herbs. Bruised or torn leaves are a statement to the consumer about your standards.

Careful harvesting and bunching of herbs for packaging is critical to offering the customer a quality product.

Food Safety and Sanitation

In recent years, the number of food-borne illnesses has been on the rise. The public's concern is driven by highly publicized cases of food contamination that caused severe illness and even death. The illnesses, most often, were caused by people consuming contaminated water, meat, dairy products, fruit, apple cider, and salad greens. Can you imagine what would happen to your business if the national media reported that someone became ill from using a fresh-cut herb that he purchased?

Food can be contaminated by viruses, such as hepatitis A, which is spread from an infected person to the food he is handling. *Cyclospora,* a parasite, has been found in fruit and occurs in drinking, irrigation, and wash water.

Many food-borne illnesses are caused by a new, more potent strain of *E. coli. E. coli* is a bacterium commonly found in human and other animal digestive tracts. Illness arises when foreign strains are ingested. *E. coli* can be introduced into food from contaminated soil and by the unsanitary practices of a worker.

The produce industry is under intense scrutiny from the public, government agencies, and itself. At this time there are no federal regulations or laws to govern the actions of the *grower* who does not process — or change it from the form in which it was harvested, such as chopping or drying — a raw agricultural commodity.

"Self-Policing"

The federal government has begun a series of food safety programs aimed at processors of fresh produce. Regulations and inspections will probably follow for growers as well. Until then, it is up to the individual grower to utilize good sanitation practices in order to prevent contamination of the crop. In light of this, several national produce organizations and industry leaders have started self-policing programs for growers in anticipation of stricter federal laws that would regulate growers. The following are some areas targeted in these programs.

Pathogens can contaminate food in all stages of production, in the field, during harvest or postharvest handling. Good sanitation habits should be practiced in all areas (see box).

Most states have their own regulations regarding growers and direct marketing to the consumer at farmer's markets, from your property, or at roadside stands. Check with the Department of Agriculture for details.

Following the Law

There are two federal statutes the grower must be aware of. The FDA administers the Good Manufacturing Practices Statute. The Department of Agriculture, food and safety division, in your state has the authority to inspect for violations of these regulations.

Section 110 of this law regulates packing, processing, and holding of human food. It contains important exclusions for the grower of raw commodities who packages his or her own produce. (Exclusions are certain parts of the statutes that would not apply to you.)

Soak your gardening and harvesting tools daily in a disinfectant to prevent the spread of diseases and possible contamination.

 critical sanitation procedures for herb harvesting

- Do not allow livestock on the field or growing beds for one year prior to planting. Several organic certification programs prohibit the use of manure or manure tea for at least 60 days prior to harvest for aboveground crops. The use of inorganic fertilizer can protect against fecal contamination except for that from pets and free-range animals.

- Water used for irrigation should be from a clean well, rather than pumped from a river or lake. For washing produce, use clean and chlorinated water to lower the bacterial count. Test the water frequently for contaminants.

- Worker hygiene and sanitation includes providing clean rest rooms with hot running water and soap for hand washing. Workers should be educated in the basics of sanitation.

- Harvesting equipment, containers, and scissors should be cleaned and disinfected daily. Make sure containers are smooth, with no rough edges, to avoid damage to the produce. Handle the produce carefully to avoid tears and bruising, which can encourage pathogens. The packing house should be clean and sanitized daily.

However, if a grower buys in a single package of herbs and repackages them for sale, he becomes a repacking house and this law would be in effect. Call the nearest office of the FDA and ask for a copy of section 110 of this statute.

The Department of Agriculture in your state administers the Food Code, which provides guidelines for the proper handling and storage of food, sanitation practices, employee practices, and much more. This a guide for retail food establishments selling to the end user, such as restaurants and grocery stores. This code is updated every two years and is adopted, in whole or in part, by the individual states. The USDA is working so that each state will adopt it as a uniform code. Until that happens, contact the Department of Agriculture in your state to see if it honors this code.

The Department of Agriculture also has the authority to inspect and take samples of fresh produce in any location, including farmer's markets and roadside stands, for pesticide analysis. They may also inspect for growers who are processing food, such as drying it, without the proper facilities.

Harvesting Fresh-Cut Herbs

The best time to harvest herbs is in the morning, after the dew has evaporated but when they are still cool from the night. It is easier to preserve the quality when the herbs are cool. This is not always possible for growers with large orders to fill, and harvesting may have to continue throughout the day.

Herbs that are harvested during the heat of the day must be chilled within minutes after cutting to preserve their quality. On large-order days, try to arrange schedules so that harvesting can be done in the morning and/or the prior evening.

It is important to know the optimum stage of growth to harvest each herb. Herbs can be harvested as soon as they are large enough, but harvesting just before the flowering stage will provide more concentrated flavor. Most herbs should not be flowering when picked for fresh-cut sales; the flavor can change or they can become bitter. Some chefs may want flowers included with their orders for use as garnishes. See part IV: Successfully Growing More Than 20 Herbs and Flowers for detailed descriptions of harvesting stages and techniques for each herb and edible flowers.

Containers and Supplies

Anything from a bucket, tub, or cooler can be use as a container in which to deposit the herbs or bunches during harvesting. The container should be lightweight and easy to clean and carry. Plastic-foam coolers work well during hot weather to protect the herbs from the sun and heat. If you use ice or freezable gel packs in the cooler, cover them with a towel to protect the herbs against damage from direct contact with the ice.

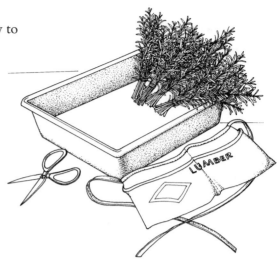

Scissors should be sharp with narrow, pointed tips for ease of cutting in small areas. They should be lightweight and comfortable to use for

A selection of harvesting tools: plastic tub, sharp scissors, and apron with pockets to hold supplies

long periods of time. Stainless-steel scissors will withstand repeated sharpening and disinfecting better than will plain metal ones. Some growers prefer to use sharp knives to cut herbs.

Use rubber bands to secure a bunch of herbs if you will be selling in bunches rather than by weight. The bands should be small, about an inch in diameter, and of the best quality you can find. Buy them in large quantities. Twist ties can also be used to secure bunches. I found that rubber bands are quicker, cheaper, and easier to use. Try them both and see which works better for you.

Wear an apron with pockets to hold the rubber bands and scissors while you work. A leather holder for the scissors attached to your belt will also work. I used canvas carpenter aprons with two pockets. These were sometimes given away at the local discount building supply superstore as free advertising. At most places they cost less than $2 apiece. Each of my employees had her own apron with her name written on it with a laundry pen. We all looked like we were employees of the same hardware store! Aprons are easy to wash and replace when needed. If you use a cloth or lightweight apron, always be aware of where the sharp end of the scissors is when you kneel or bend over.

Harvesting Techniques

Harvesting herbs for fresh-cut sales is a time-consuming process because you must do it by hand. Each stem, leaf, or flower must be examined for insects, disease, and blemishes. Only the best should be taken for sales. It is a time-saver to cut blemished parts during harvest and deposit them in a separate bucket for a later trip to the compost pile.

Diseased or insect-infested plants should be cut or pulled carefully at a later date so as not to spread the problem. Even if some stems look healthy, do not harvest diseased plants for sale. These stems can infect the cut herbs in the package and cause all the contents to rot.

If the herbs are relatively clean, and you will be bunching them for sale, it is best to bunch as you harvest. Each time the herbs are handled increases the chances of bruising or leaf loss.

It is important to develop fast and efficient techniques for harvesting and bunching. This helps to preserve the freshness of the herbs after harvest. Each person will develop a method that works best for him or her. The technique described below enabled me to pick and bunch lots of herbs quickly and efficiently.

The grab-and-cut method works well for those herbs that are clean, full, and growing approximately at the same height, such as chives, cilantro, marjoram, mint, and thyme. The technique is simple. Grab a full-size bunch of stems with one hand just above where you will cut. Cut the stems about an inch

Grab-and-Cut Technique

1. Grab a bunch of stems and cut off evenly.

2. Quickly wrap the cut stems with a rubber band about 1" from the bottom.

below your hand. Return the scissors to its holder and get a rubber band. Put the rubber band around one side of the bottom of the stems, about an inch from the bottoms. Stretch the rubber band out as you wrap it around the stems a time or two.

Don't make the rubber band too tight or pull it too hard against the stems; the stems could break. This also would make your bunch size smaller than you intended and increases the risk of disease organisms entering the wounded stems.

This is by far the fastest and most efficient way to harvest bunches. In young stages of growth many of the other herbs may be harvested this way.

Stem-by-stem cutting will have to be done with those herbs growing loosely or to different heights. This technique is a time-saver because you don't have to put down the scissors or each stem after cutting it.

Cut the first stem at the length you want the bunch to be. Hold the cut stem close to the bottom. Hold the loose stems you've already cut in the palm of your hand and with your other hand hold the next stem close to the cut point. If you hold the top of the bunch level with the top of the next stem, you will cut the stems close to the same length.

Go on to the next stem, looking as you go to make sure it is blemish-free. Move from stem to stem until your bunch is complete. Then rubber band and cut off the stem ends evenly.

Garnish cutting must be done individually to ensure that each is of top quality. Most chefs want precut garnish to be only the new growth tops of the herb. This can take time, but with practice you'll learn to work quickly. Nonetheless, some people may be fairly slow at first, so if you are paying an employee and it takes her a half hour to pick, count, and package 100 garnishes, this is not a profitable situation for you.

Look over the top of the plants and find one perfect top of the requested length. Grab the tip with two fingers and cut into the length that you need. Tuck it into the palm of your hand. Grasp the next garnish

To harvest garnish, cut only the tops to the length required. Tuck each cut stem into your palm and continue until you have the correct number for a bunch.

with two fingers, at the same time measuring the length with the previously cut ones in that same hand, and cut.

Count as you go until you have 10 (or 20 with rosemary, which does not crush as easily as other herbs) in that bunch in your hand. Do not rubber band them together. Place the bunch in a container. When you have another bunch, put it in a separate place in the container. The last bunch can stay in your hand. Go to the work area and package the garnish loose in a poly bag with all the stem ends down.

If the garnish is clean and insect-free, you can package it immediately. If it is dusty or has insects, gently swish the garnish in water a bunch at a time before packaging. Allow it to dry slightly before you package it.

The garnish should be placed neatly all in rows in the bag. This makes things simple for the chef because she can take it from the bag and place it directly on the plate. If the herbs are thrown in the bag haphazardly, the garnish may become tangled, crushed, or damaged — now in an unattractive or unusable condition.

Mist clean rosemary garnish with plain water from a spray bottle before refrigerating. This small amount of moisture helps to keep the rosemary fresh. Do not do this with other herb garnish.

Postharvest Handling

The two most important steps you can take to preserve the freshness and quality of herbs after harvest are to lower the temperature quickly and prevent moisture loss. Without moisture and cool temperatures, your herbs will quickly lose their color, flavor, and usefulness.

Washing

Washing herbs can damage the foliage and cause loss of flavor and rapid deterioration of quality. If the herbs are clean and insect-free, try to package them immediately after harvest. Many times the herbs will be dusty, have a few insects, or be muddy from rain or watering. In this case, you'll want to wash the herbs rather than deliver a dirty product to the customer.

Washing must be done with great care because most herbs are fragile. Leaves might be lost or bruised in the process. Some herbs, especially mint, actually benefit from washing. (See part IV: Successfully Growing More Than 20 Herbs and Flowers for guidelines about each herb's ability to withstand washing.)

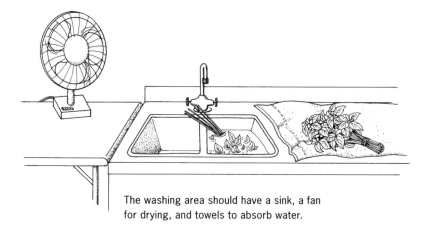

The washing area should have a sink, a fan for drying, and towels to absorb water.

Washing in very cool, but not ice cold, water has the effect of rapidly reducing the temperature of the foliage. Many large vegetable growers use a process called hydrocooling to clean and rapidly reduce the temperature of their produce. They have sophisticated and expensive equipment to accomplish this.

Your herbs can be washed in something as simple as 5-gallon plastic pails. Chlorinated city tap water is best for washing because it reduces the amount of bacteria in the water and on the leaf surface. It is important to change the water after each batch is washed and to sanitize the container daily. For this reason, a deep sink with drainage, such as a laundry tub, will be easier to use.

Bunched herbs can be washed more quickly than can individual stems. Hold the herbs by the stem ends and gently dunk them up and down in the water. This seems to be more effective at removing insects and dirt than is swishing the herbs. After washing, gently shake off the excess water. Some herbs — mint and parsley, for example — hold up well to more violent shaking if the stems are mature. Hold the stems close to the foliage to prevent breakage.

The washed herbs must be dry, or nearly dry, depending on the herb and type of packaging, before you pack them up. There are a number of spin-type dryers available for lettuce and leafy greens, but these are not recommended for fresh-cut herbs because they can damage the tender leaves.

The best method I have found is to spread the herbs on an absorbent surface or large clean towel and allow them to dry naturally. Do this in a room that is cool, well ventilated, and out of direct sunlight. A small oscillating fan kept running speeds the drying process. This allows you to do other work while they are drying. The herbs may need to be turned occasionally.

Keep a close eye on the drying process and package the herbs as soon as the excess water is evaporated from the leaf surface. If the herbs become too dry, they will lose moisture from within the leaves and start to wilt.

Preventing Moisture Loss

An herb stem begins to lose moisture to the surrounding air as soon as it is cut. This process is accelerated if harvest occurs during hot, dry weather or later in the day when the herbs have lost their nighttime coolness. Moisture loss soon results in wilting; rapid deterioration follows.

Preventing moisture loss is relatively simple. Place the cut herbs in poly bags, film wrap, or plastic containers such as those used for supermarket packaging. Beware of too much moisture, which can lead to condensation inside the closed package. This encourages the growth of molds, which cause rapid decay. Condensation can occur if you are using regular poly bags or plastic trays that are sealed tightly. The newer Modified Atmosphere Packaging (MAP) plastic film bags or trays decrease the chances for condensation to accumulate because they breathe. With this type of packaging the bags are usually sealed.

Some herbs can be held with their stem ends in water. The leaves still lose moisture, however, unless a plastic bag is placed loosely over the top. This is a fine way for the end user of the herbs to store them in the refrigerator, but it is rather time- and space-consuming for the commercial grower.

Preserving Freshness by Cooling

Cool the herbs as soon as possible after harvest to preserve the quality. The quicker the temperature is lowered, the longer the cut herb will last.

The process of lowering the temperature can begin in the field if coolers are used as containers. Herbs can be cooled before or after packaging, depending on your situation. If you are selling bunches or loose in poly bags to restaurants or wholesalers, the herbs — if they are clean and dry — can be packaged and cooled as soon as they arrive in the packing area.

Packing in supermarket trays, which generally takes more time, can be done several ways to accelerate cooling. One person can harvest while another packs. Good communication is essential to prevent harvesting more than necessary in this situation. If you are working alone, harvest small amounts, package them right away, then cool them before returning to the field to harvest more. If you would prefer to do all the packaging at one time, harvest small amounts and cool them immediately. Place the herbs in poly bags to prevent moisture loss before cooling.

The optimum temperature for preserving freshness of most herbs is between 34° and 41°F. At a temperature of 32°, cilantro, dill, marjoram, mint, oregano, parsley, rosemary, sage, savory, and tarragon should have a shelf life of three to four weeks. The higher the temperature, the shorter the shelf life. Take care when storing herbs at 32° because freeze damage is possible, especially with forced-air-type cooling.

Basil, along with watercress and perilla, should never be stored at temperatures below 45°F. Basil is chill-sensitive and probably will turn brown in two days at a temperature of 36°F. The optimum temperature for basil is between 45° and 60°F.

When packaged, basil can sometimes last longer than expected at low temperatures. This is because the respiration of the leaves results in self-heating and thus raises the temperature inside the package. This process depends on many factors, including plant age, time of day of harvest, and the cultivar of basil.

Many stores display herbs under conditions that are less than ideal. The packages are not always placed in a cooler and they are always exposed to light. Those markets that do display their herbs in coolers don't wish to separate the basil from the other herbs. Be sure to let them know that this is the reason that basil may deteriorate so quickly.

Cooling Systems

The cooling system is an important part of bringing a quality fresh-cut herb to the marketplace. There are a variety of cooling and storage

 educate your customers

Be sure to educate your customers about proper storage temperatures for the herbs they purchase from you, especially for basil. If they store the herbs at the proper temperature, the quality of the herbs — and your reputation — will be preserved. This is easier to accomplish with restaurants and distributors than with supermarkets.

systems available, from small refrigerators to large stand-alone buildings, and everything in between.

The system you choose depends on the volume of herbs that you will be holding at any given time as well as your financial resources. Most growers hold cut and packaged herbs for a short time, usually no more than two days.

Many small or beginning growers make do quite nicely using a regular home-type refrigerator or two. If you purchase a used refrigerator, monitor the temperature to make sure that it stays constant. A minimum/ maximum thermometer is helpful for this purpose: I can attest to the fact that it is very distressing to find your ready-to-deliver order frozen!

If your operation is big or you have abundant funds, you may want to purchase a larger system such as a walk-in-type refrigeration unit. There are a variety of suppliers and manufacturers of these systems. Make-it-yourself plans are available for those who have the ability and time to do it themselves. The plans should provide insulation, ventilation, cooling, and control components, as well as structure details. Check your local library for information.

For many years the function of the cooling system was to keep the produce cool at a constant temperature. It is now known that the shelf life of some vegetables, leafy greens, and herbs can be extended by modifying the atmosphere (MA) surrounding the produce. Refrigeration systems are now available that modify the air within the structure, primarily by the addition or removal of gases.

Atmosphere Modification for Optimum Shelf Life

Modifying the atmosphere (MA) or controlling the atmosphere (CA) surrounding various types of fresh produce can extend shelf life by preventing or delaying the process of decay. This is done primarily through the manipulation of carbon dioxide, oxygen, and ethylene. Controlled atmosphere is more exacting and is used for long-term storage of fruits and vegetables.

Lower levels of oxygen surrounding the produce decrease the rate of respiration and lower ethylene levels, as do lower temperatures. Higher levels of carbon dioxide have the same effect. Ethylene is caused by the respiration of fresh vegetables and speeds the decaying process. It also promotes the growth of mold, which creates ethylene as well.

Each vegetable and fruit has a different rate of respiration and therefore requires a different mix of gases to delay spoilage. Much research has been conducted on the various vegetables and fruits, but much more needs to be done on individual herbs.

Researchers at the University of California at Davis have studied many of the fresh-cut culinary herbs. It was found that most herbs produce very little ethylene but are still sensitive to it. Low levels (10 parts per million) of ethylene often cause increased respiration in the herbs. These studies report the ethylene sensitivity levels of various herbs: low sensitivity, rosemary and sage; slight sensitivity, basil, oregano, savory, and thyme; high sensitivity, marjoram, mint, and parsley. It was also found that basil increases ethylene production at temperatures over 50°F. Light, as found in most supermarket display areas, also has a detrimental effect on the shelf life of herbs.

Ethylene causes the most dramatic deterioration of fruits and vegetables, including herbs. Because all fresh produce creates ethylene, it is important not to store herbs near other produce.

Atmosphere modification for various produce can be done in the whole of the storage area or in the package. Modified atmosphere packaging (MAP) is made out of plastic film that is permeable to the different gases. It is available in bags and is used for all types of produce. Some may be used for fresh-cut herbs. Research is under way by several packaging manufacturers to develop modified atmosphere packaging of supermarket-type trays for the various fresh herbs.

Packaging Options

The type of packaging you use is determined by the kind of account you will be selling to. Wholesale distributors may want the herbs in bulk so they can repackage them with their own label. Some want them by weight or bunches in poly bags, and some request them in trays for resale to supermarkets. Most accounts want the herbs packaged by weight even if they are bunched.

Restaurants usually want the herbs in bunches or loosely packed in poly bags. Most supermarkets want the herbs in rigid plastic trays. But each herb has its own requirements for packaging (see part IV: Successfully Growing More Than 20 Herbs and Flowers for these details).

general packaging guidelines

- Package bunches with stem ends down.
- Don't overcrowd loose-packed herbs or those in trays to prevent crushing and bruising.
- Use only the best herbs that have no damage or yellowing.
- Tear or cut off leaves that are imperfect.
- Most importantly, inspect each leaf for insects.

Weighing

When selling herbs by weight, it is necessary to have a scale that is legal for trade. These are known as class 3 scales. The scale should measure down to at least a sixth of an ounce if you want to sell supermarket packages.

Try to find a scale with a digital display in the front and back of the unit for ease of use by more than one person at a time. The prices for these scales is coming down; Hobart Scales sells a class 3 scale for around $425. Most scales of this type will weigh in metric and standard American but they need to be recalibrated when you switch between the two. If you want to use both metric and American weights on your labels, do the conversion before and just measure in American or vice versa.

Poly Bags

Modified atmosphere bags for packaging herbs are the ideal. They are just beginning to make their way into the marketplace. If you can't find them yet, regular poly bags work well for packaging bunches or loose herbs. Bags with a pleat in the side (sometimes called gusseted) are good because the top expands, thus allowing more room for leafy tops.

Be sure to buy food-grade bags; this is required by law. Other types of plastic can "leach" into the food or impart off-flavors or smells. The size of the bags you buy depends on the type of accounts you have. Restaurants may want small amounts of some herbs. It would be a waste of money and resources to package three bunches of herbs in a large, dozen-size bag.

Sizes. Having three sizes of bags served my needs perfectly. The smallest bags were 4 inches wide by 12 inches tall with a 2-inch pleat in the side. These were useful for several bunches of chives or single bunches of other herbs. I also used them to hold the Veltone trays (see page 298) for supermarket sales.

The medium-size bag is 8 inches wide by 15 inches tall with a 3-inch pleat. I used these for three to six bunches and also for garnish. A hundred rosemary garnish fit well into this size bag.

The largest bag is 12 inches wide by 24 inches long with a 6-inch pleat. This is used for a dozen bunches of the bushy herbs or for loose packing.

These bags may all seem quite long but the extra length is needed for some herbs and also to allow room at the top for closure and for the herbs to breathe. It also allows you to handle the bag at the top rather than holding it right on the herbs.

Type of closure. This depends on your personal preference. I felt that it was better to leave the bag open a tiny bit for the herbs to breathe. You will probably find, as I did, that the bags end up staying open in the restaurant coolers anyway. Many growers close their bags with twist ties. Bags are available with zip tops or sticky tops; some are heat-sealed by the packer. (These may not be available with the side pleats.)

To locate a supplier, look in the telephone directory under paper supplies, restaurant supplies, or packaging materials. Be aware that most suppliers require a minimum order, so you will probably be buying large quantities at a time. Try to find a supplier near you. You'll save money in delivery costs if you pick up the order.

Part IV: Successfully Growing More Than 20 Herbs and Flowers gives details on packaging the herbs and edible flowers for supermarkets and in poly bags. In general, all herbs hold up better when packaged with the stem ends down in the bottom of the bag or tray. The packages are often stood on end in boxes for delivery and by the end user, either in the cooler or on display in the supermarket.

Package bunches and garnish with the stem ends down also. This works well with loose pack herbs as well. Some growers put loose-pack herbs in the bag any which way. This is quick and easy, but it can cause much bruising to the herbs.

Supermarket or Clamshell Trays

Herbs packaged for supermarket sales should be in rigid trays rather than poly bags. Bags do not stand upright and the herbs are easily crushed and bruised. It only takes one customer browsing through the herbs or a worker straightening up to ruin the contents of the bags. Trays provide a neat appearance, can be stood upright or stacked, and extend the shelf life of the herbs. Some trays have a tab on top with a small hole so they can also be hung.

Rigid plastic trays come in a variety of sizes and should have attached locking lids. The best dimensions for fresh-cut herbs is 3 to 4 inches wide by 6 to 7 inches tall by 1 inch deep. They should be clear to allow the customer to see the herbs. These plastic trays are expensive and usually must be ordered in very large quantities to get a price break from the supplier.

An alternative to plastic trays that provides protection for the herbs but is not as costly is a bakery tray by Veltone. Made from white paperboard, it is coated so it does not absorb moisture from the herbs. I used these for some time and received many positive comments from customers who felt they were less harmful to the environment than are plastic trays. The trays are 3½ inches by 7½ inches by 1 inch deep.

For supermarket packaging use clear, rigid plastic trays that can be stood on end to display the product.

After the bakery trays are filled with herbs, place them inside a pre-labeled bag size 4 by 12 by 2 inches. Close the bags with twist ties or tape. Do not use staples, which can cause "pokes" to consumer fingers.

Specialty Mixed Packages

Herbs can be mixed together in the same package for retail supermarket sales. Call them blends, mixes, medleys, or any other creative name you can come up with. Blends should include a recipe and the herbs should be in amounts appropriate to the recipe.

These prepackaged mixes increase sales because they are ordered in addition to the single packaged herbs. The consumer has all of the herbs at the ready, which in today's fast-paced life is most welcome.

You can blend herbs in various ways. If you have a special recipe that you have developed, by all means use it. Recipes that have been published by someone other than you cannot be used without permission. Ethnic blends that have broad appeal are usually good sellers. These include

Italian, Greek, Mexican, and Middle Eastern mixes. Soup or pasta blends are also popular. A mix of salad herbs sold well for me. It contained baby leaves of arugula, purple basil, cress, and sorrel, along with four edible flowers.

Labeling Packages

There are several ways to label an herb package. You can buy poly bags with your information printed right on them. This is expensive, though, especially if you are selling many kinds of herbs and in different-size packages. Bags can also be printed with several herbs on them and then you check off the herb that is enclosed.

Labeling for restaurant packages need not be exceptionally attractive and really must convey only the relevant information. A well-designed label does, however, make a statement about your professionalism and the quality of your product. Most executive chefs may see your labels and herbs only occasionally. It is the prep cooks and day-to-day line cooks who most often use the herbs. I once conducted an informal survey and none of these cooks cared one bit about the plain white labels or packaging; their only concern was that the herbs were clean and in good condition.

On a budget? Purchase plain white labels and use a rubber stamp, or self-inking stamp, with which to put your business name and address on them. You can then hand-write which herb is enclosed. Always use waterproof ink on the stamp and pen.

Another way to save money with labeling — and to add a bit more of a business touch — is to purchase individual rubber stamps for all of the herbs. Rubber stamps work better than the self-inking types for this purpose because they last longer and are more versatile.

The print style should be plain rather than intricate and make sure the size is large enough to be seen from a distance. Use an old-fashioned stamp pad with ink that is waterproof and safe for use near food. Your printer or packaging supplier should be able to find this for you. Use only one color of ink. Red is most noticed by customers. You can also use rubber stamps for supermarket labels.

Labels for supermarket packages should be attractive and contain more information. You can have a set of labels printed with each herb. This gets expensive when you are selling so many different herbs.

buying labels

Labels are usually sold on large rolls with 100 to 500, sometimes more, on them. The labels come in standard sizes. Call your printer to find out what sizes are available before designing your labels. A versatile size is 2 inches by 3 inches: This is good on poly bags as well as supermarket trays.

It is possible to purchase preprinted supermarket trays, but the vast majority of growers purchase the labels separately and stick them on the trays themselves.

A cost-saving way of labeling supermarket trays is to design a label with your business name and address on the outside edges, leaving the center blank. You can then use the rubber stamps to designate which herb is enclosed.

Another option to consider when planning your labels is doing it yourself on your computer. Many computers, along with the proper software programs and printer, can print labels of your own design. If you are so inclined, it would be worthwhile checking into the costs involved for the program and blank labels and comparing these costs to those of a printer.

If you apply your label to each bag and tray, do so before putting the herbs in. It is much easier to work with a flat bag or tray and you'll have one fewer opportunity to damage and bruise the herbs.

Labeling Laws

In May 1994 the federal government put into effect regulations governing nutrition labeling on fresh produce called the Nutrition Labeling and Education Act (NLEA). At this time, growers and packagers of fresh-cut herbs are exempt from these regulations because fresh herbs are not considered a major part of the diet. However, the FDA and other government agencies are taking a close look at fresh herbs and may require that these comply with nutrition labeling in the future.

Small growers of other produce are exempt from these regulations. The FDA's interpretation of a small grower is determined by the number of employees and the number of product units sold in one year. These numbers are on a sliding scale that changes annually. Contact the FDA office nearest you to find out what these numbers are currently.

If you think you are exempt, you must write to the FDA and state what your existing products are, that you are exempt, and ask for this status. If you add new products, such as salad greens, you must write again for the exempt status. A Model Small Business Food Label Exemption Notice form can be obtained by writing to the FDA. (See Resources for the address.)

Growers who do qualify for this labeling must list 14 nutritional items on the label: calories, calories from fat, total fat, saturated fat, cholesterol, sodium, total carbohydrates, dietary fiber, sugar, protein, vitamins A and C, calcium, and iron. The nutrient content of various types of fresh produce is published in the Federal Register by the FDA. At this time the nutrient content of fresh-cut herbs is not listed, but that may change. Contact the FDA office for further updates and for help in obtaining a copy of the Federal Register.

Other information that must be listed on your labels includes your business name and address. If your phone number, under your business name, is listed in the phone book, you need not include your phone number on the label. If your business phone number is not in the phone book, you *must* put your phone number on all labels.

For herbs that are packaged for retail sales, such as supermarket packs, you must declare on the label the weight of each package. Contrary to any articles you may have read, it is *not* mandatory, at this time, to list the weight of packages in metric measure. Many growers do, however, because the FDA strongly encourages that both American and metric weights be indicated on all food packages. It also is not mandatory to declare weights of bulk herbs sold to restaurants and distributors, but the FDA encourages growers to do so.

If you buy herbs for repackaging that are grown outside the United States, you must declare the country of origin on your labels. It should be stated "A product of (country name)."

Label Design and Point-of-Sale Materials

Choose an uncomplicated design for your label. It should state clearly the necessary information. The consumer wants to know certain information and to see the product; an elaborate label only serves to distract the eye from the product inside.

The herb name should be printed larger and in a color different from the other information and the design to help it stand out. A simple design might have a small decorative border with the business name and address in small letters at the bottom. The herb name should be in the center of the label. FRESH HERBS should be printed somewhere on the label to describe what your product is — the top center is a good place for this.

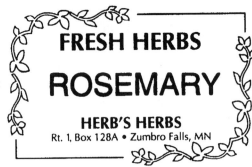

An example of label design, including company name and address, herb name, and the important words "Fresh Herbs."

There are various materials and labels that you can use at the point of sale of supermarket packs. These can increase sales, but also your expenses. For trays that contain recipes, place labels on the front of the tray for maximum exposure. Recipe labels should be small so that the consumer can still see the herbs. Labels that describe the uses of the various herbs can be put on the back of the package. You may want to insert a card inside the package so that it shows through the back. These should be printed on food-grade paper and coated so they do not absorb moisture from the herbs. They would have to be fairly large to list the most common herbs, and this can hide the herbs.

A nicely designed brochure describing the uses of the various herbs and/or recipes is a good way to increase customer awareness and sales, but it can get costly because many people will take the brochures without

 caution: labeling

A word about neatness in preparing the packaging materials: If you hand-stamp the labels, make sure the stamp is centered and straight on the label, especially on supermarket trays. Try not to smear the ink or get extra ink spots on the label. Place the labels straight on the bags or trays and all in the same place. A sloppy printing job and labels placed haphazardly present a shoddy image of you and your product.

buying the herbs. Perhaps you could share the printing costs with the store if you include the store's name in the brochure.

Bar Codes

Some supermarkets require that you have a universal product code, commonly known as bar coding, on your labels. This helps the retailer to speed up checkout, track sales, and control inventory.

You must apply to the Uniform Code Council Inc. (UCC) for your numbers. The UCC is a not-for-profit membership organization that administers Universal Product Codes (UPC). There is a rather hefty charge for obtaining your numbers but it may be necessary if your primary business is with supermarkets. You can get more information about the UCC and bar codes by accessing its Web site or by writing to the group. (See Resources.)

The Universal Product Code consists of a company prefix to identify your business and a series of numbers to identify each individual product, in this case each herb. It may be possible to use one number for all the herbs; ask your accounts if they will allow you to do this. Using one number will save you a lot of money, both in application fees to the UCC and in printing costs.

You are responsible for printing the UPC labels. Most printing companies offer this but they may send them out to be printed. You can also buy specialized printers that will print UPC codes. In the future, computer printers might have this capability. Be sure to check with your accounts about what size and type of labels their checkout equipment reads.

Organizing Orders and Harvests

This may seem like a simple process when your business is small and you only have a few orders. But as your business and your account base grow, it can be a time-consuming process to organize the orders and packaging. The more organized you are, the less time it will take to prepare your orders for delivery.

Everyone has her own way of organizing. It can be done by computer and a fax machine, or it can all be done by hand. The main objective is to have each day's orders, and the appropriate packaging materials, ready to go at the beginning of the harvest time.

If you have only one account to pick and package for, the procedure is simple: Just label the appropriate-size bags and/or the correct number of supermarket trays, write down the order on a piece of paper to take to the packing room, and now you're ready. If you have more than one account to pick for or more than one person helping, things will be a bit more complicated.

I had two, sometimes more, employees helping me harvest and package. My accounts all preferred to have their herbs first thing in the morning, so the orders were picked and packaged the day before delivery.

The night before the harvest day, all the orders were in hand. I would count how many trays and bags of each herb were needed and then mark the labels. I sold 23 different herbs, plus edible flowers and various herb garnish, so just getting the correct amount of labels done took time. The labels were placed on the bags or trays. Then each account's bags or trays were separated and marked. I made a master list with all of the accounts' orders on it. This, along with the materials, was waiting for my employees in the packing area when they arrived in the morning.

These two women were great at picking and packaging, so rather than telling them what to do, I just gave them the list and they did the work. They put a check mark on the list next to the herb they were going off to harvest to prevent duplicating an order. They would harvest and package that herb for all of the accounts. When that was completed, they drew a line through that herb for all of the accounts.

The ideal situation is to place the packaged herbs in refrigerators with each account's order in a separate area. This makes packing the boxes easier at delivery time. There were many times when the refrigerators were so full that the last few herbs packaged had to be put in wherever there was room. It took much longer to get those orders boxed and ready to go.

At delivery time, the boxes were marked with the account's name and the packing slip/invoice was ready. We packed the herbs in the appropriate box and kept the invoices separate, so as not to lose them in the bottom of the box.

successfully
growing more
than 20 herbs
and flowers

Currently there are at least 14 fresh-cut herbs that are of commercial importance. The popularity of herbs varies with where you are and the ethnic groups in your market area. Before committing large amounts of greenhouse or field space to a certain herb, make sure there is a market for it. If you have the space and time to make small plantings of herbs with little demand, of course you should do so.

Interest in a certain herb may be generated by a single magazine article or a cooking show on television. It is a definite advantage for you to be able to say, "Yes, I have that available." You never know when an herb will suddenly become popular.

It is difficult to give recommendations on how much of each herb to plant. It all depends on your market for the herb and your growing conditions. Always plant more than you think you'll need. (Ideas for ways to use unsold herbs are discussed in chapter 4, Doing Business.)

arugula

Arugula is sometimes known as roquette or rocket. (Do not confuse this with another herb, called sweet rocket.) This herb has been popular in Europe for many years and is now gaining favor here as a gourmet salad ingredient.

Arugula has a pungent, almost hot taste. People either love it or hate it. Those who have acquired a taste for it use it regularly. I have had farmer's market customers become quite upset when I was sold out of this aromatic herb.

arugula

Botanical Name:
Eruca vesicaria sativa
Type of Plant: Annual
Adult Height: 8" to 12"

USES

Arugula is used mainly as a salad herb, and the entire leaf is used. A little goes a long way due to its sharp flavor. Many people use it on sandwiches. It is also an ingredient in sauces for pasta, fish, vegetables, and stir fry. It does not make a good garnish because it wilts rather quickly when not kept cool.

Varieties

There are several varieties of arugula. 'Selvatica' *(Eruca selvatica)* and 'Sylvetta' *(Diplotaxis tenuifolia)* come from Italy and are usually referred to as "wild." These are more heat tolerant than others and are shorter and slower growing. Mix the small leaves in with your salad greens.

Arugula 'Astro' *(Eruca vesicaria,* subsp. *sativa)* is derived from the common variety rather than the "wild" types. The leaves are not as deeply lobed and have a milder taste. The best variety for growing in volume that has the most pungent flavor is regular arugula. When buying seed, make sure it is listed by the above botanical name.

Propagation

Arugula sprouts readily from seed. It does not transplant well, so sow it where it is to grow. Plant it outdoors as soon as the soil can be worked in the spring. Regular garden loam with good drainage is all that is required. Direct sow in rows 8 to 12 inches apart. Plant the seeds ¼ inch deep, about ½ inch apart and cover lightly. Keep moist until germination, which is usually six to eight days. Thin the plants to 3 or 4 inches apart. While this is not necessary, it does allow the plant to grow larger leaves. Plant as much arugula as your space allows, because only two or three leaves can be harvested from each plant a week.

The plant should reach harvestable size in 30 days. After the leaves are picked, it will take about a week to grow one or two leaves large enough to harvest again. This time length will depend on soil fertility and temperature.

Growing Arugula in the Greenhouse

Arugula grows much like loose-leaf lettuce. The leaves come up from the ground until the plant begins to mature. It then sends up a tall stalk upon which the flowers will form. Leaves may be picked from the stalk, but they are smaller and sharper in flavor, and sometimes slightly hairy.

This is a cool-weather crop that can take light frosts. Do not try to grow arugula in the greenhouse during the summer; it will produce smaller leaves, stronger flavor, and will bolt quickly in high heat. Start the greenhouse crop in late summer for fall harvest.

Arugula can be grown in large pots but it produces larger leaves if planted in beds. Continual picking will delay bolting. Depending on the quantity you sell, make succession plantings every three weeks to one month. This should guarantee a continuous supply. This herb benefits from half-strength fertilizer every month.

When the flower stalks begin to grow, cut back the plant to ground level, leaving a few of the lowest leaves. The plant will then grow another group of leaves before trying to flower again. The nearly white flowers are edible and are tangy in salads. They have a sweet aroma and the same sharp taste as the leaves.

Growing Arugula Outdoors

This herb prefers cool growing conditions. Plant it outdoors a few weeks before your last frost-free date, using the same seed depth and distance previously mentioned. Early sowings will also have more protection from flea beetle attacks.

As the weather warms, plant it in an area that receives some afternoon shade. This will prevent it from bolting prematurely.

Pests

Arugula attracts insects that like strong flavors. Aphids can be found on the back sides of the leaves away from the sun. The most bothersome pest, though, is the flea beetle. These are shiny, ¼-inch, hard-shelled, dark-colored beetles. They eat "shotgun"-type holes in the leaves, which makes them unsalable. The best way to prevent flea beetle damage, especially outdoors, is to cover plants immediately upon emergence with lightweight row covers.

Harvesting

The first harvesting of large leaves usually begins when the plants are four to six weeks old. Arugula can be cut to ground level with clippers or scissors; new leaves will grow, but the new growth leaf tips will be cut and this makes them unattractive.

Harvesting individual leaves is easier and allows more usable leaves from each plant. With your fingers, pull the leaf from the plant. Pick the large older leaves first — you'll find them on the outside of the plant. Keep the center leaves on the plant so they can produce more growth. Slide your finger down the inside of the leaf and push outward with your fingertip. This separates the leaf. Always pick perfect leaves with not even a hint of yellow. Those leaves that are not a deep green color will rapidly turn yellow in the package.

Arugula wilts quickly in the heat or sun. If the temperature is warm, pick only small amounts at a time or put the herb into a cooler immediately upon harvesting. Gently wash the leaves, if needed, and allow them to dry completely in a cool shaded area before packaging. Wet arugula rots quickly.

PACKAGING

Supermarket packages of arugula usually weigh ½ ounce. If you don't weigh your packages, put at least 12 to 15 big leaves in each. Put the stem ends down in the package. The soft leaf tips will collapse if they are facing downward and the trays are stood on end.

If you pick smaller leaves, layer them in the tray. This gives the package a fuller appearance and the leaves will stay in place if it is stood on end.

Restaurants and distributors generally want arugula in bunches even if it is sold by weight. A bunch is usually 1 to 1½ inches in diameter at the bottom where it is rubber banded together. If the leaves are large, the bunch size can be smaller, and vice versa.

basil

Sweet basil is the most important herb to grow for fresh-cut sales: Its sales far surpass those of any other culinary herbs nationwide. There are some large commercial growers who grow only basil and ship it fresh-cut worldwide.

This herb is popular not only with chefs but also with cooks in just about every ethnic group. This trend is expected to continue, so devote as much space as possible to growing basil; it will probably be your biggest seller.

Basil is quite aromatic, with flavors somewhat reminiscent of mint, cloves, and licorice mixed together. It is important to note that basil is *the* most tender herb. It will turn brown at 40° to 45°F or lower. It should be planted last in the spring and harvested first in the fall.

basil

Botanical Name:
Ocimum basilicum
Type of Plant: Annual
Adult Height: 18" to 36"

Uses

Basil is good with just about any food, especially tomatoes and Italian dishes. Pesto, a paste made with large amounts of this herb, can easily be frozen, and is popular with both chefs and home cooks. During the summer, when your basil crop is abundant, package 6 or 12 bunches in a bag. Market these as "Pesto Specials" at a reduced price and you'll probably sell out.

"Less is more" applies when cooking with most herbs, but not with basil. It is almost impossible to use too much in any dish, which helps account for its popularity. As with most herbs, you use the leaves. The top four leaves can be sold as a garnish, but they tend to wilt quickly when exposed to heat. The smaller tops of older plants hold better as a garnish.

Varieties

There are many types of basil, with more being developed. They range from the various scented types, dwarfs, and fine leafed to several varieties of purple. Listed below are the most important to grow.

Sweet basil is the most in demand with chefs and cooks. If your space is limited, this is the variety to grow. Occasionally you will see this labeled as broadleaf or large-leaf basil in seed catalogs. This is the most productive of all the varieties.

Fine-leafed or "Italian basil" is favored by some chefs and cooks. Its leaves are narrow, pointed, and smaller than those of sweet basil. It has a more spicy, delicate flavor. 'Genovese' is a larger-leafed Italian variety that has gained popularity in recent years.

Lemon basil is a delightful combination of the two flavors. Use it anywhere lemon is called for and enjoy its wonderful aroma. Chefs are beginning to request this variety and the trend should continue. Grow some if you have the space. It has a smaller habit and can be direct sown so the plants stand fairly close together.

Thai basil is used by Asian chefs and is found in small ethnic markets. It has a spicy, almost hot flavor. The stem is distinctly red or purplish. If your market area includes a substantial Asian population, do some research to determine if you could sell this variety. Most American cooks find the flavor of this variety too powerful to cook with.

Purple basils are gaining interest with cooks as garnishes and in vinegars, salads, and cooking. 'Purple Ruffles' is the most vigorous grower, with large, ruffled, thick, deeply purple leaves. It makes a fine addition to supermarket packets of salad herbs. Purple basils are difficult to grow in the greenhouse during periods of low light; these varieties require more intense light to develop their deep purple.

PROPAGATION

Plant seed indoors or in the greenhouse six weeks before transplanting outdoors or to greenhouse beds. Otherwise, you will risk having the small seed washed away by rain. Sow the seed thinly or in rows in the flats. Cover seed with fine sterilized sand or vermiculite to ¼ inch deep. Cover the flat with plastic and then a dark cover. Provide bottom heat at 70°F.

Basil seed germinates in about three days, so check the flat frequently. Remove the cover as soon as a few seeds have germinated. Basil is prone to damping-off, so it is important to provide air circulation at the first sign of germination. Continue to keep the flat moist and provide bottom heat until most of the seeds have come up. Then place the flat where it will receive full sun and warm temperatures.

When the flat is dry, and after the seedlings have 2 sets of true leaves, place it in a tub of water containing half-strength fertilizer. Watering this way prevents the tender seedlings from damage by overhead spraying. When the plants have at least two sets of true leaves, transplant them to larger pots or greenhouse beds. Do not transplant to outdoor locations until the soil is warm and the night temperatures stay above 50°F.

GROWING BASIL IN THE GREENHOUSE

Basil should have warm soil temperatures if it is to be grown during the cool season in the greenhouse. It is a fussy herb and will not grow well with cool soil temperatures. It prefers a daytime air temperature between 70° and 85°F.

Planting in greenhouse beds should be done using an odd/even spacing. This arrangement allows more sun to reach the plants, better air circulation between them, and easier harvesting.

Fertilizing basil in the greenhouse is usually necessary, depending on the texture of the soil. Lighter soil mixes have little capacity to hold nutrients. These lighter soil beds should be fertilized often. I fertilized every time I watered during spring and summer when plants are growing rapidly. During days of short light, fertilize less often.

Basil requires 80 parts per million of magnesium, which is more than most fertilizers offer. Magnesium deficiency shows up as a slight "bleaching," or light areas, between the veins of the leaves. This condition most often occurs during periods of heavy growth and when the pH is out of balance. The soil pH should be tested and adjusted every year before planting. Fertilizers specifically made for tomatoes usually contain more magnesium. Dolomitic lime, which also supplies magnesium, scratched into the soil surface, should return the pH levels to the desired range.

Magnesium, in the form of *pure* Epsom salts, has long been used by gardeners to provide a quick fix to tomatoes and peppers. Pure Epsom salts — without fragrance or other additives — is hydrated magnesium sulfate crystals. It also contains small amounts of sulfur, another nutrient needed by plants. Epsom salts are soluble in water and available at most drugstores.

Because of variations in fertilizer formulas, the nutrients in your water, and the complexities of making a complete nutrient, it is difficult to tell you the exact amount of Epsom salts to use. I used 1 or 2 teaspoons to a gallon of water to correct magnesium deficiencies in my greenhouse basil beds, but talk to a technical representative from the firm that manufactures your fertilizer for exact proportions to use. If you decide to experiment with adding Epsom salts to your fertilizer, be forewarned that too much magnesium can kill the plants.

Growing Basil Outdoors

Basil should be planted at least 18 inches apart. Planting in rows 3 to 4 feet apart allows for easy cultivation and good air circulation. It is usually not necessary to fertilize outdoor basil beds if the soil has been prepared with lots of organic matter. This herb does benefit from a sidedressing of aged compost or fertilizer after a heavy harvest.

Basil grows quite slowly after transplantation: The plants just seem to sit there for a week or two. The root system is developing at this time and little top growth appears. Then, suddenly, the plants begin to grow, and soon they are large, bushy, and demanding to be cut!

Pests

Everyone loves basil — including insects! Aphids, whiteflies, spider mites, and various worms are particular pests in the greenhouse. Outdoors, the insect problems seems to vary with location and the year. Slugs, flea beetles, and tarnished plant bugs have been problems for me in the past.

Basil is susceptible to fungal diseases, especially *Botrytis* stem rot and *Pythium*. These diseases are more likely to attack in the greenhouse during short-day-length periods due to decreased air circulation and higher humidity.

HARVESTING

The first cutting can usually begin six weeks after transplanting outdoors. In the greenhouse, during cool weather and short days, the first harvest usually begins 8 to 10 weeks after transplanting into growing beds.

During a warm summer with ideal growing conditions, you can expect a total yield of 5 pounds from three or four healthy plants. Each plant can be cut six or more times. Yields are smaller during cool, short days in the greenhouse.

Always cut basil with clean, sharp scissors. Cut the stem just above a leaf cluster even if you don't want all the stem. This forces the plant to branch and increases yield. Choose stems that have at least two leaf clusters. Don't cut more than a third of the plant at any one time or it may not recover.

As the plant begins to mature, the leaves become smaller and little flower buds appear at the tips of leaf clusters. Picking these leaves first encourages more leaf growth and delays flowering.

Check each stem, underside of the leaves, and the crevice where the leaf attaches to the stem for insects. If you find insects and they cannot be removed with your fingers, it will be necessary to wash the cuttings. Wash very gently, as this tender herb bruises easily. For insects that are stuck on, a light spray from a hose will usually wash them off. Allow to dry completely before packaging.

PACKAGING

Supermarket packages of basil should weigh ½ to 1 ounce. If you do not weigh the packages, put at least five full springs or leaf clusters in each tray. Basil leaves are large and can fill a package quickly.

When bunching basil, use long stems and rubber band six to eight of them together. The number of stems per bunch should vary depending on how many leaves are on each stem. Bunch size differs from region to region. In some areas, especially during the winter, you may see basil bunches from other growers that contain four short stems with only four little leaves per stem. Always try to make your bunches bigger than the competition's!

Pack the bunches in the appropriate-size bag and leave the top open slightly for the basil to breathe. Place the bag out of direct sunlight. *Never* store basil in the cooler. It must be held between 45° and 60°F or it will turn brown. This is a fact that you should convey to all of your accounts.

 easy way to arrange a basil package

Place one long sprig with the stem end down. Then place smaller leaf clusters around the tray to fill it up. The long sprig helps to keep the smaller clusters from sliding to the bottom of the tray when it is stood on end.

chives

Chives are a hardy perennial and easy to grow. It seems they'll always be popular with chefs and cooks, and many homes have a resident clump of chives growing somewhere in the yard.

Looking somewhat like grass, chives have rounded spikes. The lavender flowers are borne on long tough stems and are very attractive and useful in the kitchen. The sale of chive blossoms, when they are in season, can provide a bonus income for you.

As a commercial grower, I have a real dislike for chives. They are priced low nationally and yet require much work to harvest and package. It is time-consuming and can be costly (time is money) to clean the bunches and make them attractive for sale. However, if you want to supply your accounts with all of their fresh-cut-herb needs, you will have to grow them.

chives

Botanical Name:
Allium schoenoprasum
Type of Plant: Perennial
Adult Height: 10" to 12"

317

Uses

Chives have a mild onion flavor and can be used in any foods that call for onion. Many chefs use the long spikes as a garnish. Most often they are snipped or chopped finely and sprinkled on top of food dishes as a garnish. When used to flavor food, they are usually mixed with other herbs.

The whole stems are used, except the tough ones that bear the flower. To simplify chopping, bunch the chives together. This way they can be held in one hand and the tips snipped with a sharp scissors. Chefs usually lay a bunch on a cutting board and chop them with a sharp knife.

Varieties

Until just recently, there was only one variety of chives. Many chefs prefer "fine-leafed" chives, which are really first-year plants. The spikes grow thicker after the first year. Seed catalogs sometimes list a fine-leafed variety, but in the second year these grow just as thick as regular chives.

The major problem with chives for the commercial grower is that the plants stop growing vigorously after the third or fourth cutting. The best remedy is to have large amounts of plants in production.

'Grolau' is a variety developed for forcing in the greenhouse, where space is limited. 'Grolau' grows best if kept continually cut. It has performed well for me in the greenhouse over the winter. This variety is available from Nichols Garden Nursery (see Resources).

Now some new chive varieties have been introduced. These were developed for their leaf size and thickness rather than their ability to produce a continuous crop.

Propagation

Start chives by seed or division of a clump. For greenhouse use, it is best to start plants by seed. If you will be growing and cutting during winter, start with new chive plants at the beginning of your indoor cutting season. These plants should be started three or four months before you plan to take your first harvest.

Seeds. Always be sure to use only fresh chive seeds. Germination rates are uneven with seeds over a year old. Sow the seeds very thickly in flats to a depth of ¼ inch. Cover the flats with plastic and a dark cover. Provide bottom heat at 70°F. Seeds should germinate in about 10 to 14 days.

Remove the cover and place the flats in full sun. When a flat is dry, water from below with half-strength fertilizer. Transplant when the seedlings are 2 inches tall and the roots are well matted. The entire flat can be turned out on a bench. With a sharp knife that has a slightly serrated edge, cut the young chives into 2- or 3-inch squares.

Cut the roots to 1 inch in length if they are very thickly matted. Cut the tops to ½ to 1 inch in length. This promotes both top and root growth after transplantation.

Division. After the third or fourth year, chive clumps are thick and growth is less vigorous. Dig up the plants and divide to make new ones. This is best done in spring, just as the new growth is a few inches tall. You can divide and transplant outdoors in fall, but do so several weeks before your first expected frost. This will allow the plants to establish a root system before going dormant for the winter.

Prepare the beds where the new plants are going before you begin the process of division. Chives are heavy feeders, so the beds should contain organic matter and/or aged manure. If planting into greenhouse beds, where you may not want manure or compost, add lots of peat and perlite to lighten the soil mix.

Division is best accomplished when the soil is slightly moist. This allows the chive bulbs to stick together and makes the division easier. Using a sharp spade, dig all around the clump and lift it from the ground. Shake off enough soil to expose the roots. With a sharp, slightly serrated knife or a sharp heavy scissors, cut off the roots to 1 inch below the bulbs. The tops should be cut back to 1 inch.

If the clump is thick with matted roots, you may need a sharp spade to make the division. This will not hurt the plants. Or you may be able to just pull the clump apart. Divide the clump into as many sections as you wish.

If you will be replanting the divided clumps for cutting, be sure that they are at least the size of a single market-size bunch. This is 1 to 2 inches in diameter when the chives are grasped at the bottom of the bunch. If you will be potting up some of these divided chives, the size of the division will depend on the size of the pot you will transplant them into.

Replant the clumps ½ inch lower than they were in the original bed. Allow at least 8 inches between plants.

GROWING CHIVES IN THE GREENHOUSE

Chives prefer cooler temperatures, so plant them on the north side of the greenhouse. Use the odd/even method of spacing and allow 4 to

6 inches between clumps. Keep the beds absolutely free of weeds and grass; if weeds are allowed to grow within the clump, harvesting will be a real chore.

Supplemental lighting is important, as even new plants go dormant during winter without it. In addition, this herb requires a good deal of fertilizer. Growth will be slow and the color pale if chives are not fed properly. Fertilize after every cutting and you will be rewarded with rapid growth.

When growing and cutting chives from the greenhouse during winter, allow the chives to rest over the summer. Try to take all of the cuttings from outdoors during spring, summer, and fall. Cut back the greenhouse chives and fertilize them at least three weeks before you want to begin harvesting them. This will give you fresh new growth to start with.

Growing Chives Outdoors

If the soil in the chive beds was well prepared before planting, there is little that needs to be done to keep the plants happy. A side-dressing of compost or aged manure after a heavy harvest will be appreciated.

Weed frequently. If grass is allowed to grow in and around the chives, it is almost impossible to eradicate it without digging up the clumps to remove grass runners. I once had to abandon an entire 20- by 40-foot chive bed that, through neglect, had turned into lawn. It was, however, fragrant to mow!

Pests and Diseases

There are very few insects that like chives. In the greenhouse, aphids, spider mites, and leafhoppers will attack them only for lack of their favorite foods. Onion thrips can be a problem because the areas where they feed turn white and unsightly.

Various fungal diseases attack greenhouse chives if the clump is allowed to become matted down. After watering, pick up the spikes and shake off the excess water. Provide good air circulation and fungal diseases can be avoided.

Outdoors, pests should not be a problem other than an occasional "critter" hiding in a thick clump. I once found an enormous wolf spider raising young in a large clump of chives. That clump was not picked all summer!

Harvesting

It may take up to two months from transplanting to get the first cutting of a chive clump. The stems will be quite fine the first year, requiring more to make a salable bunch. Yields increase in subsequent growing seasons. It is impossible to predict the yields you can expect; they will vary with the age and size of the plants. The best advice is to have many more plants than you think you'll need.

After cutting, the plant should be ready to cut again in two or three weeks, provided conditions are right (fertility, daylight length, and temperature). Don't pick chives when the clumps are wet. Even a little moisture will cause them to rot in the package. If rain is expected on harvest day, cut them the day before. Sometimes, though, cutting them wet is unavoidable. The chives must then be taken out of the bunch, spread out away from direct sunlight, and allowed to dry.

Cutting. Tightly grab a bunch at the bottom with one hand and cut with the other. Slip a rubber band loosely around the bottom of the bunch, about 1 inch above where you cut. The bunch size should be between 1 and 2 inches in diameter where it is rubber banded together. Keep the cut bunches out of direct sunlight as you harvest. During very hot weather, put the newly cut bunches in a small cooler until you are finished.

If this is the first cutting for these plants and the clump is weed-free, the bunch should be easy to clean. Just check to make sure that there are no yellow or brown tips and debris. The flower stems, before and after flowering, must also be removed, as they are very tough. Take out any grass or weeds in the bunch. The bunch must be perfect.

Take off the rubber band, hold the bunch loosely on the bottom, and shake gently. Then grab the top of the bunch and shake. This will get rid of most of the debris and short stems. You may not get out all of the weeds and debris this way. To get the rest, hold the bunch loosely, spread it out, and pick out any more unwanted trash.

use sharp scissors

Very sharp scissors are a necessity for snipping chives. Some growers even use a knife or razor knife to cut them. Always cut chives at the bottom about 1 inch from the crown. Cutting the tips will not force the plant to regrow.

When the bunch is perfectly clean, put the rubber band back on, ½ to 1 inch from the bottom, and wrap it tightly, at least twice. Then check for any yellow or brown tips on the bunch — these must be removed also. Spread the bunched chives out with your hand. Using a small, fine-pointed, sharp scissors, cut off the bad tips. These can be numerous if the plants have had multiple cuttings. Finally, cut the bottom of the bunch evenly to make it attractive.

See why I complained about chives?

Packaging

Supermarket packages of chives usually weigh ½ ounce but 1-ounce packages look better and are a better deal for the consumer. The public perceives chives as an inexpensive plant because many people have had a plant at one time or another. They know that they are easy to grow and can't understand why they are priced the same as other herbs. Selling in larger packages gives the customer a little more for her money.

Some growers do not bunch the supermarket packs of chives. It is easier to get the weights exact if they are packaged loose. However, after a day or two spent standing on end on the shelf, the loose chives will have softened and collapsed in the package. It is also easier for the consumer to use them if they are bunched. If you don't weigh your packages, make your bunches large and attractive.

Place the bunch in the package with the bottom down. If the chives are longer than the package, twist the top of the bunch slightly and bend it downward to fit. This makes a fuller and nicer-looking package. Some growers cut the chives to fit, but after several days, the cut tips may begin to turn yellow, making the package less attractive.

Restaurant packaging is much easier. Simply place the bunches, stem ends down, in the appropriate-size bag. Some restaurants want only two or three bunches at a time. Using a smaller bag for these orders saves you money in packaging costs. A good size is 4 inches by 2 inches by 12 inches. Make sure all the chives are inside the bag — any tips left outside will dry out quickly. Leave just a small part of the bag open for the chives to breathe.

Store chives in the cooler at 40° to 50°F. They seem to soften less rapidly at this temperature; any lower, and the chives lose appeal.

Garlic Chives

garlic chives

Botanical Name:
Allium tuberosum
Type of Plant: Perennial
Adult Height: 8" to 10"

Garlic chives are sometimes known as Chinese leeks or Oriental chives. They combine the flavor of garlic and onions and are very aromatic when cut. They have gained in popularity in recent years and should be grown if space allows. This herb is very much like traditional chives in its habits. Handle garlic chives exactly as you would other chives in all respects.

There are several differences, though, that make garlic chives more useful for the commercial grower. The blades are flat and heavier than those of chives. This makes it easier to fill out bunches and it takes fewer to make the weight in packaging. Also, the clumps have a tighter growth habit, which makes it more difficult for grasses and weeds to grow within them. Finally, cleaning the bunches is easier because the color of garlic chives is lighter and the blades are wider than grass, so it is easy to spot any wayward grass in the bunch.

This herb grows much better during the winter months in the greenhouse. Many times I have substituted them for regular chives, with a chef's approval. Most chefs really liked them, and requested them after comparing the two.

Garlic chives multiply faster and can be divided sooner than regular chives. The ability to increase the number of stock plants rapidly is a benefit. When cut, the blades grow back faster and more uniformly than do those of regular chives. There is also less tip dieback after being cut. The flowers are white, fragrant, edible, and of the umbel type. Do not allow the flowers to go to seed or you'll have new plants springing up everywhere!

cilantro

Many times you will see cilantro referred to as coriander. Cilantro is the leaves; coriander is the seeds of the same plant. The fresh leaves are usually what people want; the seeds do not have the same flavor as the leaves.

Sometimes this plant is referred to as Chinese parsley. It does look somewhat like Italian or flat-leafed parsley except that cilantro leaves are more rounded. They have similar growth patterns. Inexperienced cooks may mistake the two, until the aroma gives it away.

This herb has been cultivated since ancient times for medicinal as well as culinary purposes. Its popularity continues today as a main flavoring ingredient in many ethnic dishes. Cilantro has a reputation as another "love-it-or-hate-it" herb. It has a pungent, almost soapy aroma that clings to hands even after many washings. If you find the smell of cilantro offensive, as I do, wear gloves when harvesting it.

cilantro

Botanical Name:
Coriandrum sativum
Type of Plant: Annual
Adult Height: about 18"

This herb is widely cultivated in the United States. It's found in most supermarket produce departments unpackaged in bunches alongside the display of curly parsley. These bunches of cilantro are shipped in case lots from large growers from all parts of the country and Mexico.

Because it is priced so inexpensively year-round, it may not be worthwhile taking up valuable greenhouse space to grow cilantro, especially during the winter. Do grow it, though, during the outdoor season. The demand for cilantro is high and it can be a real moneymaker if you have the space to grow it in quantity.

Uses

To take advantage of its full flavor, use cilantro when it is fresh; it loses its flavor when dried or cooked. This accounts for its abundance in the marketplace. The fresh leaves are removed from the stems and chopped. Some chefs use the leaves as a garnish, although they do not hold up well to heat.

Cilantro is used in Asian, Caribbean, East Indian, and Mexican cuisines. The recent interest in Southwestern, or Tex-Mex, cooking has added to the demand for this herb. Perhaps the most popular use for cilantro of late is in making salsa and dips.

Varieties

Cilantro is a short-lived annual and quickly goes to seed. Recently developed varieties prolong the period of harvest somewhat. 'Santo' and 'Slow-Bolt' give the grower just a week or two more to harvest before the plants bolt. 'Jantar' is a new variety that is said to have greater leaf production and is slower to bolt than 'Santo'.

The seeds are large (¼ inch) and should be purchased in 5-pound packages. The seeds remain viable for at least a year.

Propagation

This herb does not transplant well; direct-seed it where it is to grow. It does well in rich soil with good drainage and full sunlight. Each seed will send up from one to three stems upon which one or two groups of leaves will form. Because of this, cilantro seed should be planted in small clusters. Harvest will be much faster and easier if the plants are growing in a bunch.

Using a trowel or your hand, make 6-inch circles ½ inch deep. Place 10 to 15 seeds (I just grab a small handful) in the circle. Cover with soil or sand to ½ inch deep. The seeds need darkness to germinate. Water in well and keep moist until seedlings emerge, usually in 7 to 10 days.

Growing Cilantro in the Greenhouse

Cilantro grows best in full sun, so plant it on the south side of the greenhouse. Sow seed in the beds using the odd/even pattern.

Cilantro must have very fertile soil or be fertilized regularly. Without this, the leaves will have a lighter, almost yellow appearance and they won't hold as well after harvest. However, too much fertilizer results in lots of leafy growth but diminished flavor. The trick to keeping the good rich color and flavor is to watch crop growth closely. The moment you notice a lighter-colored top leaf or slowed growth pattern, fertilize!

Cilantro can be cut only twice. After the first cutting, regrowth diminishes somewhat. Usually a good second cutting can be had, although the stems may be shorter. After the second cutting the growth is sparse and low to the ground. The plant's objective in life is to produce seed and it will try to flower even on very short stems. These short stems should not be harvested, as the flavor will be bitter.

As the plant begins to mature, the new growth will have feathery leaves. It will send up a central, thicker flower stalk. Don't harvest the feathery leaves or those from the flower stalk because they are bitter and don't have the characteristic round cilantro shape.

To maintain a continuous supply of cilantro, start new plants every few weeks. This planting schedule will depend on how much you are selling, of course: The more you sell, the sooner the succession crop should be planted.

Growing Cilantro Outdoors

Planting outdoors is accomplished much the same as it is in the greenhouse. The main difference is that the clusters are planted in rows for ease of cultivating or tilling. Wait until all danger of frost has past before planting. This is a heat-loving herb and it can take only a very light frost without suffering some damage.

Pests and Diseases

Aphids can be a problem for cilantro, especially in the greenhouse. It's important to control them: Washing insects off cilantro is difficult because of the many small leaves in a bunch. Insect attacks have not been a problem for me on cilantro grown outdoors.

Fungal diseases, typically *Pythium,* are a problem in the greenhouse because of the tightly grown clusters of plants. Shake off any excess moisture from plants after watering and provide good air circulation.

HARVESTING

The first harvest can usually be taken six to eight weeks after germination. A second cutting can usually be made in 10 to 14 days. To harvest cilantro, grab a bunch-size cluster at the base of the plants, hold tight, and cut with a sharp scissors with your other hand. Place a rubber band tightly around the bottom of the bunch. Bunches should be 2 to 3 inches in size at the rubber banded bottom.

If the cilantro is clean, long, and without weeds, bunching is all you need to do. If there are weeds or yellowed leaves in the bunch, remove them. Loosen the rubber band, grab the bunch by the top leaves, and shake. The smaller stems and weeds should fall out, but this will now make your bunch smaller. This usually occurs with the second cutting because then the growth is not consistent. You may have to cut more to fill out the bunch.

After the bunch is rubber banded, cut off the bottom of the stems evenly to make it more attractive. If it is necessary to wash cilantro because of insects or dirt, be sure to allow the herb to dry completely before bunching it up. Cilantro will rot quite quickly if it is packaged wet.

PACKAGING

Cilantro is almost always sold in bunches, to both supermarket and restaurant accounts. Packaging in trays for supermarkets may be done in some areas of the United States. I once had a large supermarket account that preferred cilantro in marked packages. They said that some inexperienced cooks did not know what cilantro looked like and were too embarrassed to ask. It sold well for them this way.

Packaging is usually the same for all types of accounts. The bunches are packed in the appropriate-size poly bag and labeled. Sometimes cilantro is ordered in cases of 30 bunches. These are usually packaged in wax-lined cardboard boxes. However, some accounts may not mind if you deliver two large bags with 15 bunches apiece.

If a supermarket wishes to have this herb packaged in trays, do not make bunches. Cut off the stems so the herb fits the package with the stem ends down. The package should be full, as the small leaves are all that are used. The weight should be 1½ to 2 ounces.

dill

Dill has been cultivated since ancient times for both medicinal and culinary purposes. Its versatility in cooking has made it popular worldwide. It is an easy-to-grow herb and is much in demand in the marketplace.

USES

All parts, except the roots, are used. Most everyone is familiar with the fresh seed heads used to flavor dill pickles. These seed heads can be found in abundance in supermarkets and farmer's markets in season. You can sell them as a side crop.

Dried dill seed is commonly found in jars in markets and is used to flavor everything from breads to soups. Dried dill, often called dillweed or dill feathers, is also found on spice shelves. As a fresh-herb grower, your main

dill

Botanical Name:
Anethum graveolens
Type of Plant: Annual
Adult Height: to 48"

focus will be on the fresh leaves, which are sometimes called dill feathers. The leaves are used to flavor dips, salads, fish, and soups.

The fresh stems, stripped of leaves, can be woven into baskets when they are fresh and pliable. Even the thinnest of stems can be woven into mini baskets. The iridescence and color of the stems when dried are remarkable. The yellow flower heads dry beautifully. Pick them just when the heads begin to open in order to retain the color.

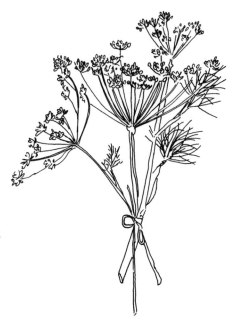

The yellow flower heads of dillweed are beautiful dried.

Varieties

For many years, the only dill seed available was for the tall-growing types, such as 'Mammoth'. 'Bouquet' is another tall variety that was developed for producing seed as well as more flavorful leaves. Be aware when purchasing this variety that some types are tall and some are dwarf. I have found some of both in the same seed packet.

These varieties are productive but can pose some problems in the garden and greenhouse. Plant them where they will not shade other plants. The larger dills also have a tendency to blow over in a strong wind. Dwarf varieties are more manageable for the commercial grower. The tetraploids produce more foliage, are slower to bolt, and are shorter in growth habit. 'Dukat' has a rich dill flavor and produces lush growth of bluish green leaves. It is of medium height.

'Fernleaf' is the best that I have found for leaf production. It is a compact dwarf growing to 18 inches tall and a good choice for the greenhouse. Fernleaf is multibranching from the base of the plant and is slow to bolt. This allows for a longer and easier harvest period. The flavor and color are exceptional.

Varieties of dill are being developed that increase and prolong leaf production. 'Hercules' is a new variety, although a tall one, that may be useful for commercial growers.

PROPAGATION

Dill is difficult to transplant and should be seeded where it is to grow; it will usually bolt shortly after transplanting.

There seems to be conflicting information about whether dill seeds need light or darkness to germinate. I have not been able to germinate dill seeds that were covered with any more than a thin coating of sand. And yet I have talked to some growers who bury dill seeds 1 inch deep with great success. Experiment to determine what works best for you.

When planting 'Fernleaf' seeds, water the greenhouse beds heavily before planting. Scatter the seeds thinly in 6-inch-wide rows, allowing 1 or 2 inches between the seeds, and the rows should be 10 to 12 inches apart. Leave the seeds uncovered except for a fine layer of sand to mark the rows and to keep the seed from moving when you water. The seeds can also be planted in clusters.

Mist or water the rows morning and night. The seeds should germinate in 10 to 14 days, but they may take as long as three weeks, especially in a hot greenhouse. Dill will germinate quicker with temperatures about 55° to 60°F.

After germination, thin the plants to at least 3 inches apart. Leaf size will be small and spindly if the plants remain crowded.

Outdoors, wait until your last frost-free date to plant dill. Seed it in wide or straight rows at least 3 feet apart. You may want to cover them with a fine layer of sand to prevent the lightweight seeds from blowing away in a strong gust. Gently water the seeds morning and night until they germinate.

Make your rows no more than 2 feet wide so you will be able to weed comfortably. Young seedlings are very delicate and don't grow well if they must compete with weeds. Dill readily reseeds itself. If seeds from the previous year were allowed to scatter, you will have bonus volunteers everywhere.

GROWING DILL IN THE GREENHOUSE

Plant dill on the south side of the greenhouse so that it receives the maximum amount of sunlight. Dwarf varieties can be planted in 2- to 4-foot tall raised beds. If you plant the taller types, ground beds are recommended; this prevents the dill from shading other herbs.

Dill, like most other herbs, grows very slowly during low light periods without supplemental lighting. It does, however, respond quite well to the additional hours of artificial lighting.

Dill seedlings are especially tender, so after a heavy watering, gently pick the seedlings up off the soil. Dill grows quickly in the first month and is a heavy feeder. Fertilize often with a high-nitrogen formula. Fertilize weekly during summer and biweekly during winter. If you notice any yellow leaves, you definitely must fertilize.

After several weeks of sprouting leaves from the base of the plant, a center stem will begin to grow. 'Fernleaf' dill takes much longer to reach this point. When the leaves have been picked from this stem, that's it — another one will *not* grow in its place. Eventually, you will have a leafless stalk. This does not stop the plant from flowering and producing seeds, though.

Because dill does not regrow new leaves as other cut plants do, succession plantings are a must. Your planting schedule will depend on the volume of dill that you sell. If this herb is a big seller in your area, plant new beds every three to four weeks for a continuous supply.

GROWING OUTDOORS

Plant tall varieties of dill in a location that is protected from strong winds because they are easily blown over. All dill should receive full sunlight. If the soil in the dill beds has been well prepared, little else needs to be done. This is a fast-growing herb that feeds heavily, so it benefits from side-dressings of aged manure or fertilizer.

PESTS AND DISEASES

Aphids love dill, especially in the greenhouse, and control of this insect is important. The foliage is so fine and tender that it is nearly impossible to remove aphids without damaging the leaves.

Outdoors, dill is attractive to a number of pests. I have seen everything from potato bugs to cabbage loopers enjoying the flavor. Usually the natural predators outdoors maintain a good balance and your crop loss will be minimal.

Fungal and viral diseases can attack greenhouse dill, especially during periods of low light. If you notice older leaves turning brown, or "white fuzz" growing on them, provide better air circulation and increase the heat. You may have to thin out some of the plants if they are crowded.

HARVESTING

Take the first leaves when the plants are 6 inches tall, usually four to five weeks after germination during the summer. In winter, the time to first harvest is longer, six to eight weeks.

The yields depend on the variety, season, and soil fertility. Dill does take a fairly long time to grow to harvestable size, and yields are small per plant. You may harvest

To harvest dill, grasp the leaf at bottom of the stem and gently break off.

only 12 or 15 big leaves from each plant. Compensate for this limited per-plant harvest by growing more and planting frequently.

Pick dill leaves by hand; don't cut with scissors. The leaves separate easily from the plant with gentle pressure. Always take the largest, oldest leaves first. If the plant is young and still growing from the crown at soil level, these will be the outermost leaves. Do not take smaller leaves from the center, as these still have some growing to do. Grab the leaf at the bottom of the stem where it is attached to the crown and gently push or pull outward with your finger.

As the center stem grows, pick the leaves from the bottom of the stem first and work your way up. These lower leaves will begin to turn brown as the plant matures. Pull the leaf downward to remove it from the stem. There may be some new growth of smaller leaves at this site.

A bunch is usually 12 to 15 larger dill leaves rubber banded together at the bottom. If the leaves are smaller, put enough in the bunch to measure 1 to 1 ½ inches in diameter at the rubber band. Be careful when putting on the rubber band, because dill stems are brittle and will break off when treated roughly.

If you are picking for supermarket packages, do not rubber band the bunches. Instead, pick enough to fill a package and lay the bunches separately in the bucket or container. This makes it easier to fill the trays.

After you have picked your bunch, grab it right below where the leaves join the stems and slap it sharply against your open hand. This will not hurt the dill but will usually knock loose any insects that are hiding. This is best done outside the greenhouse to prevent any dislodged bugs from seeking another meal.

If this does not remove insects or soil, wash the dill, gently agitating the bunch in water. Shake the bunch vigorously to remove as much water as possible. Let the bunch dry completely before packaging.

Packaging

Supermarket packs of dill usually weigh ½ ounce. Add more if the package looks bare. Place the dill in the tray with the stem ends down. If the leaves are large, you may have to cut off some of the stems to make it fit in the package. If the leaves are small, layer them in the package. This will help keep them from sliding to the bottom of a tray displayed standing on end.

To package for restaurants, place the bunches, stem ends down, in the appropriate-size bag. Leave the bag open slightly for the dill to breathe. Dill should be stored between 33° and 38°F to maintain maximum freshness.

marjoram, sweet

The aroma of this herb is wonderful — sweet and spicy at the
same time. Marjoram is a
member of the oregano family,
but it has an entirely different
flavor. The two herbs have a similar appearance, with roundish
leaves, but marjoram is lighter in
color and has smoother leaves. It
is often used in conjunction
with oregano, especially in
Italian cooking. This is one of
the herbs in the fines herbes
blend from France.

The popularity of this herb
varies from region to region. In
my marketplace, very few chefs
and cooks appreciate its complex
flavor. In some areas, though, it is
one of the most widely used
herbs. So before committing
greenhouse space to it, check the
trends in your area. Even if it is
not in demand, however, it is
important to devote some space

marjoram

Botanical Name:
Origanum majorana
Type of Plant: Tender
perennial
Adult Height: to 18"

to marjoram. Eventually, perhaps with some education from you, this herb will become more popular.

USES

Marjoram is usually combined with other herbs in a wide variety of foods, such as eggs, cheese, meats, sauces, and soups. Its flavor is intensified when dried and often is used by chefs in this form.

The leaves are stripped off the stems and chopped. The round flower buds can also be used but most chefs prefer only the leaves. The fresh sprigs don't make a good garnish because they wilt quickly when exposed to heat.

VARIETIES

Several types of marjoram are available either as seed or as plants. The varieties labeled as pot, wild, or showy marjoram do not have good flavor. Hardy sweet marjoram is a newer type, but also without good culinary flavor.

Some plants and seeds are identified only as marjoram and could be the wrong variety. Always purchase by the botanical name, *Origanum majorana,* and buy seeds from a reputable seed company. Before you buy stock plants, crush some leaves to check for the characteristic sweet aroma. A new variety, 'Erfo,' introduced in 1998, is said to produce high yields and larger leaves.

PROPAGATION

Propagate sweet marjoram by seeds, stem cuttings, or layering. Many growers prefer to do stem cuttings to be sure they are getting the exact variety they want. Sweet marjoram easily sets roots this way. See chapter 9, Starting Seeds and Plant Propagation, for information about taking stem cuttings and layering.

To start this herb from seed, first wet the soil in the flat. Sprinkle the tiny seeds thinly over the soil. *Lightly* cover them with fine sand, or do not cover them at all, and cover the flat with clear plastic. Sweet marjoram needs light and temperatures of 60° to 70°F to germinate. The seeds should be up in 7 to 14 days. This herb, as with most perennials, has a low germination rate.

When the seedlings emerge, place the flat in direct sun and away from cool drafts. Allow the soil to dry before watering. Water from below, using half-strength fertilizer. When the tiny seedlings have at least three sets of true leaves — usually six to eight weeks after germination — they are ready to transplant.

Growing Marjoram in the Greenhouse

It is best to grow this herb in the greenhouse unless you live in an area where there is no threat of freezing temperatures. It is killed by frost, growth is slowed by cool nights, and it takes considerable time to attain harvestable size.

Sweet marjoram grows well in large pots, at least 8 inches in diameter. If the demand for this herb is low, you could grow all you would need in pots. Fertilize potted herb plants weekly, especially during the rapid growth periods of spring and summer.

When transplanting the herb into raised beds, plant them in the odd/even pattern about 12 inches apart. The middle or south side of the greenhouse is the best location for this sun-loving herb. The soil mix should be rich but light; marjoram prefers dry conditions and will not grow well in heavy, moist soils.

Water carefully when the seedlings are small. Fertilize every two weeks, more often in light soil mixes. Yellowing leaves and slowed growth are indications to fertilize more often. As the stems grow longer, cut back the plant by half every two weeks. This encourages bushy new growth that will make it easier to harvest.

After many cuttings, the mature plant stops growing and rests. This usually occurs during the low light periods of winter, but not always. Only lasting a few weeks, this is not a true dormancy. Don't fertilize during this time; wait until growth begins again.

If all your plants are the same age, you could be out of marjoram during this resting time. The best way to avoid a shortage is to have younger plants just coming into production as the older ones mature.

Growing Marjoram Outdoors

Sweet marjoram should be transplanted as seedlings rather than direct-seeded. The plants are small and tender and will do better if given some protection while they are young.

In most regions, treat sweet marjoram as an annual and replant it each year. Wait until all danger of frost is past and the soil is warm before transplanting. Choose a sunny, protected location with light soil. Plant 10 inches apart in rows for ease of cultivating or tilling. If the soil has been well prepared, this herb should not need additional fertilizer.

As the nights begin to cool and frost approaches, harvest all your sweet marjoram. The leaves begin to turn brown with cooler night temperatures, and then are useless for fresh-cut sales or drying. You may dig up the plants, pot them, and bring them in for the winter. When bringing plants in from outdoors, cut off only half the foliage to allow the plant to regrow. Check carefully for insects before bringing any plant into the greenhouse.

PESTS AND DISEASES

Whiteflies can be a serious problem for marjoram in the greenhouse. It seems to be their first choice for dining. Leaf miners can also be a problem, along with aphids and spider mites. Unfortunately, insects are difficult to remove from marjoram's very tender leaves. Washing hard enough to dislodge insects often results in some leaves falling off and bruising of the remaining ones.

Marjoram is extremely sensitive to fungal diseases, especially when the plant grows older and thicker. It may be necessary to cut back thick plants to increase air circulation if they are not harvested regularly.

HARVESTING

It may take as long as five months for a single marjoram plant, started from seed, to reach harvestable size. Plants grown from cuttings will be ready sooner. Cut a few long stems from each plant to encourage new growth.

Yields, as with most herbs, depend on the age and size of the plants. It is usually possible to get five to eight bunch-size cuttings from a plant before it must rest. The ideal plant for harvesting bunches should contain 15 to 20 (or more) stems, all about 12 inches tall. When all the stems are cut the same length at the same time, the new growth will be uniform in height.

Cutting bunches for restaurants is a simple matter with the ideal plant. Grasp the size bunch that you want, cut, and put a rubber band around the bottom. Give the bundle a gentle shake to remove any debris or dead leaves. The bunches should contain at least 18 stems if they are 6 to 8 inches long; put more in the bunch if the stems are shorter. Never cut off more than two-thirds of the plant. Always leave some green leaves, or it will not grow back. Try to cut the plants before the flower buds form, as most chefs do not use them.

After being cut several times, some stems may die back. Cut these off at the bottom of the plant so they don't have to be removed from cut bunches. There will probably be times when the plants have not all grown to the same height. In this case, each stem must be cut individually, so it will take longer to make up a bunch. After doing this for a bit, you will come to appreciate those "ideal" plants and take the time to prune them.

PACKAGING

Package marjoram as you would other herbs for the restaurant trade. Place the bunches, stem ends down, in poly bags, and leave the bag open just a bit. Supermarket packages usually weigh ¾ ounce to 1 ounce. Cut enough marjoram to fill the tray but do not rubber band it together. Place the bunch in the tray with the stem ends down. Cool this herb immediately after packaging, as it wilts quickly.

mint

Who doesn't love the refreshing flavor of mint? It is used widely, in everything from mouthwash to candy. Mint is one of the most commercially cultivated herbs in this country. Much of this is peppermint, which is distilled to make the mint oil used to flavor many familiar products.

Mint has been an important herb throughout history in virtually all parts of the world. The popularity continues today for medicinal and culinary uses, and it has been used since antiquity to relieve digestive disorders.

There are more than 600 varieties of mint available today. Many are hybrids, with myriad variations in flavor, color, and taste. While many of these are fun to grow, only a few are of commercial importance to the fresh-cut-herb grower.

mint

Botanical Name: *Mentha* spp.
Type of Plant: Perennial
Adult Height: to 36"

Uses

Mint is an important culinary herb in many parts of the world. It is a significant component of Middle Eastern, Greek, European, and American dishes. It is used in everything from tabbouleh

339

(cracked wheat salad) to lamb, from tea to stir fry. In the United States, it is used mostly with sweets, vegetables, and drinks.

The leaves are usually stripped from the stems, chopped, and added to dishes at the last minute; prolonged cooking diminishes the flavor. In many restaurants, mint serves as a garnish for desserts and drinks. The top four leaves as well as a single leaf are used.

VARIETIES

Many of the multitude of mints have some culinary uses. If you have the space and time, experiment with some of these flavored varieties. You may find that some chefs like to try new products.

As a commercial grower, you should specialize in one variety in volume, and that is the one that is in demand and sells. The best type of mint for culinary purposes and garnish is spearmint *(Mentha spicata)*. Spearmint comes in numerous varieties. Some are smooth leafed, others have more "crinkly" leaves. In my region, the chefs, cooks, and distributors prefer the smooth leaves. The spearmint variety that I had most success with is 'English Mint'. This is a traditional mint with large, deep green, smooth leaves. It is a vigorous grower and has exceptional flavor. I recommend buying this variety from Bluebird Nursery (see Resources), as English mint obtained from other commercial nurseries tends not to be of the same smooth-leaved, vigorous type.

'Pineapple Mint' *(Mentha suaveolens* 'Variegata') is a large-leafed white and green mint used as a garnish. These varieties cannot be grown from seed, so you must purchase stock plants.

PROPAGATION

Mint self-propagates by sending stolons — leafless horizontal runners — over or under the soil. It is very invasive; keep it in check or it will take over your entire growing area. These runners are also the easiest means of starting new plants. Pull up a runner from the top of the soil and cut it between the nodules. (Many of these nodules will already have roots on them.) Place these pieces of runner in soil and in a week or two you'll have new plants.

Mint can also be propagated by division of the plant and by stem cuttings. Both ways will set roots and produce new plants in only a week or so. By taking stem cuttings, you avoid any possible diseases that may be harbored in the soil on the runners.

Spearmint seed can sometimes be found in catalogs. It is not advisable to grow these as your main commercial crop: The best varieties of mint do not generally produce viable seed. If seeds do grow, they are not usually true to the breed type. Buy stock plants and propagate the amount you need to be guaranteed of having one true variety.

Growing Mint in the Greenhouse

Mint likes cooler growing conditions, so plant it on the north side of the greenhouse. Planting in raised beds will help curb its invasive tendencies. Eventually the runners will find a way out of the bed, however: either over the sides, through cracks in the side boards, or down through the soil and out the bottom. Pull these out when you find them.

Space mint plants 12 inches apart in any pattern. The space between the plants will eventually fill from new plants growing from the runners. The entire bed will become one mint plant. This is actually good, because spearmint requires dormancy once a year. The young plants growing from runners will be of harvestable size while the older plants take a rest.

In time the bed will become so crowded that air circulation is inhibited. This could lead to disease problems and makes a good hideout for insects. Renew the bed at least once a year. You should have two or more beds of mint so that one is being renewed while the other is available for harvest.

All the plants in the bed should be cut off a few inches from the soil. This is a tedious but fragrant job! Pull up all the runners and dig up the smallest plants. Remove dead leaves and debris from the soil surface.

If the soil mix is very heavy and compacted, you may want to renew it too. Loosen the soil between the plants and scratch in some amendments. The alternative is to dig up all the plants, renew the bed, and replant. However, mint doesn't seem to mind heavy soil. The soil in my mint beds was not changed for six years, and I had no problems.

Fertilize and water well at this time. The mint will grow back uniformly and with increased vigor. It should be ready to harvest in three weeks. Mint planted in beds, rather than in pots, does not require constant fertilizing. A full-strength feeding once a month should be sufficient. I don't advise growing mint in pots, but if you do fertilize it weekly.

Growing Mint Outdoors

Plant mint away from other herbs or gardens because of its invasive habit. Give it space by itself where it won't matter how much it spreads.

Each variety of mint should be planted by itself to prevent possible cross-breeding. I had good luck keeping mint in check by planting it in an old lawn area that had been tilled. The runners were not very successful at setting roots in the lawn edges.

Mint likes cooler temperatures and will grow in partial shade. The soil type doesn't seem to matter as long as the plants receive plenty of moisture

PESTS AND DISEASES

Whitefly and leafhoppers are major pests, and aphids will also attack mint if their other favorite foods are not available. The white marks these insects leave after feeding cause extreme cosmetic damage and make the leaves useless for garnish. If the mint is allowed to grow behind the beds, it makes a perfect place for insects to hide, especially whitefly.

Overcrowded mint beds in the greenhouse are susceptible to a number of fungal diseases, especially rusts.

HARVESTING

A single mint plant grown from a cutting will be ready to harvest in four to five weeks following transplanting in the summer. During the winter the time is longer, usually six or seven weeks. Mint should be ready to recut in a week to 10 days.

Yields, as usual, are determined by the age of the plant and the growing conditions. My greenhouse beds totaled 120 square feet of densely packed mint. From that I harvested 12 to 15 dozen bunches every week year-round. Some were picked from outside beds during the summer while the greenhouse mint rested.

When cutting, start at the south side of the bed and cut all the mint in the area, even if you must discard some. This will have the effect of renewing the beds and allow the plants to grow back uniformly. If the stems are of many different lengths, the mint must be cut individually, stem by stem. Look for tops that are perfect — no damage or discoloration.

Wash mint, if necessary, by swishing the bunch in a bucket of cool water. Give the bunch a good shake to remove as much water as possible. It can be packaged moist for restaurants if the top of the bag is left open.

PACKAGING

Place the bunches, stem ends down, in poly bags for restaurants and distributors. Close the top of the bag loosely unless the mint is moist.

This herb dries out quickly if it has too much postharvest exposure to air. After harvesting, refrigerate mint immediately, as it wilts quickly during warm weather.

Many restaurants use mint as a garnish. Do not offer it as precut garnish; sell it in bunches and let the restaurant cut the garnish as needed. The tops do not hold up well after being cut from the stem. An exception might be a special event for which the account requests precut garnish. In this case, cut the tops right before delivery and package them in sealed, rigid plastic containers with moist foam or paper towels in the bottom. The package must be kept cool.

Supermarket packages of mint usually weigh 1 ounce. The mint is cut and packaged loose with the stem ends down. Make sure mint is dry when packaged in trays.

oregano

Oregano is the mystery herb! There are many different varieties and flavors — the dried oregano found on supermarket shelves varies in flavor from jar to jar, even when packaged by the same company. Some of the dried oregano sold is not even oregano; there are several unrelated herbs that have the same type of flavor. Mexican oregano, a variety of coleus, is often sold dried as the real thing because of its abundant growth.

The ability of oregano to cross-pollinate with other members of the species accounts for the range of flavors and varieties. Despite these variations, the flavor of oregano is very much in demand, although in your area there may not be much call for it fresh cut.

oregano

Botanical Name:
Origanum spp.
Type of Plant: Perennial
Adult Height: to 15"

Uses

Oregano is a staple in many countries. It grows wild on the hillsides in Greece, where it is used in abundance. It is also popular in Mexican dishes and, of course, Italian cooking. It is routinely paired with basil in tomato and pasta meals. Oregano is known as the "pizza herb" in the United States, as that seems to be its most common use.

The fresh leaves are stripped from the stem and chopped or snipped. It can also be sold as precut garnish but take care to prevent it from wilting during harvest and storage.

Varieties

It may take some time to find the kind of oregano that you and your customers will be happy with. The flavor of some is harsh or too strong, and others are tasteless. Before buying any plants, taste the leaves to be sure the flavor is to your liking.

Most oregano does not come true to seed, despite statements to the contrary in seed catalogs. Many times plants grown by seed from the same package will be completely different in flavor and growth habit. *Origanum vulgare*, commonly sold by seed companies, is nearly tasteless — not a good choice for the commercial grower.

Confusion abounds when you look for oregano in seed and plant catalogs. Instead, buy stock plants from a reputable grower. Taste all plants to make sure they are the same exact variety. Buy enough plants at one time so that you can propagate as many as you need yourself.

The best-flavored types are white flowering, of which there are several. One white-flowering variety, *O. heracleoticum*, is called "true Greek oregano" by some and others call it winter sweet oregano or pot marjoram. You may be able to find seeds of this variety. Johnny's Selected Seed carries a good variety of Greek oregano seed. The seeds that I tried produced all the same plants with good flavor, although the germination rate was poor.

O. heracleoticum is the variety that I grow and it was well liked by my accounts. Its flavor is quite strong — it bites you back! Your customers may find this too strong, however.

Experiment with several types until you find the one that is best for you. Plant the different types in locations far away from each other (and marjoram) and keep careful notes on the botanical name, supplier, and customer acceptance. Don't let them flower. If bees or other insects cross-pollinate these plants with another oregano or with marjoram, the efforts of your experiments might be lost.

PROPAGATION

Don't bother to start oregano from seed unless you can obtain the white-flowering variety. If you do start your oregano this way, though, be sure to take out those plants that do not have the same taste and growth habits that are desired.

To start oregano from seed, first thoroughly wet the flat because the seeds are very tiny. Scatter the seeds in the flat, sowing more than you will need to compensate for the low germination rate. Oregano seeds need light, so don't cover the seed. Cover the flat with clear plastic and keep the flat at 60°F.

The seeds should germinate in 7 to 14 days. After germination, place the flat where it will receive full sun. Transplant into pots when the seedlings have three sets of true leaves.

Plants can also be started by stem cuttings or root division. Both methods will give lots of new plants.

GROWING OREGANO IN THE GREENHOUSE

Sun- and heat-loving oregano should be planted in the middle or south side of the greenhouse. Space plants at least 18 inches apart, using the odd/even pattern.

Oregano spreads (depending on the variety) rapidly. Stems that are allowed to lie on the soil for long will take root and grow new plants. If you wish to keep your oregano beds tidy, pull up these long stems and cut them off. Stems that grow too long become woody with age. They don't always grow straight, and they make unattractive bunches, so keep them cut.

After several cuttings, new growth will come from the crown of the plant. At this time, cut off all of the old woody stems as close as possible to the new growth. This will give you all fresh growth and make bunching easier.

As with most perennials, oregano does not need frequent fertilizing. During the spring and summer, when growth is rapid, the plants will benefit from fertilizing every two weeks. Once a month during low light periods is sufficient.

Growing Oregano Outdoors

Plant oregano in regular garden soil in full sun. Allow 24 inches between plants to give them room to spread. It is even more important not to let oregano flower when planted outdoors. Bees can travel a long distance in search of nectar. If they cross-pollinate your oregano, your best culinary variety could be lost the following season.

Most oregano varieties will survive the coldest winters with good protection. Before the first frost, cut back all the plants to ground level. After three hard frosts, mulch the plants with a thick covering of straw or hay. You may want to protect the rosettes of leaves at ground level with a small wooden box or something of similar size and weight to keep the mulch from lying directly on the plant. Gradually pull the mulch off the plants as the weather begins to warm in the spring.

Pests and Diseases

Whitefly, spider mites, leaf miners, and leafhoppers are attracted to oregano. These should be controlled because, like marjoram, oregano leaves tear easily from the stems when washed.

Fungal diseases can be a problem if the clumps are allowed to grow too thickly. Be sure to keep the plants trimmed and don't let the beds become overcrowded. Cut them back periodically if you are not harvesting enough to provide good air circulation to all the plants.

Harvesting

This herb is not a rapid grower and it may take three to four months before multiple harvests can be taken from one plant. Trim the stems on young plants often to encourage new growth at the crown. Always leave green leaves on the stems when cutting; take only half the stems. Some of the old hard stems will not regrow after harvesting.

If the plant has been cut uniformly, harvesting is easy. Grasp the bunch size needed, cut, and rubber band it together. Bunches should contain 15 to 20 stems, each 8 to 10 inches long. More are needed if the stems are short. Do not include flowering stems. Cut these at crown level and discard them, unless an account requests them for a garnish.

Give the bunch a gentle shake to remove debris and dead leaves. If you must wash oregano, do so by very gently swishing the bunch in a bucket of cool water. It is tender, the leaves bruise easily, and many of the older leaves will fall off the stems with the slightest motion in water.

Precut garnish is sometimes requested. Cut only the tips of the plants, usually 3 to 4 inches long. Each one should be perfect — no insect damage or yellowed or torn leaves.

If you are cutting garnish during warm weather, immerse the cuttings in cool water before counting or packaging. When cutting from outdoors, where the plants may be dirty, wash them gently in cool water.

Packaging

Restaurant packages are the same as for other herbs. Put the bunches, stem ends down, in poly bags. Leave the top of each bag open slightly.

Package garnish according to account request, usually 100 per bag. Place the garnish in the bag with the stem ends down. Use gusseted bags so the tops of the garnish are not crushed when the bag is closed. When packaged moist, leave the top of the bag open to allow the garnish to dry.

Supermarket packages weight ¾ to 1 ounce. Do not rubber band the bunches together. Cut only the new growth and package them with the stem ends down. After packaging, oregano should be cooled immediately.

parsley

Most people are familiar with curly parsley. It is widely used as a plate garnish in restaurants throughout the country, and can be found in most supermarket produce sections unpackaged and sold by the bunch. Curly parsley is also used by cooks and chefs to flavor food, so it remains one of the most popular herbs.

It is abundant in the market-place and easy to grow; thus, the prices paid for curly parsley are quite low. It is usually sold by distributors in cases of 30 or 60 bunches. Restaurants and super-markets pay anywhere from $6 to $9 for a 30-bunch case.

It may not be profitable for you to grow curly parsley in the greenhouse during the winter, when space is at a premium. If you have the space outdoors, grow as much as you can in spring and summer.

parsley, flat-leaf

Botanical Name: *Petroselinum crispum neapolitanum*
Type of Plant: Biennial
Adult Height: to 12"

The curly parsley sold by distributors is usually shipped from a distance, and the quality and freshness can be quite poor. Your parsley should be superior to theirs, and you should be able to charge a higher price.

Flat-leaf parsley is also known as Italian parsley. This is much more in demand as a cooking ingredient, as its flavor is more pronounced. Chefs prefer it, and many order it continuously. It is gaining popularity with home gardeners and cooks as well. The prices are somewhat higher for Italian parsley. Grow it in the greenhouse over the winter if you have a market for it.

parsley, curly

Botanical Name: *Petroselinum crispum crispum*
Type of Plant: Biennial
Adult Height: to 10"

USES

The sprigs of curly parsley are commonly used as a plate garnish. It is also chopped and sprinkled on food to add color. Both curly and flat-leaf parsley are flavor blenders; that is, they are most often used in combination with other herbs. Parsley is used in every kind of cooking, from sauces to soups, and is particularly wonderful eaten fresh in salads.

VARIETIES

There are many varieties of curly parsley today, and always more being developed. Many seed companies have different names for the same types. Look for those labeled "triple curled." The types with long stiff stems are easier to bunch, although the stems may be brittle and break easily. The longer stems also help to keep the parsley clean.

Parsley labeled "dwarf" or "compact" is not a good choice because both are closer to the ground and become soiled easier. They are also harder to bunch because of the short stems. Some other qualities to look for in seed catalog descriptions are deep color, resistance to bleaching or yellow leaves, and holding quality after harvest.

Italian parsley is a more vigorous grower and the stems are longer than those of the curly type. There are only a few varieties of Italian parsley. Look for deep green color, long stems, and holding quality. 'Gigante Catalogno', available from Johnny's Seeds, has large leaves, stiff stems, and good taste.

PROPAGATION

Both types of parsley are grown from seed. Purchase seed fresh each year because the germination rate and seedling vigor diminish with older seed.

Parsley seed can take up to three weeks to germinate. Many packets recommend soaking the seed in water before planting, but this is not necessary. While this process does speed up the germination time, it also makes the seeds difficult to handle, as they stick to everything. If the seeds are kept moist in the flat and bottom heat is provided, germination usually occurs in two weeks. Italian parsley germinates a few days earlier than does curly parsley.

Plant seed six weeks before your last frost-free date. Parsley can take some frost but if the seedlings are too young and delicate, they may perish in a hard frost. Sprinkle the seeds thinly in a premoistened flat. Cover with fine sand ¼ inch deep. Cover the flat with plastic and then a dark cover. Provide bottom heat at 65°F.

Remove the coverings as soon as the first seedlings emerge. Some seeds will germinate later. Place the flat in a cool location with full sun. Fertilize the seedlings, from below, with half-strength fertilizer once a week until transplanted.

Transplant seedlings when they have no more than two sets of true leaves. Parsley has a long taproot, or main center root, and does not transplant well if it is allowed to remain too long in pots or the flat. When the roots become crowded, the plants may bolt prematurely after transplanting.

If you intend to grow parsley in the greenhouse during winter, start it five weeks before transplanting to greenhouse beds.

GROWING PARSLEY IN THE GREENHOUSE

Parsley prefers cooler temperatures and moist conditions. It does not grow well in the greenhouse during the heat of summer. For a winter crop, transplant the seedlings five weeks before you want to harvest your first crop. The first greenhouse harvest should coincide with late fall in your area, when the outdoor parsley slows its growth. This fall planting should give you a continuous supply through the winter.

Plant parsley on the north side of the greenhouse. Space the plants 10 inches apart using the odd/even pattern. The soil mix in their beds can be a little heavier (contain more soil) than other beds because this herb grows well in moister soil. Feed parsley every two weeks with full-strength fertilizer.

Because it is a biennial, parsley will flower, produce seeds, and die early in its second season of growth in normal outdoor conditions. Italian parsley planted in the fall in the greenhouse may flower by spring. This probably occurs because of the constant fertilizing and harvesting.

As the plants mature, a center stem grows, upon which flowers and seeds will form. Cut off this stem at soil level. The plant will produce a few more leaves before trying again to flower. When this occurs, pull the plants, renew and rest the beds, and start with new plants outdoors.

Growing Parsley Outdoors

The parsley beds should be well prepared with plenty of organic matter. Plant the seedlings outdoors a few weeks before your last frost-free date, unless the weather is extremely cold. Plant the seedlings 10 inches apart in rows. Weed the beds carefully when seedlings are small.

Water deeply when the beds are dry. Parsley is a heavy feeder and should be fertilized after harvesting unless the soil is very rich. Curly parsley grows low to the ground and gets muddy after a heavy rain. A thin mulch spread around the plants will help to keep them clean.

In fall, the growth will stop and the plants appear to die back. Italian parsley can take more frost and can be harvested later than the curly variety. The plants will grow again in early spring. You should be able to get one or two cuttings before the plants begin to flower. By this time, the new plants that you set out will be ready to harvest when last year's plants are finished.

Pests and Diseases

In the greenhouse, parsley sometimes attracts aphids and spider mites. Outdoors the usual array of insects can be found on parsley, including caterpillars and worms. Many of them are beneficial — spiders, for example. Insects particularly like to hide in the many folds of curly parsley. Wash the bunches to remove insects before packaging.

Parsley is susceptible to a number of diseases including gray mold *(Botrytis)*, downy mildew, bacterial soft rot, root and crown rots, various mosaic viruses, and root knot nematodes.

HARVESTING

The first leaves should be ready to pick three to four weeks after transplanting. Wait until the leaves are big enough to bunch before harvesting. You should be able to harvest from each curly parsley plant at least six or eight times in one season. Expect more harvests from Italian parsley, which grows more vigorously.

Parsley stems should be pulled from the plant rather than cut unless you take the entire plant as one bunch. Cut stem ends left on the base of the plant tend to rot, and this invites disease. It is best to take the outermost leaves and allow the crown to continue growing.

Grasp the leaf with your finger and follow the inside of the stem down to the base. Push outward with your finger using a gentle rocking motion and the stem will separate easily from the plant. You may have to steady the plant with your other hand to prevent damaging it.

Continue picking stems until you have enough for a bunch, then rubber band them together. Cut off the bottom of the stems evenly. The number of stems in a bunch will vary with the size of the leaves. Some varieties of the curly type have very large, almost top-heavy leaves. You will be able to fit only six or seven stems of this type in a bunch. When bundling this type of parsley, place the stems together in a circle. This makes a beautiful round bunch almost like a miniature maple tree. With some other types of curly parsley, and later in the season, when growth slows, more stems will be needed to make a bunch. It may take 12 to 15 stems to make a good-sized bundle.

Pick Italian parsley as you would curly parsley. The leaves are not as big, so each bunch should contain 12 to 15 stems.

When picking either types, be sure that there are no yellow or pale green leaves in the bunch. Pale-colored leaves will quickly turn yellow after harvesting. The flat-leafed types are more prone to this yellowing.

It may be necessary to wash the bunches to remove insects and mud. Grab the bunch close to the leaves and "dunk" the bunch in cool water.

Remember that the stems can be brittle and break, so hold on tightly to the stems while washing. Give the bunch a sharp shake to remove as much water as possible. Lay the bunches out of the sun to dry slightly before you package them.

Packaging

Refrigerate parsley as soon as possible after harvesting. It may be packaged slightly moist. Bunches for restaurants and distributors are usually packaged in poly bags. Leave the top of the bag open slightly to allow the herbs to breathe. If you are selling 30- or 60-bunch cases to distributors, use wax-lined cardboard boxes. Some distributors will allow you to bring the parsley in poly bags if they will repackage it in their containers. It is not advisable to sell parsley as precut garnish; it doesn't hold up well after being cut from the stem.

Some supermarkets display Italian parsley bunches unpackaged; others want it packaged in trays. Trays should weigh 1 ounce. The stems will have to be removed to fit the parsley into the package.

rosemary

Rosemary is the true treasure of herbs. It is highly prized by people around the world for its medicinal, culinary, decorative, and cosmetic uses. Rosemary has long been used as an antiseptic and astringent, and is used in cosmetics for its beneficial effect on skin and hair (especially dark hair). This herb is a symbol of loyalty and ceremony — hence the phrase "rosemary for remembrance."

It grows on single stems and its leaves resemble pine needles. The plant looks like a miniature pine tree and is often pruned and sold as a small Christmas tree. It is also used extensively in topiary. Rosemary is a favored bonsai plant.

This herb is extremely fragrant, and the aroma will cling to your clothing and hands for a very long time. The leaves release

rosemary

Botanical Name:
Rosmarinus officinalis
Type of Plant: Tender perennial
Adult Height: 36" to 60"

a sticky resin when handled that is difficult to remove, much like pine pitch. Its flavor is strong and quite unlike that of any other herb. When chopped and added fresh to food, use it sparingly.

Rosemary is a slow grower and takes considerable time to achieve a size useful to the commercial grower. If you are starting in business with small rosemary plants, you may have to buy in fresh-cut rosemary to supply your accounts until your plants are big enough to cut from repeatedly.

USES

Rosemary has long been a popular culinary herb in Europe. It is used extensively with lamb in Italy and Greece. It is gaining fans here for everything from bread to stir fry. It has an affinity for poultry and tomatoes, and is wonderful paired with garlic.

Rosemary can be used in so many ways. Strip the leaves from the stems and chop or leave them whole before adding to food. Sprigs can be used whole to add flavor in poultry and meat dishes. The long woody stems, bare or with leaves, can be used for skewers on which to broil or grill foods. Place the thicker stems on top of hot coals before grilling to impart a wonderful flavor to your feast.

If stored properly, rosemary holds its quality for a long time after cutting. It also holds better than any other herb on a hot plate. Because of this, the most profitable rosemary use for me as a commercial grower has been selling it as a garnish. I sold several thousand rosemary garnish a week for eight years! On one occasion I sold 3,000 for a special event.

VARIETIES

There are many varieties of rosemary. They are generally grouped into two classes based on growth patterns: upright and sprawling (prostrate). The upright varieties are best for the fresh-cut-herb grower because they take up less space, are easier to harvest, and make more attractive bunches and garnishes.

Prostrate rosemary *(Rosmarinus officinalis prostratus)* is good for hanging baskets, but its flavor and color are not as intense as other types. It is quite tender and the leaves are smaller than the upright types. For those who want to grow this herb outdoors in areas that have only light frosts, *R. officinalis* 'Arp' may be the best variety as it can withstand more frost than other types.

The most productive variety for me is *R. officinalis foresteri*. This is sometimes seen in plant catalogs as 'Forest Rosemary' or 'Lockwood de Forest'. This type has beautiful green foliage and is a vigorous grower. Some catalogs list this variety as having a creeping habit but for me it has always grown upright and tall. I purchased my stock plants from Bluebird Nursery (see Resources).

PROPAGATION

You can grow this herb from seed, and regular rosemary seed, *R. officinalis*, is available from catalogs. This type has rather shrubby growth with dark green foliage. Because it tends to grow to the side rather than upright, it is difficult to harvest straight tops for garnish and long, straight stems for skewers.

Rosemary seed is slow to germinate and the rate is low and uneven. Buy a number of plants of the variety you want and propagate from stem cuttings for the stock plants that you need. Take the cuttings from new tip growth. The use of rooting powder or liquid will increase the percentage of cuttings that set roots.

GROWING ROSEMARY IN THE GREENHOUSE

If you live in a climate that freezes during the winter, take care with this herb — it will not survive much frost. If it is not grown in a heated greenhouse all year, it must be grown in large pots (3-gallon size or larger) that can be brought into a heated building during cold weather. It takes at least two years of continuous growth for a rosemary plant to grow big enough to be a strong producer.

Plant rosemary in the middle or on the south side of the greenhouse. This herb prefers a dry alkaline (slightly higher pH) soil. The beds should be well prepared with a medium/light soil mix — the plants will stay in these beds for a long time. The roots grow deep and wide and the plants don't transplant well as they age and grow large.

Space the plants at least 36 inches apart using the odd/even pattern. This may seem like a big distance between plants. However, that space between the plants will fill in; as the plants mature, and you cut them often, they become very bushy. In time the entire bed will be filled in with long (4 feet tall), bushy rosemary. This arrangement can present a problem when watering and fertilizing the beds, however.

Watering. When watering overhead, it can take a long time for the water to reach the soil and for the thick foliage to dry out. To alleviate this problem, use soaker hoses to water the soil rather than the foliage. After the rosemary is planted and while plants are still small, lay soaker hoses in the beds with the hose holes facing the soil. Use enough hose so that it winds up and down on each side of the plants. Hang the connection end of the hose just over the end of the bed to make attachment easier.

Turn on the water at a fast rate at first (not too hard or the hose may rupture). This fills the hose up to the end quickly. Then decrease the rate of flow so that the water seeps out rather than spraying. Let the water run for at least a half hour to ensure that the roots get a sufficient drink. A word of caution: Don't forget that the hose is running. Many times I have forgotten and found the floor of the greenhouse to be at a minor flood stage!

This arrangement is not only better for the plants but is also a time-saver for you. The rosemary beds will be watered while you do something else. On hot days, when there is plenty of air movement in the greenhouse, water the rosemary from above occasionally. This helps to keep the foliage clean and disturbs insects that may be hiding in the foliage.

Fertilizing. Fertilize the plants at least once a month, and every two weeks during periods of rapid growth in spring. Remember that the soil mix is light and it doesn't have much capacity to hold moisture or nutrients.

Just as in watering large rosemary plants, fertilizing may also present a problem. One solution is to use a hose-end fertilizer mixer that has a removable spray-end nozzle so the water comes out in a stream. Spray the fertilizer at soil level using a slightly higher saturation rate than that called for in the instructions. Let it soak in awhile before turning on the soaker hose.

There are several other ways to be sure that the plants are getting needed nutrients. Spread dry organic (low N-P-K) or timed-release (such as Osmocote) fertilizer onto the soil surface and scratch it into the soil. Each time you turn on the hoses, the rosemary will be fed. These kinds of fertilizers must be replaced periodically.

Pruning. As the plants mature and after many cuttings, the beds will become woody, bushy, and overcrowded. Fungal disease problems may occur because of low air circulation. It also provides a good habitat for insects. Prune and cut back your plants periodically. The best time to do this pruning is early in spring, before new growth has started.

Use care when pruning rosemary. If too much of the woody stems are cut, the plant may not recover and die. Never cut back more than half the plant at a time. Always leave green leaves on the stems that you are cutting unless you do not want that stem to grow back.

Some plants can be grabbed in the middle and cut all at one time with a pruning shears. You may have to use a pruning saw for some of the thicker stems. Do this pruning plant by plant and bed by bed. Try to cut one whole bed at a time to let it regrow at an even rate. Cut only one bed at a time so you will have beds to harvest from while the other one grows back.

Some stems grow from the base trunk and are long, thin, and flexible; cut these off at the trunk. These long stems can be woven or twisted into wreaths while they are still fresh and malleable. The short trimmings can also be use to make mini-wreaths. If the stems are thicker (¼ inch or so), set them aside to dry and then sell them as rosemary skewers. Keep all the cut rosemary, as all of it can be used for something.

Yield. As previously mentioned, it takes a year or two before a rosemary plant provides any real yields. An occasional long stem can be taken from young plants. Each cutting made from young plants serves to speed up the end result by helping the plant to branch out.

Because it takes so long for rosemary to grow to good yield size, you may initially have a great many plants. As they mature, you'll probably find that you have many more plants and beds than you need. Dig up the extra mature plants and transplant them into large (3-gallon or larger) pots. They will not always survive this move, but prune those that do and sell as rosemary (or Christmas) trees at a handsome price.

GROWING ROSEMARY OUTDOORS

Growing rosemary outdoors is basically the same as growing it in the greenhouse. Plant it in a sunny location in light soil. It will survive outdoors in areas that receive only an occasional light frost. However, do mulch them and give good protection over the winter. I have heard stories of rosemary surviving very cold winters, but exercise caution. It will not grow during this time period, however. If you wish to supply your accounts year-round, you must bring the plants indoors or buy it in fresh cut to repackage.

The rosemary plants will bloom in spring with delicate blue or lavender flowers. The flowers seem to ooze the same resin that is in the leaves. They are very attractive to aphids, spider mites, and other insects. Soon you may find sticky honeydew (the excrement from insects) around the flowers and on the stems below the flowers. This honeydew provides an excellent place for black sooty mold fungus to grow. Once the mold has attached itself to the leaves, it is nearly impossible to wash off.

PESTS AND DISEASES

Rosemary appeals to those pests that like strong flavors. Leafhoppers, spider mites, and sometimes whitefly can be found on this herb. Black or green aphids occasionally are attracted to the young, tender, new growth tips. All of these insects must be controlled, especially the ones that cause the worst cosmetic damage — leafhoppers and spider mites. If the foliage is marred by white marks, it is useless as a garnish.

Watch the flowering rosemary closely. If you see any signs of insects or honeydew, cut the flowering stems off immediately and burn them. The plant may continue its mission to flower and must be monitored.

Fungal diseases, especially downy and powdery mildew, are a serious problem for rosemary. These must be controlled at the first signs of infection, because they can rapidly attack crowded beds and render all plants in them unsalable. Good air circulation and low humidity are the first line of defense against these diseases.

HARVESTING

The first thing to know about cutting rosemary is that your hands and scissors will become coated with sticky resin in a very short time. Wear rubber gloves when working with this herb; this resin is very hard to remove from skin. Some people even have an allergic reaction to it. The only thing I have found to get rid of it contains harsh chemicals. Each one of your rosemary pickers should have his own set of marked rubber (or cotton) gloves and a long-sleeved shirt!

When harvesting rosemary for any purpose, try to pick everything from one bed until the bed is picked clean. Then move to the next bed.

This has a pruning effect and each bed will have new growth uniformly.

Rosemary bunches should contain 12 to 14 stems at least 8 inches long. More stems are needed if they are shorter. If the bunches are for culinary use only (the leaves stripped from the stems and chopped up), the stems need not contain the new-growth tips. It is easier to strip off the leaves when the stems are more mature and thicker.

The length, width, and dryness of rosemary skewers should be determined by your accounts, depending on how they use them. They are usually straight, dried, with leaves, and at least ¼ inch in thickness. Some chefs prefer a slight curve to the skewer to enhance its presentation on the plate.

Rosemary garnish should be the top of the stem with new-growth tips. Do not use stems that have been cut or are branching. The tops that have matured a bit hold better than will soft new growth. Each garnish should be perfect, with no yellowing or insect damage. Garnish length is dictated by your accounts but is usually 3 or 4 inches long. Anything longer is in the "sprig" category, and you should adjust prices accordingly.

Cutting large amounts of rosemary garnish in a short time takes a bit of practice. I was able to pick 100 rosemary garnish in just 4 minutes! This was, of course, under ideal growing conditions and with years of practice. The technique that I used is described in chapter 13, Harvesting, Handling & Packaging.

PACKAGING

Rosemary holds fairly well after picking if it is kept cold. It may even be packaged slightly wet, with the bag open a bit, and it will not rot as quickly as will other herbs.

Restaurant packages of rosemary are the same as for other herbs. Place the bunches in the poly bag, stem ends down. Close the bag unless the rosemary is moist. Supermarket packages usually weigh ½ ounce. Cut the sprigs to fit the length of the tray. Rosemary should be packaged dry with the stem ends down.

sage

Sage is a respected ancient herb used for culinary, medicinal, cosmetic, and decorative purposes worldwide. The astringent properties of sage have been proven by scientific studies. In cosmetics, it is said to have a stimulating and cleansing effect to the scalp and hair. Sage was reputed to provide a person with improved health and a long life. Today it does not seem to be valued as much, either by cooks or by herbalists, but it is certainly a very pretty and decorative herb, especially the various colored types.

There are hundreds of varieties of *Salvia*. Most are flowering ornamentals, but some are ornamental culinary herbs as well. Garden sage is a handsome plant with oval gray-green leaves and attractive lavender flowers. Its flavor is strong and fresh.

sage

Botanical Name:
Salvia officinalis
Type of Plant: Perennial
Adult Height: to 36"

CULINARY USES

Many people use sage dried, which is unfortunate because the flavor of it fresh is far superior. As an ingredient in the Thanksgiving turkey stuffing, it is legendary; for some, that may be the only time it is used all year.

Sage is good with all poultry, breads, sausages, cheese, and herb butters. Many chefs are creating new ways to make use of this refreshing flavor. One chef whom I know places fresh whole sage leaves on top of pizza before baking. When crisp, the leaves impart a wonderful flavor to the pizza.

The thick leaves are usually stripped from the stem and chopped fine. The tops or whole leaves can also be used as a garnish. There are three varieties of colorful sage that have gained wide acceptance in my market area as garnish.

VARIETIES

For cooking, the regular garden sage is the best. 'Holt's Mammoth Sage' is a steady producer. The leaves are larger and have good culinary taste. It can be propagated only by cuttings. Dwarf sage, *Salvia officinalis* 'Nana', has good flavor but is not a good choice for the commercial grower. In this instance, bigger is better.

The flavor of the colorful sages is not as intense as that of garden sage. They make excellent garnishes, however, and their holding quality after cutting is better. All of the colorful or fruit-flavored sages, although vigorous growers, are not hardy and will not withstand frost.

'Golden sage', *S. aurea,* has green with yellow leaves. *S. officinalis* 'Tricolor' has green, gold (sometimes white), and purple leaves and is definitely one of the most attractive for a garnish. 'Purple sage', *S. officinalis* 'Purpurea', is also striking as a garnish, since the new-growth leaves are deep purple.

Pineapple-scented sage, *S. elegans,* has a rich pineapple aroma and flavor. It does not hold well after cutting for use as a garnish. The scarlet, trumpet-shaped flowers, however, are very sweet, and are excellent in salad herb mixes. Look inside the flowers carefully before using them because insects love them too. The plant flowers late in the season, so start plants early in the greenhouse. Pineapple sage roots readily from stem cuttings.

PROPAGATION

Start sage from seed, by stem cuttings, or by layering. As with most perennial herbs, this one can be difficult to start from seed. Begin with a premoistened flat. Scatter the seeds thinly and cover them ¼ to ½ inch deep. Cover the flat with plastic and then a dark cover. Provide bottom heat of 70°F. The seeds should germinate in 7 to 21 days, but expect a low germination rate. When the seedlings have two sets of true leaves, transplant them into pots or the greenhouse beds.

A quicker and more dependable way to get the needed number of stock plants is to propagate by stem cuttings. This is particularly easy if you dip cuttings in a rooting hormone. Sage is another herb that has variations in flavor and leaf color, so propagating your own from stem cuttings will ensure that all of your stock plants are the same. The colorful and fruit-flavored sages must be started from cuttings.

Growing Sage in the Greenhouse

Before committing large amounts of greenhouse space to sage, you should determine what the market for this herb will be. In some areas the demand will be quite low; in others, it could be a big seller.

Plant sage in the middle or south side of the greenhouse. The soil mix should be medium light. Sage grows better in dry soil but the mix should be dense enough to support heavy plants.

Space the plants 24 inches apart in the odd/even pattern. Sage grows outward as much as upward, so avoid overcrowding. Three-year-old plants could easily reach 36 inches tall by 36 inches wide if left unpruned.

Sage also can be planted in large (3-gallon) pots, but production then is limited. The colorful sages are more suited to container growing. They are shorter, more rounded, and they don't sprawl quite so much. Container growing these types causes leaves to be smaller, making them more useful as garnish.

Fertilize sage in beds every two weeks during periods of rapid growth. At other times, once a month should be sufficient. Fertilize container-grown sage weekly.

Sage flowers early in the growing season. This flowering time may be erratic with sage grown in the greenhouse year-round, however. Cut off the flowering stems as they form. This will stimulate leaf growth. Eventually, the plant should be allowed to flower, because leaf growth stops at this time and the plant will put all of its energy into flowering. When flowering is nearly completed, prune back the plants. Don't cut back all of the plants at the same time.

Cut the stems back by a third. Cutting above a leaf junction stimulates new growth and causes the plant to branch. You may want to cut off many of the lateral stems at the main trunk. This helps increase air circulation in the bed. Pruning this way may cause new growth to come from the base of the plant. If this happens, allow it to grow to several inches and then cut off the main old-growth stems. Fertilize well after pruning — the sage will grow back with vigor.

Do very little pruning, if any, in the fall if the greenhouse is allowed to freeze during winter. The new growth, and possibly the whole plant, may perish in the cold. Prune the old growth when the plants begin to grow in early spring.

Growing Sage Outdoors

Do not plant sage seedlings outdoors until all danger of frost is past. If the plants have been hardened off, they can go into the garden a few weeks earlier. Space them 24 to 30 inches apart in rows. Plant in the odd/even pattern in wide rows if you will be weeding by hand. Sage can grow quite big, and reaching the middle of wide rows for weeding and harvesting could be difficult after the first year of growth.

Always plant sage in soil that has good drainage. Take care not to plant it near other herbs that require a lot of watering; excessive moisture promotes fungal diseases. Because of this, sage is tricky to grow in hot, humid areas of the country.

In northern climates, sage is hardy and should survive harsh winter conditions. After three or four hard frosts, mulch the plants with a thick layer of straw or hay. Gradually remove the mulch as the weather warms in late winter or early spring. As growth resumes, you may want to cut back any dead branches. A side-dressing of fertilizer or compost will help the plant to grow back vigorously.

Pests and Diseases

Whitefly, spider mites, and leafhoppers love sage. Leafhoppers especially like the plant and can do a great deal of cosmetic damage. Aphids will sometimes attack the soft, new growth. As always, control any insects: Sage does not hold up well to strong washing.

This herb is susceptible to many fungal infections. A plant may survive a fungal disease but yields will be severely curtailed. Keep the growing conditions dry and don't let the beds become overcrowded. Allow the beds to dry before rewatering.

Harvesting

In the first year of growth, the plant may not reach harvestable size for 10 or 12 weeks. This period will be even longer if the plant was started from seed. In the next year (from outdoor-grown plants), they will be ready to cut four weeks after new growth appears in the spring. Greenhouse plants will give you continuous cuts after they reach harvestable size except during the flowering stage.

Sage bunches should contain 10 to 12 stems (fewer if the stems are very bushy) 7 to 8 inches long. There should be no yellow or brown on the lower leaves. Do not include any flowering stems.

These bunches usually have to be cut stem by stem because sage does not grow to a uniform height. Cut the stems just above a leaf junction to encourage branching. This should leave enough stem on the cut branches to place the rubber band.

If you must wash sage, do so very gently by swishing the bunches in cool water. The lower leaves may come off the stem, so be careful. Gently shake out excess water and lay the cuttings on a towel or other absorbent surface in a cool area to dry before packaging.

Garnish of the colored sages should be cut individually. The tops are very full and could be crushed if more than a few are held in the hand at the same time. The garnish is usually 2 inches long and contains at least the top four leaves.

Packaging

Package bunches in poly bags with the stem ends down for restaurant and distributor accounts. Garnish should be placed very loosely in the bag. Don't bother to try to put all of the stem ends down in the bag — they'll probably be crushed this way. Try not to let the garnish pack down, and don't put more than 100 in a bag.

Supermarket packages of sage usually weigh ¾ to 1 ounce. Use only the new-growth tips cut to fit the tray. Package with the stem ends down. Sage leaves sometimes curl inward, so be sure to check inside for insects.

sorrel

This herb has long been a culinary staple in Europe, especially in Russia and France. In the early days it was used for medicinal purposes and as a meat tenderizer. Scientific research has shown that compounds in sorrel have antiseptic and laxative effects. Sorrel contains oxalic acid, extremely large amounts of which can be toxic, especially to those in ill health.

The tart, lemony flavor of sorrel is not appreciated in much of this country, although there are some devotees in most areas. This is unfortunate because sorrel is easy to grow and has so many uses. If sorrel is not popular in your area, perhaps you can generate some interest in it with free samples and recipes.

If you do not have a large demand for sorrel, start with only six or eight plants. The

sorrel

Botanical Name: *Rumex scutatus* or *R. acetosa*
Type of Plant: Perennial
Adult Height: 18"

plants grow larger and spread somewhat, and the beds will eventually fill in.

CULINARY USES

Sorrel grows from a single crown, much like spinach and loose-leaf lettuce. Its broad leaves are used in much the same way as lettuce. Use the leaves — minus the tough stem — fresh. Sorrel does not retain its flavor when dried.

It is a wonderful addition to salads, sandwiches, and omelets. Wrap meats and seafood in the leaves before broiling for a tenderizing effect. Sorrel is widely used in sauces and butters with seafood. Perhaps the most popular use is in the wonderful cream of sorrel soup.

Sorrel is not suitable as a garnish because it wilts quickly after harvest when exposed to warm temperatures. The small leaves, however, can be used whole to adorn salads and cold foods.

VARIETIES

French sorrel, *Rumex scutatus,* with its arrow-shaped leaves, is the variety preferred by chefs. This type is said to contain less acid and is better for use fresh. Most people, myself included, find that the other main culinary type of sorrel, *R. acetosa,* is just as good in all respects.

PROPAGATION

Sorrel can be started from seed or by root division. The seeds germinate readily and the plant will self-sow if allowed to go to seed. Start seed outdoors in beds after the danger of frost has past. Plant ¼ inch deep in rows 3 feet apart. Cover and keep moist until germination, which outdoors is usually in 10 to 14 days. Thin the plants to 6 to 8 inches apart.

You may prefer to start the seed indoors and then transplant seedlings to beds. Sprinkle the seeds thinly in a premoistened flat. Cover them with fine sand to ¼ inch deep. Cover the flat with clear plastic and provide bottom heat between 65° and 70°F. The seeds should germinate in 5 to 10 days.

When the plants are 2 to 3 inches high and have at least three leaves, they are ready for transplanting. During the summer they will be ready to transplant in three to four weeks after germination; in winter, this time will be extended to five to six weeks.

Growing Sorrel in the Greenhouse

Sorrel is a cool-weather herb that does well in partial shade, so plant it on the north side of the greenhouse. Space the plants 12 inches apart using the odd/even pattern. The plants will grow larger and fill in any empty areas. The soil mix should be medium to heavy, with good drainage. Sorrel also can be planted in large pots. Because this herb has a strong, deep root system, growth from container sorrel will be limited.

Sorrel beds must be weeded and cleaned regularly. As the plants mature, the large outer leaves have a tendency to fall over. If these are left on top of the soil, they will begin to rot, resulting in a slimy mess. This also provides a fine breeding ground for fungal diseases and insects. The large, older leaves should be removed to encourage new growth from the crown and to keep the beds clean.

If the beds become overgrown, some of the plants can be cut off right at the crown. The plant will grow back with tender new leaves that will be ready to harvest in a month. In order to maintain a continuous supply, don't cut back all of the plants at the same time.

Fertilize the beds once a month during low light periods and every two weeks in spring and summer. Container-grown sorrel should be fertilized every two weeks year-round.

During the summer, the long (3 to 4 feet) flower stalks begin to grow. Cut off the stalks where they grow from the crown. They can sometimes be pulled out with a strong tug while you stabilize the plant with your other hand. After the flower stalks are removed, the plant will continue to grow new leaves.

Sorrel beds may become thick and overcrowded after several years. Dig up the smallest plants and replant them elsewhere. The larger plants should be dug up and divided.

Growing Sorrel Outdoors

Plant sorrel in any well-drained garden soil in a sunny location. French sorrel (R. scutatus) prefers drier and sunnier conditions than does R. acetosa. Make sure these beds are where you will want them to be for a long time. It can be difficult to eradicate this herb — I have on occasion deep-tilled sorrel beds, only to have them reappear with renewed vigor!

This herb requires little care when grown outdoors. The beds should be cleaned of fallen leaves and weeded periodically. Remove the flower stalks when they form; if it is allowed to produce seeds, the beds and adjacent areas will be full of little sorrel plants the next season.

In the fall the leaves and stems take on a reddish hue and become a little thicker. This does not affect the flavor or production. Sorrel is hardy and can take light frosts. It will be one of the first herbs to appear in the spring.

Pests and Diseases

Aphids are attracted to sorrel and can be found on the back side of the leaves, usually along the stem. Slugs and snails are also fond of sorrel but they are usually more of a problem outdoors. Greenhouse beds overcrowded with rotting leaves on the soil are an invitation to slugs. Flea beetles will sometimes eat sorrel, and the small holes they create in the leaves may be mistaken for slug damage.

Sorrel leaves are tender and brittle, and they also bruise easily. It is quite common to tear a leaf while trying to remove insects with your fingers. Washing the leaves, by dunking or swishing in water, will often tear or bruise the foliage or break the stems. This is why it is important to control insects on this herb.

Sorrel is not particularly prone to fungal diseases unless leaves are allowed to rot on the soil surface; this may pave the way for bacterial soft rot to get a foothold.

Harvesting

Sorrel should be ready to harvest four weeks after transplanting. You can expect to harvest continually from the plants without a slowdown in growth. Harvesting is usually done stem by stem because the leaves do not grow to a uniform height. The entire plant can be cut at one time and bunched, but that is rather unattractive. If done this way, shake out the small leaves before you rubber band.

When harvesting sorrel for bunches, look for the larger outer leaves that have no holes or insect damage. Slide your finger down the inside of the stem to the base of the plant and gently push outward. You may have to rock the stem back and forth to pull it from the plant. This way there is no short stem left on the plant to rot. Bunch size is usually 12 to 15 larger leaves, more if the leaves are small.

Line up the lower edges of the leaves even with each other and place the rubber band 2 or 3 inches from the bottom of the stems. You may have to cut off the stems evenly before rubber banding. As always, be gentle.

When harvesting for supermarket packages, pick leaves that will fit into the trays. If the leaves are too long for the trays they'll have to be bent over to fit. This usually causes the leaf to break. Initially, this won't look bad, but after several days on the supermarket shelf this break line begins to turn a rather unappetizing brown.

When picking sorrel, clean the plants as you go. Pull the damaged or yellowed leaves off the plant and put them in a compost bucket. This can be a time-saver later. If the bunches must be washed before packaging, do so very gently. Grasp the bunch right next to the leaves and hold tightly. Swish, rather than dunk, the bunch in water. Shake out the excess water gently.

If you must wash leaves for supermarket packages, do them individually. Use a hose-end sprayer on the gentle spray setting to wash the undersides of the leaves. Support the leaves on your hand as you spray them to prevent breakage.

After washing, allow the leaves to dry completely in a cool area out of the sun before packaging. Sorrel rots quickly when packaged wet.

PACKAGING

For restaurant packages, place the dry bunches in a poly bag, stem ends down. Leave the top of the bag open slightly for the herb to breathe.

Supermarket trays of sorrel usually weigh 1 ounce. If you include recipes for cream of sorrel soup or sauces on the package, you may wish to make different-size packages. The soup recipes usually call for 2 to 4 cups of shredded sorrel. This is much more than the standard package contains. This increased size and weight package should be discussed and decided upon jointly with your accounts.

The leaves, minus the stems, should fit into the trays. Package with the stem ends down and layer one leaf on top of the other. Make sure each leaf is dry. Sorrel should be cooled as soon as possible after harvesting.

tarragon, french

This herb grows like a small shrub, and its stems become woody rather quickly. It has long, narrow, shiny leaves that have very little fragrance unless they are bruised. True French tarragon is one of the most desirable — and scarce — culinary herbs. Its sweet, slight licorice flavor is an indispensable ingredient in French cuisine. The flavor concentrates considerably when dried.

Tarragon's growth habits are such that it is difficult to maintain a continuous abundant supply of tarragon for fresh-cut use. It also takes at least two years for a plant to spread and mature enough to provide good yields. This accounts for the fact that it is one of the most expensive herbs. The financial rewards are definitely worth the extra work and time to grow this herb.

tarragon, french

Botanical Name: *Artemisia dracunculus sativa*
Type of Plant: Perennial
Adult Height: to 30"

You may have to buy in tarragon for resale during your first year in business. As you will learn, there are ways to manage tarragon so that you can maintain a continuous supply, even in the warmer climates. Because this herb requires dormancy, it can be tough to grow where freezing temperatures do not occur for an extended time.

CULINARY USES

Tarragon is used extensively in French sauces such as hollandaise and béarnaise. It is one of the ingredients in the blend known as fines herbes, which is used fresh or dried. It is commonly used to make herb vinegars. Eggs, fish, poultry, and vegetables all become "gourmet" dishes when flavored with French tarragon.

This is one of the few fresh herbs whose flavor does not diminish drastically with prolonged cooking. The leaves are stripped off the stem before being chopped up. Tarragon can be used as a garnish but take care when harvesting and cooling; it bruises easily and wilts quickly under warm conditions.

VARIETIES

True French tarragon *(Artemisia dracunculus sativa)*, does not produce viable seed. It can be propagated only by root division or rooting cuttings. Some seed companies offer tarragon seed that is really Russian tarragon *(A. dracunculus)*. The flavor of this variety is mild and very much unlike French tarragon; it is useless as a culinary herb. Although the leaves look similar, they are rather dull in appearance and the plant's growth is weedy and lanky.

Purchase stock plants of true French tarragon from a reliable grower or nursery. Taste the leaves, if possible, to make sure the licorice or anise flavor is there. Some unwitting or new nurseries plant Russian tarragon seeds and sell the plants as tarragon. You must be sure you are getting the real thing.

PROPAGATION

After purchasing your stock plants, put them in the ground as soon as possible. If the plants have more than 3 or 4 inches of growth on them, cut the stems down to 1 inch above the crown before planting. This will encourage new root and top growth.

Because tarragon is a slow grower, you won't be able to propagate it by root division until the second or third year. You should be able to root some cuttings during the first year, however. Tarragon does not set roots easily, but, with care, it can be done. Always take the new soft growth for cuttings — woody stems will not set roots easily. The cuttings should be at least 3 inches long.

Rooting hormone of medium strength helps increase the number of cuttings that set roots. Usually 50 percent to 75 percent of these cuttings will grow roots, usually in about two or three weeks. When the roots are 1 to 2 inches long, the little plants should be transplanted into 4-inch pots to grow on for several more weeks. It is easier to coddle the plants in pots until they grow up a bit. Fertilize them every week with half-strength fertilizer.

They can be transplanted directly into greenhouse or outdoor beds when they are 4 to 5 inches tall. However, take care to avoid temperature extremes until the plants become established. Tiny tarragon plants do not compete well and will die if crowded by weeds.

Propagate by root division as the plants begin their second year of growth. At this still young stage there may be only one extra root to take from each plant. It is better to wait until the third year to make extensive divisions. Root divisions in subsequent years will provide you with many new tarragon plants.

Division can be done in the spring, just as the new shoots break through the soil, or in the fall, after the plants have started to die back but before frequent frosts. Cut the woody stems off 1 inch above soil level before digging up the root ball if you are doing fall divisions.

Tarragon roots grow laterally as well as downward. When digging up the plants, make sure you place the shovel well away from the main crown to avoid cutting too many roots. Dig up the entire root ball; this can be quite large with mature plants. With water, spray the soil off the root ball to allow you to see where the roots intertwine.

The creamy, tan-colored roots are thick at the top and taper to a thin point at the bottom. They will be wound around and between each other. Many will have a whitish growing tip at the top; some already may be sprouting some green. These are the roots that will make new plants.

Gently separate the roots. You will see how the roots can be divided without a problem. Some roots will be attached to the mother plant just below the crown. Pull these off the main stem. They will separate easily and this does not harm the plant. If you are dividing to make potted plants, use the smallest single roots. If your purpose is to increase your stock plants for fresh cutting, leave several roots together; this quickly will give you bigger, more productive, plants.

Before replanting the mother plants in their same location, renew their beds. Dig up any weeds and grasses. Add whatever soil amendments are necessary to provide good drainage, and add some nutrients or compost as well. The mother plants should be replanted at the same height as before. Plant bare roots so that the new growth tip is at or just below soil level.

GROWING TARRAGON IN THE GREENHOUSE

First-year plants from stem cuttings may not grow enough to harvest more than once. These should be grown outdoors so that valuable greenhouse space is not wasted with an herb that cannot be harvested. In the spring of their second year, when the roots have matured, they can be dug up and replanted in the greenhouse.

Plant tarragon in the middle or the south side of the greenhouse. The soil mix should be medium-heavy with good drainage. Space the plants 12 inches apart in the odd/even pattern. If bare roots are being planted, dig a hole wide and deep enough to accommodate the entire root. Plant it so the growing tips are just below soil level. Fill in the soil and water well.

When planting potted plants, cut back the tops to 1 inch above the crown. Plant them at the same height that they were in the pots.

In three weeks or so, the plants (from both bare roots and plants cut back) will begin to send up new shoots. This growth will be bright green and uniform. Fertilize when the new growth is 2 inches tall. Fertilize each time the plants are cut back to soil level or harvested heavily. If you notice any yellowing leaves early in a plant's growth, it needs some fertilizing.

As the plants mature, the stems become long and woody. Soon the older leaves on the woody stems turn yellow, then brown, and then die. This dieback will continue all the way to the ends of the stems. This is a normal part of the plant's growth and does not mean the plant is dying. The timing of this dieback will depend on the age and growing conditions of the plant. It usually begins when the plant has been growing for 8 to 10 weeks.

When this dieback starts, cut off all the stems at 1 inch above soil level. In a few weeks the plant will again send up some shoots, usually more than before, and begin a whole new period of growth. The plant seems to grow with more vigor each time it is cut back and allowed to regrow. This dieback, cutting, and regrowth pattern can occur three or four times before the plant finally goes dormant.

Obviously, you don't want to cut back all of the plants at the same time; you must control the pattern of growth in order to maintain a continuous supply. Successive plantings made four weeks apart should provide a good yield.

Another method for keeping a continuous supply is to cut back half the plants prematurely, at least four weeks before the dieback starts. This seems like such a waste when they are growing so well, but perhaps this cutback will coincide with a large order!

Tarragon requires dormancy in order to recoup its energy. After three or four cutbacks it simply will stop growing, no matter what you do. If you wish to grow year-round, you can provide the dormancy for this herb yourself. In colder climates the greenhouse tarragon should be dormant during the summer while you harvest from the outdoor beds.

Creating a dormancy period. When the outdoor plants become of harvestable size, cut back the greenhouse plants to 1 inch above soil level. Dig up the root balls and wash them thoroughly. Divide them if you wish and allow the roots to dry completely in a cool, shaded area.

Package the roots in plastic bags that have several holes punched in them. Seal and date the bags. Put the bags in a cooler or refrigerator set between 33° and 38°F. Don't freeze the bare roots. They should be kept cool for at least a month, but two months is better.

Open the bags every week to let in some air and to check the roots for any signs of mold. Move the roots around a little to let the moist air out of the bags. Checking for mold or fungal growth is extremely important. I once lost 100 mature tarragon roots because I had forgotten to check them. When I finally did give them a look, they were covered with a thick layer of gray fuzz!

 weed well and often

It is important to keep tarragon beds weed-free. This herb does not grow well with competition. If weeds are allowed to grow within the root structure, it is almost impossible to eradicate them. The root ball must be dug up and separated and the weeds removed. This process takes a valuable plant out of production for some time.

Do not let the roots get too warm when doing this or the dormancy may break and the roots begin to grow prematurely. After the roots have completed the dormancy period, remove them from the bags and allow them to dry and warm in a cool, shaded room. When they have warmed enough, plant them again in the greenhouse. This should be done 10 weeks before you want to harvest from them. The same technique can be used in warm-climate areas to provide the cool dormancy period. It may be necessary to vary the timing, depending on the climate in your location.

Growing Tarragon Outdoors

Tarragon can take a light frost without damage and can be planted a few weeks before your last frost-free date. Young plants will do better if planted when the soil is warm. Plant them in well-drained soil in full sun. Space the plants at least 18 inches apart. In areas with hot, humid climates, choose a bed that will provide partial shade during the hottest part of the day.

Watch the plants closely for signs of dieback. When this begins, cut back all the stems and allow the plant to regrow. Even in the northernmost climates, this cutback and regrowth pattern should be instituted at least once during the growing season.

Pests and Diseases

Aphids are sometimes attracted to tarragon, especially the new growth tips. Spider mites can also be a problem. Spittle bugs have been a minor nuisance both in my greenhouse and outdoors. They seem to be more of a problem in the North. Young spittle bugs do suck the plant juices but it seems to be the frothy white foam they produce that causes the most headaches. The foam, if allowed to remain on the plant, encourages fungal diseases on the stems. The bugs are easily removed by running your fingers up the stem through the foam and squashing them. After removing the bugs, wash the remaining foam off the stem with a fine spray of water.

Tarragon does not seem to be susceptible to fungal diseases unless the plants are overcrowded or sprawling stems are allowed to lie on the soil surface for long periods of time.

Harvesting

As previously mentioned, first-year plants may not yield any more than a sprig or two. Plants in their second and subsequent years should be ready for harvesting six to eight weeks after new growth begins. With winter-greenhouse-grown plants, plan on 8 to 10 weeks before the first harvest.

The yields you can expect depend on the age and stage of growth. The older the plant, when it was cut back, and how many roots are in the clump will all determine how much and often it can be cut. Multiple harvests can be made on each plant.

When the plant is in its initial growth or after being cut back, harvesting is simple. All the stems are growing at nearly the same height. Bunch stems with one grasp of the hand, cut with a scissors, and rubber band together. With plants that are older and whose growth is not uniform, bunching will have to be done stem by stem. Bundles should contain 8 to 12 stems that are 8 inches long. Too many stems in a bunch will crush and cause bruising of this tender herb.

If washing is necessary, gently dunk or swish the bunches in cool water. New-growth leaves will not separate from the stems as easily as the older growth will. Work delicately with this herb.

Tarragon garnish is usually only 2 or 3 inches long. The tender new-growth tips look the prettiest but do not hold up well after harvesting. If you have plants that are beginning to die back and have woody stems, take the garnish from them. This is a way to make use of the old growth before cutting back, and the garnish will hold better after harvesting.

Cut the garnish individually, or in groups of 10, and gently swish them in cool water. This helps to preserve the quality after cutting. Allow them to dry almost completely before packaging.

Packaging

Restaurant bunches should be packaged, stem ends down, in poly bags. Leave the top of the bag open slightly and refrigerate quickly. Place the garnish very loosely in a larger-than-necessary bag to prevent it from packing down and bruising. Leave the top of the bag partly open and refrigerate immediately.

Supermarket packages usually weigh ½ ounce. Cut the tarragon to fit the tray. If shorter stems are used, layer them in the tray to prevent them from sliding to the bottom when it is stood on end.

thyme

Thyme is another herb with a history that dates back to biblical times. It has been used for ceremonial, decorative, medicinal, and culinary purposes. Thyme contains thymol, a disinfectant that is effective against bacteria and fungi, which made this herb a favorite of the ancient cultures as a remedy for relieving coughs. Its tiny narrow leaves are very aromatic. The many types of thyme are also quite popular for decoration, ranging from bonsai to ground covers in rock gardens.

This shrubby plant grows wild in southeast Europe. In the United States it is hardy in all but the northernmost regions. Thyme is not a vigorous grower and it takes a long time for a plant to reach harvestable size. Because of its small size and slow growth habits, more than the usual amount of plants are required to

thyme

Botanical Name:
Thymus vulgaris
Type of Plant: Perennial
Adult Height: to 12"

produce adequate quantities for your markets. It is an herb very much in demand in most parts of the country and should be on the herb list of every commercial grower. Be sure to allocate ample space for your thyme beds.

CULINARY USES

Thyme is used with many types of food but seems to be most popular with beef and poultry. Lemon thyme has a wonderful lemon flavor and is used wherever that taste is desired. It is especially useful with salads, fish, and poultry.

The small leaves are stripped from the stems before using. The easiest method to remove the leaves is to hold the stem at the tip and, with two fingers, strip the leaves downward. If the stems are branched, this is a tedious chore, but is well worth the effort. The tender new growth stems may tear using this method, but they can be chopped up fine and used with the leaves. Sometimes the leaves are used whole because they are so tiny. If the stems are older and woody, hold the bunch firmly and strip off the leaves with a fork.

Thyme foliage and the flowering tops make a good garnish because they hold well after harvesting. Try to avoid offering it as precut garnish; because of its small size, it is very time-consuming (and not cost-effective) for you to cut it into garnish.

VARIETIES

There are hundreds of varieties of thyme. Many of them are decorative and have little or no culinary value. The best variety for fresh-cut sales is known by several names: common, garden, English, or winter thyme (Thymus vulgaris). This type has small, pointed green leaves and is winter hardy. There are some variations in this thyme. Some types will have larger but sparser leaves and others have smaller but denser leaves.

French or summer thyme (T. vulgaris var.) has very small, grayish leaves. Its growth is upright, shorter, more woody, and it is not as hardy and vigorous as English thyme. This variety is preferred by French chefs. Unless you have specific requests for French thyme, grow it only in small quantities, if at all.

Lemon thyme (T. x citriodorus) is gaining in popularity with chefs and cooks for its wonderful lemon aroma and taste. The leaves are larger and more rounded and it makes an attractive garnish. Grow as much lemon thyme as your space will allow.

There are several types of variegated thyme that are useful to sell as garnish material. *T. x citriodorus* 'Doone Valley' and golden lemon thyme *(T. x citriodorus aureus)* are both lemon thyme but their flavor is not as pronounced as that of the green

lemon thyme

Botanical Name:
Thymus x citriodorus
Type of Plant: Perennial
Adult Height: to 14"

lemon thyme. Their leaves are green and gold and they have a creeping habit. This makes them a little harder to bunch. A popular variety for garnish is 'Silver Posey', or silver-edge thyme (*T. vulgaris* 'Argenteus'). It has a beautiful silver color, upright growth, and is easy to bunch.

Some thymes have diverse flavors, such as oregano, nutmeg, and caraway. These may be interesting to grow but so far they do not seem to be very popular with chefs; thus, they have little value for the commercial grower. You may want to try a few different types and see if you can generate some interest among the chefs in your market.

PROPAGATION

Thyme can be propagated from seed, stem cuttings, layering, and root division. Lemon thyme and the flavored and variegated thymes can be propagated only by stem cuttings and root division.

Thyme is difficult to start from seed and the germination rate is low. Seed should be planted at least eight weeks before your last frost-free date. For best results, scatter the tiny seeds thinly in a premoistened flat. Make sure the soil mix provides good drainage. The seeds should be barely covered with fine sand.

Cover the flat with plastic and then a dark cover. Provide bottom heat at 70° to 75°F for 14 to 20 days. You can expect only 50 percent of the seeds to germinate. The tiny plants will be ready to transplant into pots when they have four sets of true leaves, usually four or five weeks after germination.

Because of the variations in thyme grown from seed, it is best to purchase stock plants and propagate by stem cuttings. This method will give you larger plants faster and they will all be of the type that you want. Thyme and lemon thyme cuttings are easy to root. Take cuttings from the new-growth tips — woody stems will not set roots well. The cuttings should be 3 to 4 inches long. Use a medium-strength rooting hormone. The cuttings should set roots and be ready to transplant in three to four weeks.

GROWING THYME IN THE GREENHOUSE

Thyme grows best in full sun but it has done well in both north and south locations in my greenhouses. The soil mix should be light to medium with very good drainage. Space the plants 10 inches apart using the odd/even pattern.

The small plants grow quite slowly. When they are 5 inches tall, pinch an inch or so off the tops. This encourages branching and new growth from the crown.

Be sure to keep the beds free of weeds. This is especially important when the plants are small because they can be lost when crowded out. Weed control is essential as the plants grow larger. If weeds are allowed to grow within the crown, they are nearly impossible to eradicate. Weeds are also a time-consuming nuisance to remove from the bunches because thyme stems are so thin.

As the plants mature, the crown becomes thicker and some spreading occurs. Some stems may grow long and fall over. If they are allowed to lie on the soil for long, they may grow roots. Cut these from the mother plant and pot or move them to another bed. This will prevent overcrowding and keep the beds tidy. Fertilize thyme every three to four weeks in spring and summer. During short day and low light periods, fertilizing every five to six weeks should be adequate.

Aging plants. The stems of thyme become woody and thicker as the plants age. This usually happens around the third year. If the plants are cut frequently, the new growth will remain pliable and suitable for harvesting for a longer period. Eventually, the plants will become thick and very woody. The growth slows as if the plant is "spent."

Replace these plants with vigorous seedlings before this stage of growth goes too far. The ideal is to replant thyme and lemon thyme beds every two or three years. By having more than one bed of each type, you can replant one bed each year and maintain a continuous supply. Replanting should be done in the spring, just as your outdoor thyme beds reach harvestable size.

Lemon thyme grows much the same as does English thyme. The stems do not become as woody and thick, however, and growth is less upright and more sprawling. If the plants are not cut frequently, the stems may lie on the soil in a circular pattern. These will take root and the main mother plant will die.

The beds can become untidy and hard to maintain if the plants are allowed to spread this way. To prevent spreading, pull up the long stems all around the mother plant and cut them off 4 inches from the crown. Make sure there are green leaves below your cutting point. Sometimes lemon thyme will regrow from the crown, but not always. If you cut off the stems too short, the plant may die. This is also true of English thyme.

Growing Thyme Outdoors

Plant English thyme 14 to 16 inches apart. Lemon thyme should be spaced 18 to 24 inches apart because of its sprawling habit. Choose a sunny location in cooler climates, or partial shade in very hot climates. The beds should be well prepared and provide good drainage.

You may wish to prune the thymes after the first frost. Pruning is not necessary if you want to continue cutting from the plants. This is especially true if the plants are covered with snow or mulch. Many times I have harvested beautiful thyme and lemon thyme in December after brushing away the thick snow cover.

The stems and leaves of lemon thyme may take on a red hue as fall approaches. This does not alter the flavor and is quite attractive. Both thymes are winter hardy but may die if not given some protection from severe cold. If you live in a cold climate, mulch your thyme beds as soon as the ground freezes. Pull off the mulch in late winter before warm days approach. This will prevent fungal diseases from attacking the plants.

Pests and Diseases

The thymes are not particularly attractive to most insects pests. Leafhoppers, whitefly, and spider mites feed on thyme when they lack their favorite foods. These pests can cause considerable cosmetic damage.

If the beds are allowed to become overgrown and thick, they provide a perfect hiding place for everything from mice to grasshoppers. Fungal diseases are more likely to grow in crowded beds as well.

Harvesting

The thymes are very slow growers and the yields per plant are small. Plants may be cut when they are 5 to 6 inches tall. It generally takes 14 to 16 weeks before a few sprigs can be harvested from plants started from seed. If plants are started from cuttings, the time to first harvest is usually 10 to 12 weeks.

Plants in their first season of growth yield only four to six bunches. Regrowth time may be as long as three to four weeks in summer and five or six weeks in winter. As the plants mature, the rate of growth is faster and more cuttings can be made. Yields will be higher, also, because each time the plant is cut, more stems will grow. There may even be new growth from the crown.

Thymes should never be cut any lower than 3 inches from the crown. Always make sure there are green leaves below the cut point or the plant may die. Older plants should not be cut too far into the woody part; try to take only the new growth.

It is especially important with thyme to maintain a uniform rate of growth for ease of harvesting. The stems are thin, and it takes many stems to make a bunch. If the stems are nearly all the same length, it is simple to grasp a bunch, cut, and rubber band it. If the bunches need to be washed, dunk or swish them in cool water. Shake out the excess water and allow to partially dry before packaging. Thyme holds up well under this washing process; lemon thyme is slightly more fragile.

PACKAGING

If the stems are all of different lengths, cut them individually so the bunch size and length are fairly uniform. This can be a time-consuming process. The bunch size should be 1 inch in diameter at the rubber band if the stems are 6 inches or longer; shorter stems require a thicker bunch. Package the bunches, stem ends down, in poly bags, leaving the top open slightly.

Supermarket packages of thyme usually weigh ¼ ounce. Some growers package thyme in ½-ounce weights. Place the sprigs loose in the tray with the stem ends down. Layer shorter stems in the tray.

Thyme should be cooled immediately after packaging, at temperatures between 33° and 38°F.

more herbs with commercial potential

The herbs in the previous chapters are the best-selling ones currently. What follows is a list of herbs that may have some limited value for the commercial grower of fresh-cut herbs. In some markets there will be no demand; in others they may sell very well. Determine your market needs before committing large areas to these herbs.

Bay

Bay leaves were once thought to be useful only when dried. People nowadays are learning to use these flavorful leaves fresh, as do European cooks. The market for fresh bay leaves, especially in supermarket packages, is increasing.

Bay trees are sometimes known as sweet bay or bay laurel. When purchasing bay plants or

bay

Botanical Name:
Laurus nobilis
Type of Plant: Tender perennial
Adult Height: to 25'

seeds, be sure to get the variety *Laurus nobilis*. Many plants look, smell, taste, or sound alike, but not all are safe for culinary purposes. *Umbellularia californica,* California bay, for instance, is toxic, yet its leaves are dried to make wreaths.

Growing and Harvesting

Bay is difficult to propagate by stem cuttings and sometimes hard to find (and expensive) as started plants. Viable seed has recently been made available to growers, but is still scarce. If you wish to buy young bay trees or seeds, begin your search during the winter and prepay your order to guarantee delivery.

Laurel is an evergreen, shrubby tree, hardy in zones 7 and 8, where it can be planted outdoors. In other areas, grow it in containers so you can bring it indoors for winter protection. It will overwinter in a warm environment or at cooler temperatures, just above freezing. Mature plants can withstand intermittent light frosts. Either way, new growth will not occur during low light periods.

This slow grower does well in containers. Repot young plants into increasingly larger containers as they grow. When the tree is 24 inches tall and has many side branches, pot it into a large tub (18 inches or larger), where it can remain indefinitely. Fertilize container-grown bays every few weeks with a medium-strength formula. During the summer, when you water frequently, timed-released fertilizer granules are the most convenient for bay, as well as all container-grown plants.

In summer, place the pots outside in full sun. If bay is wintered indoors, gradually move it into full sun outdoors and provide some protection until it is hardened off. After this, it will need some shade during the hottest part of the day.

You can help the plant to branch out by cutting off 3 or 4 inches from the tips of a few stems. Harvest the leaves by pulling them downward off the stem rather than cutting them. The plant will usually grow a new branch from that site. Take only the stiff older leaves to sell.

Packaging

Supermarket packages of bay leaves usually weigh ¼ ounce. Package the leaves loose with no moisture on the leaves. They will last a long time packaged this way.

Chervil

Chervil has been a mainstay in French cooking for many years, and is a major part of the fresh version of fines herbes. It is used in salads and is a good addition to a salad herb mix for supermarket sales. It should always be used fresh, as its flavor diminishes when dried.

This herb is a moderate- to cool-weather crop and it grows well in the greenhouse over the winter. It can be grown in any greenhouse location and doesn't require much fertilizer.

Growing and Harvesting

In summer, grow chervil outdoors in the shade in a rich, moist soil. Direct sun may cause the leaves to bleach. In very warm climates, it may not grow well during the summer.

chervil

Botanical Name:
Anthriscus cerefolium
Type of Plant: Annual
Adult Height: to 12"

Always use fresh seeds. Sow them directly in the ground where the plants are to grow — chervil does not transplant well. (It can also be grown in large pots.) Plant seed in rows or clusters. The seeds need light to germinate. Keep them moist until seedlings emerge, usually in 7 to 14 days.

Chervil is a short-lived plant. It will bolt quickly during warm weather. Succession planting, every three to six weeks, depending on the quantities that you sell, will ensure a steady supply. During the winter, the plants don't go to seed as quickly and succession planting can be done every six to eight weeks.

Aphids love chervil, and they can be a real problem in the greenhouse. The delicate foliage does not hold up well to vigorous washing, so try to control these pests.

Harvest the first leaves about five to six weeks after germination. Pick the large outside leaves first to allow the crown to continue growing. Pull the leaves from the plant; don't cut them. Create your bundles stem by stem instead of cutting the whole plant. Expect three to six harvests from each plant.

Packaging

Supermarket trays usually weigh ½ ounce. Although chervil makes a pretty garnish, it is not often used this way because it wilts quickly upon exposure to warmth.

Epazote

This herb is commonly used in Mexican cooking. Depending on your market, it could be a big seller. It grows wild in Mexico and has become naturalized in parts of the South. However, many prefer to buy it packaged even there. It is generally used fresh.

Propagating

Epazote is easy to grow from seed. It does not transplant well, so direct-sow it where it is to grow. It prefers a warm sunny location and medium to dry soil, but will grow just about anywhere. Plant the seeds at least ¼ inch deep and cover them only lightly — they need light to germinate. Seedlings emerge in 7 to 14 days. The rate of germination of purchased seeds is usually about 50 percent, so plan accordingly.

epazote

Botanical Name:
Chenopodium ambrosioides
Type of Plant: Annual
Adult Height: to 4'

Growing and Harvesting

Epazote grows tall and lanky. Choose a location in the greenhouse where it won't shade other plants and has enough headroom to grow without leaning into the glazing. Don't fertilize this herb unless the soil mix is very poor. Too much nitrogen results in abundant leaf production but diminished flavor. If the lower leaves begin to turn yellow, use a half-strength fertilizer.

This plant readily self-sows, so it is important to cut off the seed heads when they develop in the fall. Once this plant becomes established, it can be difficult to eradicate.

When the plant grows to 18 inches, begin harvesting. Cut the tall center stem first to encourage the side branches to grow. Don't harvest more than half the plant at a time.

Packaging

Make your bunches large because the leaves grow sparsely on the stems. Supermarket trays usually weigh 1 ounce.

Fennel

There are three types of fennel, each with its own uses and culture. All possess the sweet anise/licorice flavor and aroma that are so interesting in a variety of cuisines. The flavorful foliage especially enhances fish, salads, and soups. The seeds are commonly used in breads. In the northern United States, the plant may not have enough time to produce seeds. All three varieties have lacy foliage, which is attractive both in the garden and as a garnish. The demand for fennel may be light in your area.

Growing and Harvesting

Sweet fennel (*F. vulgare dulce*) is a tender perennial. This is the variety most often grown for its flavorful, light green foliage. It is the tallest-growing of the fennels, sometimes reaching a height of 5 feet.

Bronze fennel (*F. vulgare dulce nigra* 'Purpurascens' or 'Rubrum') is also classified as a tender perennial. Its foliage has a red, almost copper color that is as striking in the garden as it is on the plate. The flavor does not seem to be as intense as that of sweet fennel. It grows to a height of 3 feet. Its flower and seed heads form later in the season.

Florence fennel *(F. vulgare azoricum)* is also known as bulb or finocchio fennel. This is an annual grown for the bulb that forms at the base of the plant. The bulb is used as a vegetable and is very popular in Europe. It grows to a height of 2 to 3 feet. Cut off the flower stalks when they form to concentrate energy in the bulb.

This variety requires a long growing season to form a large bulb. It is also prone to bolting during hot weather. A new variety from Switzerland, 'Zefa Fino', has a 65-day (from transplant) growing season and may be more suitable for those in northern climates.

All fennel is easy to grow from seed. Direct-sow it into the garden when all danger of frost is past. It can also be started indoors and transplanted to the garden. Be careful not to disturb the roots when transplanting.

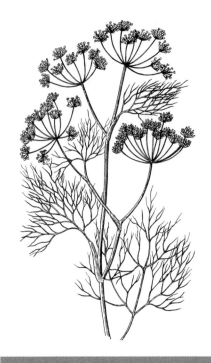

fennel

Botanical Name:
Foeniculum vulgare
Type of Plant: Annual; perennial
Adult Height: to 5'

Seeds started indoors should be given bottom heat between 70° and 80°F. Germination takes 7 to 14 days; bronze fennel usually needs 14 days. Transplant when the seedlings have three branches, which is usually three weeks after germination. The soil should be rich, well drained, and receive full sun. Plants should stand 18 inches apart. Make sure the plants receive plenty of moisture.

Harvesting of the leaves and stems can begin when the plants are 12 inches tall. Don't take more than a few stems at a time from any plant or it may not recover. To harvest the leaves, gently pull them downward off the main stem while holding the stem for support.

Packaging

It is best to package fennel loose (by weight) rather than by bunching. The stems are tender and have a tendency to break when rubber banded together. Fennel can be offered as precut garnish but exercise care in handling to avoid tearing the fragile foliage. Supermarket trays usually weigh ½ ounce.

Lemon Balm

Lemon balm looks and grows like mint, but without mint's invasive tendencies. The wonderful lemon flavor and aroma make this herb a treat in everything from salads to tea. Sadly, this easy-to-grow herb does not seem to be much in demand in most areas of the country.

Growing and Harvesting

This plant can be propagated by seed, stem cuttings, or division. Start seeds indoors at least two months before transplanting. Light is needed for germination, so do not cover the seeds. Expect only half the seeds to germinate.

Choose a location in full sun or partial shade with a rich, moist soil. Space the plants at least 36 inches apart. Lemon balm does not spread by runners, but the clumps do get

lemon balm

Botanical Name:
Melissa officinalis
Type of Plant: Perennial
Adult Height: to 30"

larger each year. It can be grown in big pots, but production will be limited. Fertilize container-grown plants every two weeks. In the greenhouse, watch for attacks by whitefly.

Flower spikes form in midsummer, which stops leaf production. Cut off these flowering stems 6 inches from the crown. The cut stems will regrow and new growth will emerge from the crown. Lemon balm is very hardy and will come back year after year even after the most severe winter. It will grow from the crown early in spring.

The first cutting from a new plant can usually begin 10 weeks after transplanting. Harvests will be small the first year, but plants in the second year and after will yield multiple harvests. Do not cut more than half the plant at a time. Bunch size should be the same as mint. Although lemon balm makes a pretty garnish, do not offer it as precut garnish unless special care is taken to cool the cuttings immediately after harvest. It has a tendency to wilt quickly after cutting when exposed to warm temperatures.

Packaging

Supermarket trays usually weigh ½ ounce. Lemon balm should be packaged with the stem ends down.

Lemongrass

Lemongrass has long been used for tea, medicine, and in Oriental cooking. Dried, it is excellent in potpourris because of its intense lemon aroma. Its recent gain in popularity may make it worthwhile to grow, even if you want only a clump or two.

The long blades of lemongrass are used fresh to flavor dishes and make teas. They are stringy and chefs usually remove them after cooking. The lighter-colored tender base of the stalk is cut off and used in stir fry and other cooked dishes, much as you would use green onions and leeks.

Growing and Harvesting

Lemongrass can be propagated only by root division, so you must purchase stock plants. Another variety of lemongrass, *Cymbopogon flexuosus* 'East Indian', can be grown from seed, but the germination rate is low.

The clumps increase in size each year and division can begin with second-year plants. When dividing clumps, make sure that each stalk has a least 1 inch of root attached or it probably will not grow. Cut down the blades to 2 inches above the base before dividing a clump.

This very tender herb can be grown year-round outdoors in the southern United States. Plant it in full sun in rich, moist soil. This plant will die when frozen, so in cold climates dig it up in the early fall, pot it, and bring it inside to overwinter. It should survive this procedure but will not grow until spring.

If you have year-round heated greenhouses, plant lemongrass in raised or ground beds. It can be container-grown in large pots, but because of the extensive root system, division should be done yearly. Fertilize every two weeks during the growing season. Growth will be limited during low light periods without supplemental lighting.

Cutting can begin when the blades are 12 inches tall. Where the plant is cut depends on how the herb is to be used. If the blades will be used dry or to flavor food, the cut should be made 1 inch above the crown. This method ensures that the plant will regrow. Use a very sharp, heavy scissors to cut this herb because the stalks are quite tough. I actually once had to use a saw to cut the tough stems of older established plants.

If the tender base of the stalks is what's needed, cut the stems as close as possible to the roots. In many ethnic markets lemongrass is sold with a small amount of the roots attached. Cutting the stalks this way usually means the stems will not regrow.

lemongrass

Botanical Name:
Cymbopogon citratus
Type of Plant: Tender perennial
Adult Height: to 36"

Packaging

When packaging lemongrass, leave the blades as long as possible (unless your accounts request otherwise). The blades may have to be gently folded to fit into the bag or tray. Supermarket trays usually weigh 1 ounce.

Lovage

Lovage has a strong, yet sweet, celery aroma and flavor. The leaves and tender stems are used to flavor stews, soups, poultry, and vegetable dishes. Lovage is almost always used fresh, as the flavor diminishes when the herb is dried. Although interest in this herb is growing, there still is not much demand. Because it is so easy to grow, everyone should have a few plants, at least during the summer.

Growing and Harvesting

Propagate lovage by seed or root division. Always use fresh seed; old seed does not germinate well. You can expect only 50 percent of fresh seed to germinate. Start seed eight to nine weeks before transplanting to the beds. Sow the seed in a premoistened flat, then cover with plastic and a

lovage

Botanical Name:
Levisticum officinale
Type of Plant: Perennial
Adult Height: to 72"

dark cover. Provide bottom heat between 65° and 70°F. The seeds should germinate in 8 to 12 days.

In three weeks or so, when the seedlings are a few inches tall, transplant them into 4-inch pots to allow them to grow on. Fertilize them weekly for five weeks. When the plants are at least 4 inches tall, transplant them to beds. Lovage is a heavy feeder, so prepare the beds with lots of composted manure. Additional fertilizing may be necessary to maintain adequate production. Choose a location that receives full sun or partial shade with rich, moist soil. Because it grows so tall, make sure that lovage is planted on the north side so it won't shade other plants. Space the plants at least 36 inches apart.

For year-round production, lovage can be planted in the greenhouse. Because of its size, plant it in ground beds; it is not suitable for container growing. Use a heavy soil mix and fertilize every two weeks.

Lovage is very hardy and will survive even severe winters. In mid-summer, long flower stalks form. Cut these back so the plant will produce more leaves. In the third year, divide the roots — the plant will then grow with more vigor.

First-year plants will grow only about a foot tall, and only small harvests can be made. In the second year the plants can grow up to 6 feet tall, and yields will be greatly increased. Harvest by cutting the side branches off the main stem. Bunching is easier if the stems are long. Bunch size is usually 1 to 1½ inches in diameter at the rubber band.

Packaging

Supermarket trays usually weigh ½ ounce. Remove the stems so the leaves will fit into the trays.

Salad Burnet

The round, serrated leaves of this herb have a sweet cucumber taste and aroma. As the name implies, it is used mostly in salads, but it also makes a good vinegar. Perhaps few people in your area are familiar with this herb: You can introduce this easy-to-grow herb by giving free samples to gourmet restaurants and including it in salad-herb mixes.

salad burnet

Botanical Name: *Poterium sanguisorba*

Type of Plant: Perennial

Adult Height: 18"

Growing and Harvesting

Salad burnet can be propagated by seed or root division. It is very slow to germinate unless it is started indoors. Sow the tiny seeds in premoistened soil, do not cover with soil or sand, and then cover with plastic. The seeds germinate best between 70° and 75°F. Expect only 50 percent of the seeds to germinate, in 5 to 10 days.

The plants should be ready to transplant into beds about six weeks after germination. Choose a spot in full sun or partial shade. The soil should be dry to average with good drainage. Space the plants 18 inches apart. When the plants become established, growth will be vigorous.

This herb can be grown in large pots but you must fertilize every two weeks. It will produce good yields when grown in the greenhouse during the winter. Aphids and rabbits are attracted to this herb.

Growth is slow the first year, with the plants reaching a height of only 8 inches or so. Salad burnet is quite hardy and will survive cold winters. Divide each clump every three years, as growth becomes sluggish and tough if you omit this chore.

You can begin small harvests when the plant reaches 6 inches in height, about four to six weeks after transplanting. The whole stem should be cut; don't simply remove the individual leaves. The outer stems are usually tough, so take the tender new growth toward the center. Cut the stems or gently pull them off.

Packaging

Bunches should contain at least 20 long stems; cut the stems to fit the tray. Supermarket packages usually weigh ½ ounce.

Summer Savory

This quick-to-grow, short-lived herb has been a culinary staple in Europe for years. It is gaining popularity here because its peppery flavor has a natural affinity for beans. Used fresh or dried, it can act as a salt replacement and a flavor blender.

Growing and Harvesting

Savory is quick to germinate, usually in four or five days, so it is best to direct-sow. It can be transplanted, with care, if that is preferable to you. Choose a location in full sun with medium-rich, but light, soil.

After the danger of frost, sow the seeds in rows ¼ inch deep. Cover them with fine sand so that light can reach the seeds. Keep moist until the seedlings are up, then thin to 6 inches apart. Plants can be closer together but picking bunches is easier with some space between the branches. Make succession sowings every three or four weeks because by midsummer your crop will begin to flower.

When savory flowers, it seems that the entire plant is flowers, with no leaves left. Nothing can be done to stop this process. It is the leaves that are used in cooking, so keep them available by succession sowing. The entire plant should be harvested just as flowering begins.

The first harvest can usually be taken six weeks after germination, when the plants are 6 inches tall. Cut only select branches and never cut more than a third of the plant at a time. Harvest the entire plant before the first frost — it will not withstand even the briefest freeze. A very light mulch of grass clippings helps to keep the plant clean. Savory seems to attract mud from splashing rain. It does not hold up well to vigorous washing.

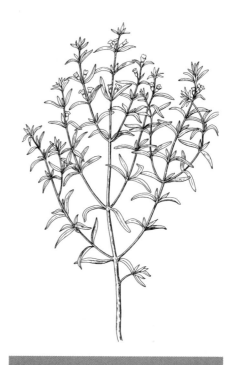

summer savory

Botanical Name:
Satureja hortensis
Type of Plant: Annual
Adult Height: 18"

Packaging

Bunches have to be made stem by stem because the woody stems spread out and grow upright. The bunch should contain 12 to 15 long stems. Supermarket trays usually weigh 1 ounce.

Winter Savory

Winter savory has basically the same flavor properties of summer savory, only a bit stronger. Many growers substitute this for the summer type during winter. Being a perennial, it is easier to grow in the greenhouse during winter than is the summer variety. The growth is woodier and a little more spread out, but chefs who love savory don't seem to mind the difference.

Growing and Harvesting

Winter savory can be propagated by seed, stem cuttings, division, or layering. Seed is said to be slow to germinate but in my experience it has not been that difficult. Be sure to allow at least three or four months from sowing to transplanting to the final beds or pots.

Sow the seeds in premoistened soil, cover with clear plastic, and provide bottom heat at 70°F. The seeds should begin to germinate in 7 to 14 days. Expect only 50 percent of the seeds to germinate.

The little seedlings are slow growers and should be fertilized every week. It may take up to a month before they are ready to

winter savory

Botanical Name:
Satureja montana
Type of Plant: Perennial
Adult Height: to 12"

transplant. Transplant to 4-inch pots and give them at least another month to grow before setting them into their final beds.

Space the plants 18 inches apart in average to well-drained soil. This herb grows best in full sun, although it can be grown in large pots in the greenhouse. It is quite hardy, but during especially cold winters it is best to cover the plants with a thick layer of mulch after the ground is frozen. (It is not the cold that kills this herb over winter; rather, moisture is the villain.)

Harvest, bunching, and packaging are the same as for summer savory. This herb is easier to bunch because it does not spread out, the stems grow closer together, and growth is more uniform.

edible flowers

Cooking with edible flowers is not a new trend. During medieval and Elizabethan times, the use of blossoms in food was commonplace. Today this nearly lost art is enjoying a revival. Although there is some speculation today about edible flowers being a passing fad, it doesn't really matter. As long as people buy and use them, you should grow them. Because they require little care, many of them are easy to grow — and they provide a good profit margin.

It is important that you educate your chefs about the flowers and that they pass on this information to their service staff. Customers can then be assured that the flowers are edible. Also, some decorative flowers have little culinary merit. Fuchsia, for example, makes a striking garnish but it just plain tastes bad. If you sell any of these, be sure to inform the buyer about the taste.

The petals of many flowers must be removed from the calyx (a group of small leaves, usually green, that enclose a flower bud) before being eaten. The calyx may be very bitter or too large and tough to eat. This information, too, should be passed along to the consumer before selling the flowers.

Most flowers sold by florists have been chemically treated and should not be consumed. Many

Nasturtiums (left) and pansies (right) are two popular edible flowers.

nurseries treat young flower plants with systemic fungicides and pesticides, and these are not safe to eat. These chemicals can remain within the plant throughout its life cycle. Exercise extreme care when purchasing plants to ensure that they, or the soil they grow in, has not been sprayed or treated with *any* chemicals. It is better to start your own plants from seed than to risk the health of your customers and your business.

Not all flowers are edible, and indeed some are poisonous. Be careful that the flowers you choose to sell really are edible and not on the American Medical Association's list of poisonous plants. Also be aware that if a flower is listed as edible, it does not necessarily mean that the rest of the plant is safe to eat. There are many sources for lists of edible flowers. The most comprehensive I have found is published by The Herbfarm (see Resources).

Boosting Business with Edible Flowers

There are numerous varieties of flowers to choose from and it can be tempting to grow as many types as possible. As a commercial grower, though, you should have two or three varieties as your main crop. You can still grow some of the other flowers if you have the time, space, and the demand for them. The best edible flowers for your business are nasturtiums and pansies. They are the most requested, are easy to grow, and have good holding quality after harvest.

Growing

Flowers can be grown in hanging baskets in the greenhouse during winter. Without supplemental lighting, however, few plants will bloom.

Flowering plants need some care periodically. Yellow leaves, blemished flowers, and seed heads should be removed to keep the plants blooming. If seeds are allowed to develop, most annual plants — having completed their mission in life — stop flowering.

Controlling Pests and Diseases

Insects love flowers as much as we do. Aphids, spider mites, thrips, and a multitude of others can create serious problems, especially in the greenhouse. There are very few insecticides labeled (legal to use) for use on edible flowers. Synthetic chemicals should not be used to suppress insects.

Insecticidal soaps, some pyrethrum sprays, and predatory insects are all good methods of control. However, soap and pyrethrum can damage flower petals if sprayed directly onto the blossoms or buds.

Downy mildew and other fungal diseases also infect flowering-plant foliage, especially during a winter spent in the greenhouse. These should be controlled because they can suppress flowering on many plants. By providing good air circulation, warm temperatures, and a clean soil surface, you can help prevent the spread of fungal diseases.

Harvesting

Harvest when the flowers are dry, usually in the afternoon. Newly opened blossoms will have more staying power after harvesting and less insect damage than will older blossoms. Take the entire flower stem, so nothing is left to rot on the plant. This also encourages the plant to branch and ultimately produce more flowers. Gently pull off the stems or cut them with scissors.

Transport flowers to a cool area immediately after harvest. This cooling is of the utmost importance for good holding quality. Either package the flowers right then or leave them in the container to cool for a while.

Each flower that you sell must be absolutely perfect. The longer stems make this examination process easier. Each perfect-looking flower must then be inspected carefully for insects. Pull back each petal, look in front and behind, inspect underneath the calyx and inside the center of the flower. Most insects can be removed with your fingers.

It is crucial that there be no insects on these flowers — no one wants to have bugs crawling around his food. The owner/chef of an expensive gourmet restaurant once told me that a small worm had crawled out of my nasturtium onto the plate of one of his valued customers. He explained to the patron that these flowers were organically grown and handled with the utmost care. The customer was grateful for that fact and nothing more was said. Not all chefs will be that understanding, however; nor will all customers!

Packaging

Plastic trays with clear domed lids are the best to use for flowers. They protect the flowers from damage and look attractive as well, but these can be expensive unless you purchase in case lots. The size you want depends on how many flowers are packaged for your markets. The usual package is

50 flowers in a tray. Some growers use two sizes of trays, a larger one for restaurant sales and a smaller one for gourmet market sales.

Because of the expense of the trays, small growers may want to ask some of their accounts to save the empty used trays and return them. They can be washed, sanitized, and reused. This works well with the heavier-weight, two-piece types. It is time-consuming to sanitize, but it does save money.

The flowers need some moisture in the trays to stay at their freshest. A damp paper towel folded to fit the bottom of the tray works just as well as the more expensive florist's foam.

To package edible flowers, use plastic trays with clear, domed lids.

You can expect the flowers to last a week packaged this way if they are stored between 34° and 38°F. If it is possible to store the packages somewhere between 42° and 50°, the flowers will last even longer. Most restaurant coolers are set for the colder range.

Another way to make the flowers last longer is to use dense florist's foam — ½ inch thick — in the bottom of the trays. Stem-size holes are poked in the foam and the foam is then soaked with water. The flowers, two to a hole, stand upright in the package. Leave enough stem on the flowers so the petals do not touch the foam. This method takes time and is expensive, so use it only with really "picky," select accounts.

Regular packaging is simple. After the flower has passed inspection, remove the stem at the base of the flower and place it in the tray. Arrange the flowers any way you like, but only one layer deep. This prevents crushing of the fragile petals. Refrigerate the full trays immediately after packaging.

Here are the most commonly used edible flowers, along with cultural recommendations of the two most popular types. Some flowers you may be familiar with have been omitted from this list because of bad flavor, tough texture, or difficulty in maintaining quality after harvesting.

Herb Flowers

All culinary herb flowers are edible. Not all taste good, not all are attractive, and not all hold up well after harvest. Many are flavorings as well as garnish. Most herb flowers are quite small and are usually sold on the stem. You can sell them by the stem, perhaps charging as you would for garnish. This is a list of the most frequently used herb flowers.

- **Anise hyssop.** Small blue to purple flowers carry a strong anise flavor.
- **Arugula.** Medium-size, creamy white flowers have the mild flavor of arugula. Use the blossoms in salads.
- **Basil.** All of the flavored, purple, and scented basils have aromatic and tasty flowers. They are quite small and don't hold well after harvest.
- **Chervil.** These tiny white flowers are sold in a cluster and have the delicate flavor and aroma of chervil.
- **Chives.** Both onion and garlic chives have very useful flowers. The flowers of onion chives are lavender-colored and round. Garlic chive flowers are tiny, white, and sweetly aromatic. Both can be cut from the stem or sold on long stems in water, as is often done with cut flowers.
- **Oregano.** The white or lavender flowers are small but they make an attractive garnish for oregano-flavored dishes.
- **Rosemary.** These delicate flowers are blue to purple. The flowering stems make an attractive garnish. They are sometimes used in cosmetics and have some medicinal value.
- **Thyme.** These tiny flower clusters should be sold on the stem. They are served mainly as a garnish.

Popular Edible Flowers

The following flowers differ in size but are mostly larger than the herb flowers you sell by the stem. Most of these flowers are somewhat unusual for chefs to find; many wholesale distributors do not carry a wide variety of edible flowers. Package them as you would pansies or nasturtiums. If you offer any of these for sale, you should charge a price per flower plus perhaps a little more than you do for pansies or nasturtiums.

- **Bergamot** *(Monarda didyma).* Also known as bee balm, these citrus-scented flowers are now available in many pastel colors. They are a little larger than the optimum garnish size, which is usually 1 to 1½ inches.

They must be picked when the flowers just begin to open because they fade rapidly. Bee balm does not hold well after harvest.

• **Borage** *(Borago officinalis)*. These blue, star-shaped flowers add a cucumber taste and aroma to salads and other dishes. They are easy to grow and reseed readily. The petals are delicate and don't hold well after harvest.

• **Calendula** *(Calendula officinalis)*. The petals from these orange flowers must be removed from the calyx before eating.

• **Chamomile** *(Matricaria recutita)*. These small, white, daisylike flowers are best known for making a soothing tea. They also make a pretty garnish and are a flavorful addition to salads.

• **Chrysanthemum** *(Chrysanthemum x morifolium)*. The flowers and leaves have a mild cauliflower flavor and have been used for many years in Oriental cuisines. The petals are usually removed from the calyx before eating.

• **Citrus blossoms** *(Citrus limon, C. auranstiifolia, C. sinensis)*. The blossoms of lemon, lime, and orange trees are all edible and aromatic. If you live in an area where these trees grow, be sure the trees have not been chemically treated before harvesting their flowers.

• **Daisy,** English *(Bellis perennis)*. These white, sometimes yellow flowers are used for garnish and desserts. Remove the petals from the calyx before eating.

• **Daylily** *(Hemerocallis fulva)*. These large orange flowers have been naturalized over many parts of the country. New sizes and hybrids are commonly introduced to the marketplace. Daylilies can be deep-fried, sautéed, stuffed, or used as a garnish.

• **Fuchsia** *(Fuchsia spp.)*. These multicolored flowers are very exotic looking. They have a tart flavor that many people dislike.

• **Lavender** *(Lavandula spp.)*. The flowers have a long history of being used for their fragrance. They are equally good for flavoring food and as a garnish. The top 3 or 4 inches of the flowering stem is usually all that is used. Some chefs will request longer stems.

• **Marigold** *(Tagetes erecta, T. tenuifolia, T. lucida)*. While all varieties are edible, the flavor can be bitter. The best tasting are the 'Lemon Gem' and 'Tangerine Gem' types. The petals are removed from the calyx before eating.

• **Rose** *(Rosa spp.)*. Most varieties of roses have large flowers. They are striking as a garnish used whole. Most often the petals are removed and used in cooking and as garnish. Take care to use roses that have not been chemically treated.

• **Scarlet runner bean** *(Phaseolus coccineus)*. These flowers are as tasty as they are beautiful. 'Painted Lady' is a good choice.

• **Scented geraniums** *(Pelargonium* spp.). There are many, many varieties of scented geraniums, which are grown primarily for their interesting foliage and unusual scents. Most either have inconspicuous flowers or rarely flower. Some types do produce wonderfully colored and scented flowers large enough to separate and sell. Because there is so much cross-breeding of scented geraniums, you can never be sure a purchased plant will produce large flowers.

These are the varieties that have been useful for me: apple, apricot, old-fashioned rose, 'Attar of Rose', eucalyptus, 'Mrs. Taylor', 'Prince Rupert', 'Joy Lucille', 'Lady Mary', crowfoot, peppermint rose, and staghorn oak.

• **Squash blossoms.** All varieties of squash have edible blossoms. The summer squashes are more prolific. These are usually stuffed with all manner of gourmet delicacies before serving. The female blossoms have small fuzzy fruits attached; the males do not. Most chefs will have a preference as to type. These blossoms are very deep and delicate and often have insects hiding inside.

Nasturtium

Nasturtiums come in numerous types and colors. All are edible and have a wonderful peppery taste, much like that of watercress. The leaves and unripe seed heads are also edible and have nearly the same taste as the flowers. Nasturtium seeds can be purchased in mixed or single color lots. Always use new seed because old seed does not germinate well.

'Dwarf Jewel' has double-petaled flowers that are 1 to 1½ inches in size. It has a height of 10 to 12 inches and comes in red, orange, and yellow. The flowers

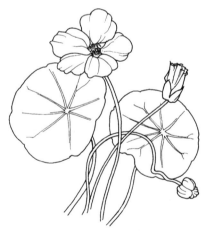

nasturtium

Botanical Name: *Tropaeolum majus; T. minus; T. tuberosum*
Type of Plant: Annual
Adult Height: to 30"

have a 1-inch spur on the back, which can prevent them from lying flat on the plate.

'Whirlybird' grows 12 to 14 inches tall and has semidouble 2-inch flowers. This is the best commercial variety because the flowers are spurless and lie flat on the plate. They produce a nice wide variety of colors, such as rose, orange, and even multicolors. Both of the compact types do well in hanging baskets.

Growing and Harvesting

When the soil is warm, plant seeds in groups of three or four, with 12 inches between groups. If they will be grown in hanging baskets, plant three or four seeds in one 10- or 12-inch basket. Plant the seeds 1 inch deep and keep moist until seedlings emerge, which usually occurs in 7 to 12 days. If sowing in flats for transplanting later, leave 1 inch between seeds. Provide a dark cover and bottom heat of 70°F. Transplant very carefully when the seedlings have two sets of leaves.

If you are selling foliage as well as flowers, the plants should be fertilized every month — more often if they are container grown. If the flowers are your only concern, less or no fertilizer will result in more flower production but somewhat sickly looking foliage.

The first flowers should be ready to harvest in 8 to 9 weeks after germination, 11 to 12 weeks during winter greenhouse conditions. After three or four months of producing flowers, the plants seem to run out of steam and gradually stop flowering. To ensure a steady supply, make new plantings every two or three months.

Always begin a winter greenhouse season with new plants. Succession plantings should be sown every two months during periods of low light.

Packaging

With nasturtiums, very close inspection for insects is necessary before packaging. The center of the flower is deep and the many overlapping petals make excellent hiding places. Try to remove the insects by hand; the petals are delicate and will not hold up to washing.

Nasturtium foliage is also edible. Some chefs may want the leaves to use as garnish or for floating in soups. In addition, the leaves are useful in the packaging itself to line the bottom of the tray before layering in the blossoms.

Pansy

Pansies are the ideal edible flower. They are the perfect size, come in a wide variety of colors, are easy to grow, and, of course, taste good.

There is a multitude of pansies to choose from. Each seed company has its own series and names for them. You can choose the colors, face types, aroma, size of the flowers, and growth patterns that you like.

Some types do better in warm climates. In southern parts of the country, the F_1 hybrids are more heat tolerant than others. In the North most varieties will flourish, as pansies are a cool-weather plant.

pansy

Botanical Name:
Viola x *wittrockiana*
Type of Plant: Perennial, grown as annual
Adult Height: to 12"

Growing a mix of colors, sizes, and face types can have advantages. Some chefs like to use certain colors for special gourmet meals. It is in your favor to be able to accommodate them. It is also nice to have a few plants of the giant variety, whose flowers can be as big as 5 inches across.

Viola *(Viola cornuta)* and Johnny-jump-up *(V. tricolor)* produce smaller flowers than pansies. Johnny-jump-ups are especially abundant producers. Occasionally chefs may request a large number of the same color flower for a special event. Plant a big bed of Johnny-jump-ups and you should be able to fill this type of order.

Generally, look for varieties that offer a mix of colors and sizes and an extended flowering time. Grimes Seed Company offers a Gourmet Brand Salad Mixture that has performed very well for me, both outdoors and in the greenhouse during winter.

Growing and Harvesting

Pansies can be started by stem cuttings and from seed. They will survive most winters but produce fewer flowers as the plants age. For the commercial grower, it is best to start new plants each season to ensure maximum yields.

Always start a winter greenhouse growing season with new plants. Start the seeds 16 weeks before your indoor growing season. For outdoor beds, sow seeds at least 14 weeks before flowers are desired. For early-spring blossoms in cool climates, seed can be started in August and overwintered in a cold frame or in the ground (with good winter protection). In warm climates, new pansy plants will bloom in the fall and early spring.

Pansy seeds require a little work to germinate. They must be chilled for a week or two before they are exposed to warmer temperatures. Sow them in a flat, then cover with plastic and a dark cover for two weeks at 34°F. The seed packet can also be prechilled at 40° for one week before sowing. After the chilling period, place the flat where the temperatures are approximately 65° during the day and 45° at night. The seeds should germinate in two to three weeks.

As an alternative, primed seed of some varieties is available from some suppliers. This more expensive seed has been preconditioned and the grower can simply sow it without prechilling.

Pansies are a cool-weather crop and like a moist, well-drained soil. In the North, plant them in full sun outdoors. In warmer climates, the plants like partial shade during the hottest part of the day. They won't need fertilizer unless the soil is very poor.

In the greenhouse, they do well in 10-inch hanging baskets during the winter. Fertilize every month with the same fertilizer used for herbs. If blossoming slows during the winter, apply a fertilizer specifically for flower production, such as 10-20-10. Moving the plants to a location where they will receive more light should also help production.

Harvest the flowers often; the more they are picked, the more they will produce.

Packaging

Newly opened pansy blossoms are durable and will withstand some dunking and washing should you need to remove insects. Allow the petals to dry before packaging. See page 295 for information on packaging.

-A-

acclimate. To adjust to another condition, such as temperature or climate. The process achieved when hardening off plants.

acicular. Having needlelike leaves, such as those of rosemary.

acuminate. Having leaves that taper abruptly to a point.

aeration. The porous nature of the soil and its ability to hold air spaces, which are necessary for good plant growth. Also, the act of improving the porosity of the soil by adding soil amendments or by tilling.

alkaloid. A nitrogen-containing organic compound found especially in seed-bearing plants.

alternate. Leaves that arise singly along the stem, in contrast to opposite.

amendment. Material, usually organic or minerals, that is added to soil to improve it.

annual. Completing the life cycle from seed to seed production in one growing season.

aromatic. Having a pleasant odor that is easily released from a plant.

axil. The upper angle formed by a leaf or branch as it grows from a main axis or stem. An axillary bud is usually present at the axil, which may develop into a new shoot.

axillary. Located or growing in the axil.

axis. A plant stem or branches growing from the main stem.

-B-

backfill. To return soil to a planting hole after the roots have been placed.

basal. Growing at the base of a stem.

base temperature. The temperature below which a particular species of plant will not grow.

bed. A specific area where plants are grown either in the ground or raised.

bedding plants. Ornamental plants, usually annuals, that are sold by nurseries for transplanting into growing beds.

bench. A modern term meaning greenhouse tables or shelves where containers of plants are grown.

biennial. A plant that completes its life cycle in two years.

biocontrols. Biological controls. Pest management using natural methods such as predatory insects and pathogens.

bolt. The rapid flowering of a plant that usually occurs after the plant has been stressed.

botanical. Pertaining to plants. Commonly used to refer to pesticides that are derived from plants.

botanical name. The scientific name of a plant, listed by family, genus, and species. Most commonly seen are genus and species. Variety and/or cultivar (in single quotation marks) may also be given. Names in Latin are italicized.

bottom heat. Heat supplied by artificial means on which to place flats or containers of planted seeds or cuttings to induce rapid germination or rooting.

bottom watering. Allowing a plant or plants to soak up water from below rather than watering from above.

bract. A small leaf usually found outside the calyx or at the base of a flower stalk.

broadcast. To scatter seeds rather than planting them in rows.

bud. A small growth on the end or side of a stem that usually develops into a flower, leaf, or branch. A terminal bud is one growing at the tip of the main stem.

buffering capacity. The ability of water or soil to minimize fluctuations of pH.

bunch. A group of plant stems held together by a rubber band or by other means.

-C-

callus. A growth around a wound on a plant or cutting. On cuttings, the callus often precedes the growth of roots.

calyx. A group of small leaves, usually green, that enclose a flower bud.

canopy. The stage of plant growth when the upper leaves provide nearly complete shade to the soil below. The tops of a stand of plants.

chlorophyll. The substance within a plant that causes it to be green.

chlorosis. Yellowing of leaves, usually caused by a nutrient deficiency.

cold frame. An outside framed-in growing bed with a removable glass or plastic cover. Usually used without artificial heat.

come true. A term meaning that a plant, grown from seed, is almost exactly like its parents.

commercial grower. A person who grows plants for the purpose of selling them or plant products for profit.

compost. Decayed, rich organic matter that is used to enrich soil.

compound leaf. A leaf divided into two or more leaflets.

corolla. All of the petals of a flower.

cotyledon. The first leaves of a sprouted seed.

crenate. A leaf with scalloped edges, such as on most mints.

crop scheduling. A method of determining when to plant seeds or propagate plants to achieve your desired results.

crown. The part of the plant above the roots and stem that produces the leaves and branches.

culinary. Pertaining to being used in the kitchen for food or flavoring.

cultivar. A "named" plant variety, different from others of its species, that is usually propagated asexually.

culture. A certain set of environmental conditions that are most beneficial for a particular type of plant to encourage optimum growth.

cutting. A plant part that will grow new individual plants, if placed in a favorable environment. Stem cuttings are also known as *slips*.

-D-

damping-off. A fungal disease that affects seedlings or seeds, causing them to die quickly.

deadheading. Removing old and faded flowers to encourage new flowers and prevent the formation of seeds.

deciduous. Describing a plant or tree that sheds its leaves at the end of the growing season.

desiccant. A material, usually silica, that absorbs moisture from the air. Sometimes comes in small packets.

diatomaceous earth. Skeletal remains of particular algae that are ground into dust. The particles are very sharp, causing abrasions on an insect's body that will often result in its death.

dibble. A narrow, pointed tool used to lift young plants out of the soil; a device to make holes in the soil to receive young plants.

dichotomous. Describing a stem that divides or that grows side branches.

direct seeding. Sowing seeds where they are to grow rather than transplanting.

division. A propagation technique whereby a plant is separated into two or more plants, each with its own roots and stems.

dolomitic lime. Lime for field use composed of calcium and magnesium carbonates.

dormancy. A resting period in perennial plants during which there is no growth and few signs of life, yet the plant is alive. Also a period of inactivity in seeds when the ability to germinate stops.

drainage. The movement of water down through the soil.

drip irrigation. A type of watering system that uses tubes to carry and release water to plants in individual pots, or in raised or flat beds.

-E-

ebb-and-flow system. A watering system for potted plants in which bottom water is provided by flooding the benches or molded concrete floor. The water, sometimes containing fertilizer, is then drained off, collected, and reused.

electroconductivity (EC). A measure of soluble salts in water or a growing medium determined by its ability to conduct electricity. EC meters are used to determine that the correct amount of fertilizer is being applied to the plants.

embryo. A plantlet developing inside a seed.

emergence. The embryo of a new plant breaking through the seed. Also, new sprouts just breaking through the soil.

enhanced seed. Seed that has been treated or coated to promote germination or to make the use of mechanical seeders easier.

epicalyx. Additional whorl of bracts growing outside the calyx on some flowers.

ethylene. A natural plant hormone caused by the respiration of plant material. It causes, in part, the process of decay. Also a by-product of incomplete combustion of fossil fuels.

evaporation. Loss of water from soil or other surfaces as it is converted into water vapor.

-F-

family. A plant classification category, below an order and above a genus. A botanical family can include more than one genus. Plants in the same family share characteristics, but not as many as do those in the same genus.

farmer's market. An area where growers and farmers gather on a regular basis to sell their produce and products to the general public.

fertile. Characterizes soil that has abundant nutrients needed by plants. To fertilize is to supply these nutrients.

fertilizer. Compounds, organic and inorganic, that contain measurable amounts of nutrients necessary for plant growth.

flat. A shallow tray used to grow seedlings and to hold pots or divided cell trays.

flora. The plants growing wild in their natural habitat.

flower. A plant organ used for sexual reproduction. Usually the most colorful part of a plant.

foot-candles (FC). A unit of illuminance equal to the illumination on one square foot of surface produced by one candle (called an international candle) at a foot away from the source.

forcing. A process that causes plants to grow when they normally wouldn't.

formal garden. A garden where plants are laid out in regular patterns with defined pathways.

friable. Soil texture that is crumbly, loose, and easily penetrated by plant roots.

frond. A large compound leaf, such as those of fennel.

full sun. Describes an area that receives a minimum of six hours of unfiltered sun per day.

fungicide. An agent that kills fungal growth.

-G-

garnish. A plant part, or other edible item, used to decorate a plate of food to make it more attractive.

genus. A plant classification category, below a family but above a species. Plants in the same genus share more characteristics than those of the same family. A genus may include many species. The genus name is the first word of a plant's scientific (botanical) name and is always capitalized.

germination. The sprouting and initial growth of seeds.

germination chamber. A small room or enclosed area used specifically to sprout seeds. Usually equipped with heat, ventilation, moisture, and supplemental lighting.

grade. The direction and degree of a slope of land.

grandiflora. Unusually large flowers for a particular species.

growing on. Allowing a small plant to grow to a larger size and maturity, usually after transplanting.

-H-

habit. The growth pattern of a plant, such as upright, spreading, or rounded.

habitat. The area where a plant grows naturally.

half-hardy. Characterizes a plant able to withstand a few degrees of cold but not extreme cold temperatures.

harden off. A process of gradually conditioning a plant to the outdoor growing conditions such as temperature, sun, and wind.

hardiness. The ability of a plant to survive a certain low temperature.

heel cutting. A cutting of new growth with a piece of older stem attached to its base.

heel in. To temporarily place plants roots in a trench to prevent them from drying out until they can be planted permanently.

herb. Edible plant that is used for flavoring food or for medicinal or aromatic purposes.

herbaceous plant. A plant whose stem aboveground does not become woody but stays soft. Many annual herbs are herbaceous.

herbicide. An agent, usually chemical, used to kill plants and weeds. Selective herbicides kill only certain plants; while nonselective herbicides kill all plants they touch.

high-intensity discharge lights (HID). High-output supplemental lights that efficiently produce light in various spectrums beneficial for plant growth.

horizontal air flow (HAF). Continuous movement of air inside the greenhouse caused by the placement of small fans.

horticultural grade. Describes chemicals or substances that are free of most contaminants that may harm plants.

horticultural oil. A highly refined, light mineral oil that, when mixed with water, can be used as an insecticide.

humidity. The amount of water vapor in the atmosphere.

humus. Beneficial decomposed organic matter in the soil.

hybrid. The offspring of crossbreeding varieties of the same species.

hygrometer. A gauge for measuring relative humidity.

-I-

inflation fan. A small fan used to blow air in between two layers of polyethylene on a greenhouse to keep them separated.

injector. A device attached to a water system that introduces a set amount of material, usually fertilizer, into the irrigation lines.

inorganic. Material that is not derived from living organisms.

insect growth regulator (IGR). A type of insecticide, usually hormone-based, that interrupts the growth cycle of certain insects, which results in their death.

insecticidal soap. An agent made from the fatty acids of plants and other ingredients that is mixed with water and used as an insecticide.

insecticide. An agent that kills insects.

instar. A stage of an insect's larval growth between molts.

interplant. To put a variety of plants together in the same area.

-J-

juvenile. A young plant that has not reached the flowering stage.

-L-

layering. The process of covering a certain part of a plant with a growing medium to cause it to grow roots and eventually be separated from the mother plant to grow a new plant.

leaching. Loss of nutrients, salts, and other water-soluble compounds from the soil, usually caused by excessive water draining through it.

lesion. A wound, off-color area, or damaged area on a leaf or stem.

liners. A small young plant that is ready for transplanting.

loam. An ideal type of balanced soil that contains much organic matter and many mineral particles.

lobe. A distinct rounded shape on the margin of a leaf.

-M-

macronutrients. Six essential elements required by plants in larger quantities: N (nitrogen), P (phosphorus), K (potassium), Ca (calcium), Mg (magnesium), S (sulfur).

margin. The outside edge of a leaf.

mature plant. One that is old enough to produce flowers and seeds.

medium. A material in which to grow plants. It can contain many ingredients and may or may not contain soil.

microclimate. Local conditions of light, temperature, drainage, air circulation, and other factors in a certain area.

micronutrients. Elements, also called trace elements, that are needed for plant growth but in small amounts: Fe (iron), Mn (manganese), B (boron), Zn (zinc), Cu (copper), Mo (molybdenum).

miticide. An agent formulated for the purpose of killing small arachnids in the mite family.

moisture-holding capacity. The ability of the soil to retain moisture, thus making moisture available for plants to utilize.

moisture stress. A condition whereby the plant is losing water faster than it can absorb it. This often causes wilting.

mother plant. A stock plant, usually a perennial, that is grown for the purpose of taking cuttings or layering to propagate new plants.

mound layering. The process of mounding soil around the center stems of plants to cause roots to form. Also known as *stooling*. Useful for propagating thyme and winter savory.

mucilage. A thick, sticky substance found in some plants. It is also formed when seeds of some plants, such as basil, become moist.

mulch. Material spread on top of the soil to keep it moist, provide winter protection, or suppress weed growth. Mulch can be organic material or artificial, such as plastic.

multiflora. Plants with mixed flower sizes.

-N-

naturalized. Plants from other areas that become established and grow wild along with local plants.

necrosis. Dead, brown, or scorched areas on leaves that are caused by disease or nutrient deficiencies.

needle. A rigid, narrow, long leaf such as those on pines and rosemary.

nematode. A microscopic, soil-dwelling roundworm. Some are beneficial and some are pests.

nit. The egg or young adult of some parasitic insects.

node. A point on a stem at which a leaf grows. The part between the nodes is the internode.

nutrients. Nitrogen, phosphorus, potassium, and other elements needed by plants to grow. These are supplied by minerals and organic matter in the soil and by fertilizers.

-O-

oblong. Describes a leaf that is more long than wide, with nearly parallel sides and rounded ends.

opposite. Characterizes leaves arranged in pairs on opposite sides of the stem.

organic. Refers to crops that are grown without the use of chemical or synthetic agents and are grown in accordance with the National Organic Program. Also refers to material that is derived from living organisms.

orientation. The location of plants or greenhouses in relationship to the points on a compass.

ornamental. In horticulture, referring to plants that are grown only for their visual appeal, such as flowers.

oval. Describes a leaf that is egg-shaped, with the narrower end opposite the base.

-P-

packs. Lightweight plastic container in which two or more plants are growing. A number of packs usually fit inside a flat.

pathogen. A living organism that causes diseases in plants. Pathogens can be bacteria, fungi, viruses, or nematodes.

peat moss. Sedges, mosses, and other organic material that is partially decomposed. Peat is mined from bog areas and used to improve garden soil and in potting mixes.

pelleted seed. Very small or irregular seed that is coated with a clay-type material to increase the seed size and make it easier to handle or use with mechanical seeders.

perennial. A plant that has a life cycle of more than two years.

perlite. A growing medium amendment made from a volcanic mineral that is heated, causing it to expand into a lightweight white aggregate.

pesticide. An agent, botanical or chemical, used to kill unwanted insects.

petal. One of the modified leaves, often colorful, that form the corolla of a flower.

petiole. The stalk of a leaf.

pH. A measure of acidity or alkalinity that ranges between 0 and 14. A reading of 7 is neutral; below 7 is acid and above 7 is alkaline.

photoperiod. The amount of time a plant is exposed to light and darkness.

photosynthesis. The process plants use to convert light into carbohydrates using carbon dioxide, water, oxygen, and minerals.

phytotoxicity. The capacity of a pesticide or other agent to poison plants, causing damage to foliage or flowers.

pinch back. To remove the tips of new shoots to encourage the plant to branch out. Also to remove the flower buds to inhibit flower formation and promote leaf growth.

plug. A small plant grown from seed to transplant stage in trays with small individual compartments for each plant.

plug stages. 1. radicle emergence. 2. stem and cotyledon emergence. 3. development of the true leaves. 4. ready to transplant.

pollen. Minute particles produced by the male stamens of a flower.

porosity. The amount of pore spaces in a growing medium.

postharvest. Referring to the treatment and care given to plant material after it has been separated from the plant.

ppm. Parts per million.

prick out. To remove small plants from their growing beds or containers in order to transplant them.

primed seed. Seed that has been partially germinated and then stopped and dried again. When it is sown again, it germinates quickly with better performance.

propagate. To cause a plant to multiply through planting of seeds or rooting cuttings.

puddle in. To transplant into a hole that has been filled with water and allowed to drain.

-R-

radicle. The first rootlet emerging from a seed.

resin. A sticky, gummy, and usually aromatic substance secreted by such plants as rosemary.

respiration. A plant's process of breaking down carbohydrates or sugars to release energy for other processes.

respirator. A protective mask worn by a person spraying agents to prevent him from inhaling the vapors.

rhizome. A horizontal underground stem, often with thick parts, that store nutrients for the plant's growth. They can usually send roots into the air or deep in the ground.

root. The part of a plant, usually belowground, that holds it and absorbs water and nutrients for its growth.

root-bound. A condition of a container-grown plant that has outgrown the pot. The roots have grown in a circular mass and may not recover when the plant is transplanted.

root cutting. Small pieces of a plant's root cut to produce new plants.

rooting hormone. A liquid or powder containing plant hormones and other ingredients that stimulate root formation in cuttings.

root-zone heating. Heating below the bench or in growing beds that maintains soil temperature rather than air temperature.

rosette. A circle of leaves around the stem of a plant. Basal rosettes occur at the base of the stem.

runner. A long, slender stolon growing above or below soil level.

-S-

sap. The fluid that circulates through a plant carrying nutrients and water to the tissues.

scale. Small insects that suck plant juices. They appear as tiny dots clustered on leaf surfaces.

scarification. A process of wounding hard seed coatings to break their dormancy. They may be scratched, nicked, or soaked briefly in hot water, acid, or another agent to aid germination.

seed. A fertilized and mature offspring of a plant that is usually dormant but will sprout when given favorable conditions.

self-sowing. Refers to plants that drop their seeds, which then germinate and grow.

sepal. One of the set of floral leaves that occur outside the petals.

shade house. An area, usually outdoors, covered with a material to provide shade to the plants that are placed inside or beneath it.

sheath. In some plants, such as dill, the lower part of a leaf enveloping the stem.

shoot. New plant growth, usually an aerial stem growing upward out of the soil.

shrub. Usually a woody plant with multiple stems and growing no more than a few feet in height.

simple layering. The process of covering a plant stem with soil to cause it to form roots and eventually be separated from the mother plant. Useful for propagating oregano, rosemary, sage, thyme, and winter savory.

softwood cutting. Cutting taken from new growth early in the growing season.

soluble salts. Salts that are dissolved in water. These include most inorganic fertilizers and mineral salts dissolved in irrigation water.

species. A plant classification category that is below genus and contains individuals capable of interbreeding. Varieties of a species may look very different.

sphagnum peat moss. A type of peat moss that has the lowest pH (3.0 to 4.5) and has the best qualities for a growing medium.

sport. An atypical stem on a plant that sometimes contains genetic mutations. These are often used to create new cultivars.

spur. A tubelike projection extending out from the flower petals.

starter charge. A small amount of fertilizer added to a growing medium before it is used.

stem. The main axis of a plant on which buds, leaves, and branches grow.

sterile. Describes a plant that is unable to produce seeds.

stippling. Tiny marks on a leaf surface caused by feeding insects.

stock plants. A plant that is grown for the purpose of taking cuttings or for layering to propagate new plants.

stolon. A branch or stem that grows out horizontally above or just below the ground from the base of a plant for the purpose of producing new plants. Common to the mint family.

stratify. To treat certain seeds with cold temperatures, darkness, and moist conditions for a period of time so they will germinate.

subsoil. The layer of earth below the topsoil.

sucker shoot. A shoot growing on the lower part of the stem.

supplemental lighting. Additional lighting used to increase the footcandles of light reaching the plant surfaces or to extend the daylight hours. Usually used in greenhouses or germination areas.

systemic. Describes agents, usually chemicals, that are absorbed by the plant. They usually stay within the plant for its lifetime.

-T-

taproot. The main root of a plant.

tempered water. Water that is heated before irrigation.

tender. Characterizes a plant that will not survive or will be damaged by freezing temperatures. Also foliage that is easily bruised or damaged.

terminal bud. The bud on the top of the plant or the tip of a stem.

tetraploid. A plant that has twice the number of chromosomes and usually is larger, has more leaves, and is more vigorous than others.

thinning. Removing some plants, usually seedlings, from an overcrowded bed.

tissue culture. An asexual method of propagating plants in a laboratory using careful, well-controlled cell growth.

top-dress. To place compost, more soil, or fertilizer on the surface of the soil around a plant without mixing it in.

toxic. Poisonous.

trace elements. Micronutrients that are needed for plant growth but in very small amounts.

transpiration. The evaporation of moisture from a plant, mostly through the leaves. This helps to cool the plant.

transplanting. Removing a plant or seedling from an area or container and planting it in another area or container.

-U-

umbel. A flower cluster with a flat top from which individual stalks radiate at a central point, as with dill and garlic chive flowers.

-V-

variegated. Characterizes foliage that is striped, marked, or blotched with a color other than green.

variety. A member of a species that has characteristics that are slightly different from other members of a species. Properly called a *cultivar.*

vascular. Pertaining to the circulatory system, which carries water and nutrients around the plant.

vegetative. Refers to the propagation of new plants by means other than seeds. Also pertains to the growing parts of a plant rather than the flowers.

vein. A channel that carries water and nutrients to and from a leaf.

vermiculite. An inorganic growing medium amendment made from a micalike ore that is heated to cause the layers to expand.

viable. In seeds, being alive and able to germinate and grow.

vigorous. Healthy, strong, able to withstand stress.

volatile oil. A general term describing the compounds found within a plant that give the plant its characteristic flavor, taste, and aroma.

volunteer. A plant that grows from self-sown seed, often in an area away from others of its kind.

-W-

water-holding capacity. The ability of the growing medium to hold water.

water-soluble fertilizer. Fertilizers that completely dissolve in water.

weed. A native plant growing in an area where it is not wanted.

wetting agent. A compound used with a peat moss–based growing medium to allow it to absorb water easily.

whorl. A ring of leaves, bracts, or flowers radiating horizontally from a common point on a stem.

wilting. The drooping of a plant that does not contain enough moisture.

woody plant. A plant with tough, thick tissue rather than the soft growth of herbaceous stems.

Wholesale Herb Seeds, Plants, Plugs, and Liners

Bluebird Nursery
519 Bryan Street
Clarkson, NE 68629
(800) 356-9164
Fax (402) 892-3738
e-mail: bluebrd@megavision.com
Plants, plugs, and liners

Flowery Branch Seed Company
PO Box 1330
Flowery Branch, GA 30542
(770) 532-7825
Fax (770) 532-7825

Gari's Greenhouses
1018 Rose Dale Lane
Dover, DE 19904
(302) 734-4606
Fax (302) 653-7166
Plugs and liners

Grimes Seeds and Plants
11335 Concord-Hambden Road
Concord, OH 44077
(440) 352-3333 / (800) 241-7333
Fax (440) 352-1800
e-mail: grimesseed@aol.com
Wholesale seeds, plugs, liners

Johnny's Selected Seeds
RR 1 Box 2580
Albion, ME 04910
(207) 437-9294
Fax (800) 437-4290
Fax outside U.S. (207) 437-2165
e-mail: commercial@
  johnnyseeds.com
Seeds and plants

Meadow View Growers, Inc.
755 N. Dayton Lakeview Road
New Carlisle, OH 45344
(800) 276-6841
Fax (937) 845-4082
Herb liners

Parks Seed Wholesale
1 Parkton Avenue
Greenwood, SC 29647-0002
(800) 845-3366
Fax (800) 209-0360
e-mail: wholesale@parkseed.com
Seed, plugs, liners

Richter's Herbs
357 Highway 47
Goodwood, ONT
Canada L0C 1A0
(905) 640-6677
Fax (905) 640-6641
e-mail: catalog@richters.com
Wholesale seeds, plugs

Herb Nurseries

Companion Plants
7247 North Coolville Ridge Road
Athens, OH 45701
(740) 592-4643
Fax (740) 593-3092
e-mail: complants@frognet.net

Good Scents
1308 N. Meridian Road
Meridian, ID 83642
(208) 887-1784
e-mail: basil@micron.net

Nichols Garden Nursery
1190 N. Pacific Highway NE
Albany, OR 97321-4580
(541) 928-9280
Fax (541) 967-8406
e-mail: nichols@
   gardennursery.com
Seeds, plants, wholesale seeds

Rasland Farm
RR 1 Box 65
Godwin, NC 28344-9712
(910) 567-2705
Fax (910) 567-6716
e-mail: rasland@intrstar.net

Sandy Mush Herb Nursery
316 Surrett Cove Road
Leicester, NC 28748-5517
(828) 683-2014

Shady Acres Herb Farm
7815 Highway 212
Chaska, MN 55318
(612) 466-3391
Fax (612) 466-4739
e-mail: herbs@shadyacres.com

Southern Perennials & Herbs
98 Bridges Road
Tylertown, MS 39667-9338
(601) 684-1769
Fax (601) 684-3729
e-mail: sph@neosost.cpm

Well-Sweep Herb Farm
205 Mt. Bethel Road
Port Murray, NJ 07865
(908) 852-5390
Fax (908) 852-1649

Sources for Buying Bulk Fresh-Cut Herbs

California Specialty Farms
2420 Modoc Street
Los Angeles, CA 90021
(800) 437-2702
Fax (213) 587-0050
e-mail: specfarms@aol.com

Fresh Foods Hawaii, Inc.
PO Box 30867
Honolulu, HI 96820
(808) 833-3664
Fax (808) 839-6689

Generation Farms
1109 N. McKinney Street
Rice, TX 75155
(903) 326-4263
Fax (903) 326-6511

Quail Mountain Herbs
PO Box 1049
Watsonville, CA 95077-1049
(408) 722-8456
Fax (408) 722-9472

Riddle Mill Organic Farm
1098 Riddle Mill Road
Lake Wylie, SC 29710-9618
(803) 831-2506
Fax (803) 831-2506
e-mail: wiccalady@aol.com

Taylor's Garden, Inc.
PO Box 362
Congress, AZ 85332
(520) 427-3201
Fax (520) 427-3660

Grower's Associations

Herb Growing and Marketing
   Network
PO Box 245
Silver Spring, PA 17575-0245
(717) 393-3293
Fax (717) 393-9261
e-mail: HERBWORLD@aol.com

International Herb Association
PO Box 317
Mundelein, IL 60060-0317
(847) 949-4372
Fax (847) 949-5896
e-mail: IHAoffice@aol.com

Periodicals and Trade Publications

*The American Vegetable Grower
   Greenhouse Grower*
Meister Publishing Company
37733 Euclid Avenue
Willoughby, OH 44094
(440) 942-2000
Fax (440) 942-0662
e-mail: avg_circ@meisterpubl.com
or gg_circ@meisterpubl.com
For commercial growers

Business of Herbs
439 Ponderosa Way
Jemez Spring, NM 87025-8025
(505) 829-3448
Fax (505) 829-3449
e-mail: HerbBiz@aol.com

*Greenhouse Management and
   Production*
Branch-Smith Publishing
PO Box 1868
Fort Worth, TX 76101
(817) 882-4120 / (800) 434-6776
Fax (817) 882-4121
e-mail: circulation@
   bsipublishing.com
For commercial growers

Grower Talks Magazine
PO Box 9
Batavia, IL 60510
(630) 208-9080
Fax (630) 208-9350
For commercial growers

Growing for Market
PO Box 3747
Lawrence, KS 66046
(800) 307-8949
Fax (785) 748-0609

Herb Companion
201 East Fourth Street
Loveland, CO 80537
(800) 272-2193
Fax (970) 667-8317
e-mail: HC@iwp.ccmail.
   compuserve.com

Herb Quarterly
PO Box 689
San Anselmo, CA 94979
(415) 455-9540
Fax (415) 455-9541
e-mail: HerbQuart@aol.com

The Herbal Connection
PO Box 245
Silver Spring, PA 17575-0245
(717) 393-3295
Fax (717) 393-9261
e-mail: HERBWORLD@aol.com

Organic Food Business News
PO Box 161132
Altamonte Springs, FL 32716
(407) 628-1377
Fax (407) 628-9935
e-mail: 74562774@
   compuserve.com

Government Agencies

United States Department of
   Agriculture (USDA)
Washington, DC 20250
(202) 720-2791

Environmental Protection Agency
   (EPA)
Worker Protection Standards
401 M Street SW,
   Mail Code 7506C
Washington, DC 20460
(703) 305-7666

Food and Drug Administration
   (FDA)
409 3rd Street SW
Washington, DC 20416
(202) 205-6600

Immigration & Naturalization
   Service (INS)
200 Constitution Avenue NW
Washington, D.C. 20210
(202) 219-9098
www.ins.gov

Internal Revenue Service
Kansas City, MO 64999
(800) 829-1040
www.irs.gov

Small Business Administration
5600 Fishers Lane
Rockville, MD 20857
(301) 443-1544 / (800) 368-5855

Biological Pest and Disease Controls

Arbico
PO Box 4247 CRB
Tucson, AZ 85738
(800) 827-2847
Fax (520) 825-2038
e-mail: Arbico@aol.com
Beneficial insects and other supplies

Beneficial Insectary
14751 Oak Run Road
Oak Run, CA 96069
(530) 472-3715
Fax (530) 472-3523
e-mail: bi@insectary.co

Gardens Alive!
PO Box 149
Sunman, IN 47041
(812) 537-8650
Fax (812) 537-5108
e-mail: gardener@
     gardens-alive.com
Beneficial insects and other supplies

IPM Laboratories, Inc.
PO Box 300
Locke, NY 13092-0300
(315) 497-2063
Fax (315) 497-3129
e-mail: ipmlabs@ipmlabs.com

M & R Durango, Inc.
PO Box 886
Bayfield, CO 81122
(800) 526-4075
Fax (970) 259-3857
e-mail: sales@goodbug.com
Beneficial insects

Peaceful Valley Farm Supply
PO Box 2209
Grass Valley, CA 95945
(530) 272-4769
Fax (530) 272-4794
Beneficial insects and other supplies

Planet Natural
PO Box 3146
Bozeman, MT 59772
(800) 289-6656
Fax (406) 587-0223
e-mail: ecostore@mcn.net

Rincon-Vitova Insectaries, Inc
PO Box 1555
Ventura, CA 93002
(800) 248-2847
Fax (805) 643-6267
e-mail: bugnet@west.net
Beneficial insects

Greenhouses, Equipment, Growing Supplies

Agra Tech
2131 Piedmont Way
Pittsburg, CA 94565
(925) 432-3399
Fax (925) 432-3521
e-mail: ati@slip.net
Greenhouses, equipment

CropKing, Inc.
5050 Greenwich Road
Seville, OH 44273
(330) 769-2002
Fax (330) 769-2616
e-mail: cropking@cropking.com
Greenhouses, equipment

Hydro-Gardens, Inc.
PO Box 25845
Colorado Springs, CO 80936
(800) 634-6362
e-mail: hgi@usa.net
Greenhouses, equipment, supplies

J.R. Johnson Supply
2582 Long Lake Road
St. Paul, MN 55113
(800) 652-9022 MN Wats
(800) 328-9221 nationwide
Greenhouses, equipment, supplies

P.L. Lights Systems
PO Box 206
Grimsby, ONT Canada L3M 4G3
(800) 263-0213
Fax (905) 945-0444
Supplemental lighting

Poly-tex, Inc
27725 Danville Avenue
Castle Rock, MN 55010
(800) 852-3443
Fax (612) 463-2479
e-mail: polytex@compuserve.com
Greenhouses, equipment

Schaefer Ventilation Equipment
PO Box 647
Waite Park, MN 56387
(320) 251-8696
(800) 779-3267
Fax (320) 251-2922
e-mail: sales@schaffan.com
Equipment

Packaging Supplies

Action Bag & Display
501 N. Edgewood Avenue
Wood Dale, IL 60191-1410
(800) 824-2247
Fax (800) 400-4451
e-mail: actionbg@ix.netcom.com
Poly bags

Monte Package Company
Riverside, MI 49084
(616) 849-1722
Fax (616) 849-0185
e-mail: montepkg@atm.net
Clamshell trays, poly bags

Schilling Paper Company
PO Box 369
LaCrosse, WI 54602
(800) 888-1885
Fax (608) 781-2344
e-mail: info@schillingpaper.com
Clamshell trays, poly bags

Ultra Pac, Inc.
21925 Industrial Boulevard
Rogers, MN 55374-9575
(800) 999-9001
Fax (612) 428-3462
e-mail: upac@ultrapac.com.
Clamshell trays for herbs

Horticultural Industry Insurance

Butler Florists and Growers
  Insurance
20 South Street
Westborough, MA 01581-1696
(800) 288-5373
Fax (800) 866-2884

Florists' Mutual Insurance
  Company
500 St. Louis Street
Edwardsville, IL 62025
(800) 851-7740
Fax (800) 233-3642
e-mail: feedback@plantnet.com

Hylant MacLean, Inc.
PO Box 1687
Toledo, OH 43603-1687
(800) 249-5368
Fax (419) 255-7557
e-mail: jeannie.hylant@hylant.com

Businesses Profiled in This Book

Farmer's Daughter Herbs
4220 W. 2100 South, Suite E
Salt Lake City, UT 84120
(801) 973-7875

Maggie's Herbs
1400 County Road 13N
St. Augustine, FL 32092
(904) 829-0722

O'Toole's Herb Farm
Rocky Ford Road
PO Box 268
Madison, FL 32341
(904) 973-3629

Quail Mountain Herbs
PO Box 1049
Watsonville, CA 95077-1049
(408) 722-8456
Fax (408) 722-9472

Riddle Mill Organic Farm
1098 Riddle Mill Road
Lake Wylie, SC 29710-9618
(803) 831-2506
Fax (803) 831-2506
e-mail: wiccalady@aol.com

Farmer's Market Information

Dynamic Farmer's Marketing,
by Jeff Ishee available through
the Web site:
www.farmersmarketonline.com/
dynamic.htm

Two other Web sites focus on
farmer's markets:
www.farmersmarketonline.com/
openmark.htm, *and* openair.org/

Miscellaneous

Uniform Code Council, Inc.
8163 Old Yankee Street, Suite J
Dayton, OH 45458
(937) 435-3870
Fax (937) 435-7317
www.uc-council.org
Bar or UPC codes

The Herbfarm
32804 Iss-Fall City Road
Fall City, WA 98024
(800) 866-4372
Fax (206) 789-2279
e-mail: HERBORDER@aol.com
Edible flowers list

*USDA Fruit and Vegetable
Market News*
(National Wholesale Herb Market
Report)
230 South Dearborn Street,
Room 512
Chicago, IL 60604-1503
(312) 352-0111
Fax (312) 886-3766

Illustrators and photographers whose work appears in this book:

Agra Tech, Inc.
photo on page 89 (left)

Cathy Baker
illustrations on pages 223, 235 (top), 236, 238 (bottom), 239 (top), 240, and 282

Beverly Duncan
illustrations on pages 10, 15, 26, 34, 38, 41, 78, 84, 85, 86, 94 (top), 96, 110, 111, 112, 115, 116, 117, 118, 119, 122, 124, 126, 128 (top), 138, 139, 150, 164, 167, 169, 171, 172, 182, 183, 190, 196, 198, 204, 206, 207, 208, 212, 225 (middle, bottom), 233, 244, 257, 262, 268, 271, 273, 277, 278, 280, 287, 289, 291, 332, 388, 405, and 407

Judy Eliason
illustrations on pages 199, 227, 229 (bottom), 231, 235 (bottom), 237, 238 (top), 239 (middle, bottom), 260, and 263

LaVonne Francis
illustrations on pages 184, 283, 329, and 399

Brigita Fuhrmann
illustration on page 209

Deb House–Finlay
photos on pages 128 (bottom), and 288

Jack Huhnerkoch
photos on part opening pages; also pages 298, 308, 311, 317, 324, 328, 334, 339, 344, 349, 350, 355, 362, 367, 372, 379, and 402

Charles Joslin
illustrations on pages 175, 176, 177, 385, 387, 390, 391, 394, 395, 397, and 398

Alison Kolesar
illustration on page 162

Doug Paisley
illustrations on pages 285 and 393

P.L Light Systems Canada, Inc.
photo on page 99

Poly-tex, Inc.
photo on page 89 (right)

Schaefer Fan Company, Inc.
photo on page 93, 94 (middle), and 95

Elayne Sears
illustration on page 174

Brian Whitehurst
illustrations on pages 25, 129, 130, 181, 221, 225 (top), 229 (top), 238 (middle), 254, and 269

index

Note: *Page numbers in italics refer to illustrations or photos; those in* **boldface** *refer to charts.*

Hardiness of herbs, **137**
Harvesting, 62, 203, 283,
    286–290, *288,* 303–304.
    See also specific herbs
Health food stores, 46
Heating cable, 130–131, *130,*
    163–165, *164*
Heating systems, greenhouse,
    79–80, 140–141, 219. *See
    also* Root-zone heating
    boilers, hot-water and steam,
    90
    furnaces, 90–91, 138, 141
    propane, use of, 69, 91
    for seed germination,
    163–165, *164*
    thermostats, 96–97, 141, 142
Heating systems, packaging area,
    120
Height of herbs, **137**
Herb garnishes
    container growing, 127
    cutting, 289–290, *289*
    mint, 340
    packaging, 297
    pest damage and, 244
    pricing, 59
    rosemary, 360–361
Herb Growing & Marketing
    Network, 60
Herbal Connection, The, 60
Herbal products, 70, 71
Herbicides, 147, 208, 209
Herbs
    cultural information, **136–137**
    flavor, consistency of, 75
    handling of (*See* Handling,
    postharvest)

production time, **160,178**
    trends in, 66
Honeydew, 221, 359, 360
Hoop greenhouses, 83, *84,* 85,
    88, 103
    assembling, 111–114,
    111–112
    foundation, 110–111,
    110–111
    painting, 115, *115,* 118
Horizontal air flow (HAF) fans,
    95–96, *95,139,* 140
Horsetail tea, 272, 276
Horticultural oils, 260
    for aphid control, 224
    for fungal disease control,
    271–272
    for leaf miner control, 238
    for leafhopper control, 237
    for mealybug control, 238
    for mite control, 228
    phytotoxic damage from, 254
    for scale control, 239
    for thrip control, 231
    for whitefly control, 234
Hoses, 125, 149–150, 201
Hours of work, 19, 44, 53, 76
Humidity, 138–139
Hybrid seeds, 157–159

-I-
Ideal bookkeeping system, 51
Immigration and Naturalization
    Service (INS), 55
Independent contractors, 54–55
Indicator plants, 246
Information sources, ix, 2, 12–14,
    60, 222, 425–427, 429–431

Other Storey Titles You Will Enjoy

The Big Book of Gardening Secrets, by Charles W.G. Smith. 352 pages. Paperback. ISBN 1-58017-000-5.

The Big Book of Gardening Skills, by the Editors of Garden Way Publishing. 352 pages. Paperback. ISBN 0-88266-795-5.

The Big Book of Preserving the Harvest, by Carol W. Costenbader. 352 pages. Paperback. ISBN 0-88266-978-8.

Cordials from Your Kitchen, by Pattie Vargas and Rich Gulling. 176 pages. Paperback. ISBN 0-88266-986-9.

The Gardener's Bug Book: Earth-Safe Insect Control, by Barbara Pleasant. 160 pages. Paperback. ISBN 0-88266-609-6.

The Gardener's Guide to Plant Diseases: Earth-Safe Remedies, by Barbara Pleasant. 192 pages. Paperback. ISBN 0-88266-274-0.

Growing Your Herb Business, by Bertha Reppert. 192 pages. Paperback. ISBN 0-88266-612-6.

Herb Mixtures and Spicy Blends, edited by Deborah Balmuth with an introduction by Maggie Oster. 160 pages. Paperback. ISBN 0-88266-918-4.

The Herbal Home Remedy Book: Simple Recipes for Tinctures, Teas, Salves, Tonics and Syrups, by Joyce Wardwell. 176 pages. Paperback. ISBN 1-58017-016-1.

The Herbal Palate Cookbook, by Maggie Oster and Sal Gilbertie. 176 pages. Paperback. ISBN 1-58017-025-0.

Herbal Vinegar, by Maggie Oster. 176 pages. Paperback. ISBN 0-88266-843-9.

Herbed Wine Cuisine: Creating & Cooking with Herb-Infused Wines, by Janie Therese Mancuso. 160 pages. Hardcover. ISBN 0-88266-967-2.

The Herb Gardener: A Guide for All Seasons, by Susan McClure. 240 pages. Paperback. ISBN 0-88266-873-0.

The Pleasure of Herbs: A Month-by-Month Guide to Growing, Using, and Enjoying Herbs, by Phyllis Shaudys. 288 pages. Paperback. ISBN 0-88266-423-9.

These and other Storey books are available at your bookstore, farm store, garden center, or directly from Storey Books, Schoolhouse Road, Pownal, Vermont 05261, or by calling 1-800-441-5700. Or visit our Web site at www.storey.com.